Bound and Conjugated Pesticide Residues

Bound and Conjugated Pesticide Residues

Donald D. Kaufman, EDITOR

Gerald G. Still, EDITOR

Gaylord D. Paulson, EDITOR
U.S. Department of Agriculture

Suresh K. Bandal, EDITOR
3M Company

A symposium sponsored by
the Division of Pesticide
Chemistry at Division
Workshop, Vail, Colo.,
June 22–26, 1975

ACS SYMPOSIUM SERIES **29**

AMERICAN CHEMICAL SOCIETY
WASHINGTON, D. C. 1976

*ad
12/7/76*

Library of Congress CIP Data

Bound and conjugated pesticide residues.
  (ACS symposium series; 29 ISSN 0097-6156)

  Includes bibliographical references and index.

  1. Pesticides—Environmental aspects—Congresses.
  I. Kaufman, Donald DeVere, 1933-   . II. American
Chemical Society. Division of Pesticide Chemistry. III.
Series: American Chemical Society. ACS symposium series;
29.

QH545.P4B63   632'.95'042   76-13011
ISBN 0-8412-0334-2   ACSMC8 29 1–396

Copyright © 1976

American Chemical Society

PRINTED IN THE UNITED STATES OF AMERICA

ACS Symposium Series

Robert F. Gould, *Editor*

C8400

FOREWORD

The ACS SYMPOSIUM SERIES was founded in 1974 to provide a medium for publishing symposia quickly in book form. The format of the SERIES parallels that of the continuing ADVANCES IN CHEMISTRY SERIES except that in order to save time the papers are not typeset but are reproduced as they are submitted by the authors in camera-ready form. As a further means of saving time, the papers are not edited or reviewed except by the symposium chairman, who becomes editor of the book. Papers published in the ACS SYMPOSIUM SERIES are original contributions not published elsewhere in whole or major part and include reports of research as well as reviews since symposia may embrace both types of presentation.

CONTENTS

Preface xi

1. Bound and Conjugated Pesticide Residues 1
   Donald D. Kaufman

2. Biological Activity of Pesticide Conjugates 11
   H. W. Dorough

3. Pesticide Conjugates—Glycosides 35
   D. S. Frear

4. Recent Advances in the Isolation and Identification of Glucuronide
   Conjugates 55
   Jerome E. Bakke

5. Amino Acid Conjugates 68
   Ralph O. Mumma and Robert H. Hamilton

6. Sulfate Ester Conjugates—Their Synthesis, Purification, Hydrolysis,
   and Chemical and Spectral Properties 86
   G. D. Paulson

7. Glutathione Conjugates 103
   D. H. Hutson

8. Miscellaneous Conjugates—Acylation and Alkylation of Xenobiotics
   in Physiologically Active Systems 132
   Jorg Iwan

9. Nature of Propanil Bound Residues in Rice Plants as Measured by
   Plant Fractionation and Animal Bioavailability Experiments 153
   M. L. Sutherland

10. Solubiliztion of Bound Residues from 3,4-Dichloroaniline-^{14}C and
    Propanil–Phenyl–^{14}C Treated Rice Root Tissues 156
    Gerald G. Still, Frank A. Norris, and Jorg Iwan

11. Classification and Analysis of Pesticides Bound to Plant Material .. 166
    J. Wieneke

12. Bound Residues of Nitrofen in Cereal Grain and Straw 170
    R. C. Honeycutt, J. P. Wargo, and I. L. Adler

13. Use of Radiotracer Studies in the Estimation of Conjugated and
    Bound Metabolites of Dichlobenil in Field Crops 173
    A. Verloop

14. Metabolite Fate of *p*-Toluoyl Chloride Phenylhydrazone (TCPH) in Sheep. The Nature of Bound Residues in Erythrocytes 178
Prem S. Jaglan, Ronald E. Gosline, and A. William Neff

15. Organic Matter Reactions Involving Pesticides in Soil 180
F. J. Stevenson

16. Clay–Pesticide Interactions 208
Joe L. White

17. Turnover of Pesticide Residues in Soil 219
John W. Hamaker and Cleve A. I. Goring

18. Microbial Synthesis of Humic Materials 244
K. Haider

19. Spectroscopic Characterization of Soil Organic Matter 258
R. Bartha and T.-S. Hsu

20. Classification of Bound Residues in Soil Organic Matter: Polymeric Nature of Residues in Humic Substance 272
R. W. Meikle, A. J. Regoli, N. H. Kurihara, and D. A. Laskowski

21. Chemical Extraction of Certain Trifluoromethanesulfonanilide Pesticides and Related Compounds from the Soil 285
Suresh K. Bandal, Henry B. Clark, and Jay T. Hewitt

22. Biological Unavailability of Bound Paraquat Residues in Soil 301
D. Riley, B. V. Tucker, and W. Wilkinson

23. Fixed and Biologically Available Soil Bound Pesticides 354
J. B. Weber

24. An Experimental Approach to the Study of the Plant Availability of Soil Bound Pesticide Residues 356
F. Fuhr

25. Degradation of the Insecticide Pirimicarb in Soil—Characterization of "Bound" Residues 358
I. R. Hill

26. Chloroaniline–Humus Complexes—Formation, Persistence, and Problems in Monitoring 362
R. Bartha and T.-S. Hsu

27. Determination of the Release of Bound Fluchloralin Residues from Soil into Water 364
Gary M. Booth, R. Ward Rhees, Duane Ferrell, and J. R. Larsen

28. Dinitroaniline Herbicide Bound Residues in Soils 366
Charles S. Helling

29. Summary of Conjugate Papers 368
Hans Geissbuhler

30. Summary of Soil Bound Residues Discussion Session 378
    P. C. Kearney

Index 385

PREFACE

Production of adequate supplies of food and fiber presently requires the use of pesticides. Pesticides are used deliberately to alter the ecology, that is, to eliminate or restrict undesirable species in favor of species considered necessary for man's continued existence. The ubiquitous nature of many biological and biochemical processes makes it likely that even highly specific pesticides will affect some nontarget organisms. It is therefore imperative to determine what ecological changes pesticides may produce, which changes are permanent or temporary, and to decide which are acceptable or unacceptable.

In the past, the inability to reisolate a pesticide or its degradation products enabled us to conclude glibly that it was detoxified, degraded, metabolized, or eliminated by some unknown mechanism from any need for further concern. Radiolabeled pesticides and more sophisticated analytical technology, however, have brought a halt to such practices. We now recognize that our inability to isolate a chemical does not constitute metabolism or complete detoxication to innocuous products, but rather it constitutes a complex environmental research problem requiring the most sophisticated inputs of a multitude of scientific disciplines.

This workshop was organized because of current interest and concern for bound and conjugated pesticide residues in animals, plants, and soils. The objective of the conference was to bring together scientists with biological, chemical, and physical expertise in environmental fate of pesticides so that they could examine in some detail the formation and fate, synthesis, extraction, and methods of characterization of such pesticide residues, and if possible, provide scientific insights to future considerations regarding their overall significance. In the absence of definitive conclusions it was thought that the information discussed would provide the foundation for further research toward such a goal.

U.S. Department of Agriculture
Beltsville, Md.

DONALD D. KAUFMAN

U.S. Department of Agriculture
Fargo, N. Dak.

GERALD G. STILL
GAYLORD D. PAULSON

3M Company
St. Paul, Minn.
January 20, 1976

SURESH K. BANDAL

Bound and Conjugated Pesticide Residues

DONALD D. KAUFMAN

Pesticide Degradation Laboratory, Agricultural Environmental Quality Institute, ARS, U. S. Department of Agriculture, Beltsville, Md. 20705

Man has developed the capacity to manufacture and use on a vast scale, organic compounds which are initially foreign to our own body and to our environment. Such synthetic organic compounds are used as drugs for sickness, pesticides of various kinds for agriculture and health purposes, coloring matters, emulsifiers and stabilizing agents for food and drink, dyes for clothes, plasticizers, lubricants, coolants, and cleansing agents for all sorts of purposes, flame retardants, beauty preparations, spermicides and ovicides for population control, explosives and poison gases for military use and so on. Past research experience has clearly demonstrated that certain of these compounds, or their degradation products can enter into almost every phase of our environment and civilized existence. It is therefore essential that we know what happens to these compounds if and when they enter our bodies, our foodstuffs, and our environment so as to avoid any damaging effects or, if they cause damage, their use can be avoided in favor of less harmful compounds, or the risks associated with their continued use can be adequately evaluated.

Almost as rapidly as man has learned to generate complex new organic chemicals our environment has adjusted to cope with not all, but the majority of these chemicals. Investigations of the environmental fate, behavior, and metabolism of synthetic organic chemicals have revealed many new and fascinating environmental processes, metabolic pathways, and chemical and physical reactions which heretofore were either not recognized, or their significance not fully understood or appreciated. We all quickly recognize where man would be without the development of synthetic and natural organic chemicals in medical science. Similarly, where would environmental science and food production technology be today if it were not for the advent of agricultural chemicals. We know that agricultural chemicals are dissipated or utilized by many mechanisms and biochemical processes including oxidation, reduction, hydrolysis, dehalogenation, dehydrohalogenation, ring cleavage, etc.

In the past metabolic changes of foreign organic compounds were referred to as "detoxication." In the past the inability to isolate a chemical enabled us to glibly conclude that it was degraded, metabolized, or by some unknown mechanism eliminated from any need for further concern. The advent of radiolabeled pesticides and more sophisticated analytical technology, however, gradually called a halt to what many have referred to as "bathtub chemistry." We now recognize that our inability to isolate a chemical does not consitute metabolism or complete detoxication to inocuous products, but rather, it constitutes a complex environmental research problem requiring the most sophisticated inputs of not only a multitude of scientific disciplines, but economic, resource management, and development techniques, as well.

We know now that all chemical or metabolic changes which occur with a pesticide do not necessarily lead to complete degradation or mineralization of the pesticide molecule. In some instances a parent pesticide molecule may be converted to a more toxic substance. Indeed, a few pesticides actually require molecular changes for activation. Other pesticides or their metabolites enter into synthetic reactions which frequently result in the formation of molecules far more complex than the parent pesticide molecule. It is this latter phenomenon which brings us to our present conference regarding bound and conjugated pesticide residues.

A review of radiolabeled pesticide degradation or dissipation studies reveals numerous basic similarities. Briefly, these similarities can be characterized as three experimental fractions:

A.   Volatilized or eliminated products.
B.   Extractable products.
C.   Unextractable (or residual) products.

Volatilized materials would include respired materials, i.e., parent or intermediate products, and CO_2, as well as materials lost by the physical processes of volatilization. Other products may be eliminated in wastes or excreta. Generally speaking, the products lost by these mechanisms are readily trapped and characterized once the process and the environmental factors affecting the process are recognized.

For purposes of this presentation extractable products are considered as those products readily removed from the treated material by any one or more of a variety of solvents and extraction techniques. Although characterization of extractable products has challenged our very best minds and technology, they too represent at present a more easily workable and identifiable fraction. Included in this fraction are some of the parent compound and its many metabolic products resulting from both synthetic and degradative reactions. Generally, but not always, degradative reactions lead toward

more polar products, hence the tendency to use solvent extraction
techniques designed for maximum polar product removal. Included
in this polar solvent extract are products resulting from synthetic reactions occurring in the treated material. Experience
has demonstrated that these synthetic products are frequently
the result of various conjugative processes which occur within
the treated material. Part of the purpose of this conference is
to examine in some detail the synthesis, extraction, and methods
of characterization of these products, i.e., conjugates, and if
possible provide scientific insights to future considerations
regarding their overall significance.

Isolation and removal of the extractable products from the
original pesticide treated material invariably involves sacrific-
ing the natural physical state of the treated material. While
the resulting mass of unextractable materials is generally known
and describable in plants as: lipids, proteins, and structural
carbohydrates; in animals as: lipids, fats, proteins, and skel-
etal materials; and in soil as: sand, silt, clay, and organic
matter; its precise chemical nature has generally defied completely meaningful and accurate description. It is frequently
within this portion of the pesticide treated material that any-
where from a few percent to nearly 100% of some pesticide residues will remain in an extractable form. It is this unextractable and heretofore largely undescribable pesticide residue which
has been loosely characterized by many scientists as the "bound
residue." As with many such general quasi-scientific terms, however, this terminology is not fully understood or appreciated by
all pesticide scientists, regulatory agencies, or their administrators. Thus, a second part of our conference is to examine
what is known about the "bound residue": What is it? How can
it be characterized, defined, or identified? and if not, What is
its significance? How far must we go in characterizing it? If
it is truly "bound", and not readily available to significant
biological systems, must it be fully characterized? What does
the qualification of "readily available" mean? These are but a
few of the questions we hope to either answer or lay the foundation for further research to find the answer for during this
conference.

Conjugates

There are very few chemicals which enter into biological
systems that are not subject to chemical changes. A few "biochemically inert" compounds remain unchanged, although they may
be toxicologically active. The type of change which occurs depends primarily upon the structure of the compound, but other
factors such as species, route of entrance, and nutritional balance may also be important. There are several types of synthetic
reactions common to pesticides. Conjugate type reactions which
have been observed include (1):
   A. Reactions with carbohydrates (glycoside formation):

    1. Glucuronic acid conjugation.
    2. Glucoside formation.
    3. Other (riboside, gentiobioside).
 B. Reactions with aminoacids:
    1. Simple amino acids (glycine, alanine, etc.)
    2. Complex amino acids (glutathione, cysteine, etc.)
 C. Reactions involving sulfur (sulfate conjugation): and
 D. Reactions involving alkylation and acylation.

Reactions of pesticides and pesticide metabolites with carbohydrates and amino acids are common in plant and animal metabolism. Few, if any, carbohydrate-pesticide conjugates have been isolated from soil. This is somewhat surprising in view of the relative ease with which simple sugars can be reacted with aromatic amines in the laboratory. On the other hand such conjugates may be quickly metabolized by soil microorganisms and therefore seldom isolated.

Sulfate conjugation is a common reaction in animal metabolism. Reports have also indicated the occurrence of sulfate conjugation in plants and microorganisms. Alkylation and acylation reactions are common to plants, animals, and soil microorganisms.

Other types of synthetic reactions involving pesticide residues and metabolites include condensation type reactions yielding dimeric and polymeric compounds such as have been observed in soils. These will be discussed further under soil bound residues.

Whether or not a given compound will undergo any of the above syntheses depends upon its possessing particular chemical groups or reactive sites. If the compound does not initially contain such a group, it may aquire one by oxidation or reduction or some other process. Perhaps the simplest example of such a reaction is the hydroxylation of benzene to phenol which is subsequently conjugated through the hydroxyl group.

Several hypotheses have been put forward regarding the purpose or significance of conjugative reactions in biological systems. These hypotheses include:
    A. Chemical defense (2).
    B. Surface tension (3).
    C. Increased acidity (4).

Briefly, the chemical defense hypothesis is based on the assumption that metabolic products of foreign compounds are less toxic and more soluble than their precursors. Such is not always the case, however. The surface tension hypothesis observes that compounds which lower the surface tension of water tend to accumulate at cell surfaces and thereby attain toxic concentrations. Conjugated products do not lower surface tensions appreciably, and thus do not accumulate to toxic concentrations at surfaces. The hypothesis of increased acidity notes that conjugation generally changes a weak acid

which the body can not eliminate to a strong acid which it
can eliminate. While all three hypotheses seem to account
for certain aspects of the problem, none of them provides
a general explanation of how and why such processes occur.
It is not the primary objective of this conference to determine
why such conjugates form but rather, how are they formed,
how are they isolated and characterized, and what is their
possible role in the formation of the unextractable or bound
residues observed in various biological systems. Also, what
is their possible significance in terms of biological availability and toxicity in subsequent food chain organisms.
In general conjugates are considerably more polar and less
lipophlic than the parent pesticide molecule, and as such
are therefore more readily eliminated from animals. Plants,
however, do not have efficient excretory mechanisms, thus
while conjugation may lead to detoxication of the pesticide
it does not necessarily lead to elimination. The role of
such conjugates in catabolism and ultimate binding of the
pesticide molecule is not clear. An alternative mechanism
to binding, however, may be the direct interaction of the
pesticide with functional groups on protein or complex carbohydrate molecules.

Bound Residues

Perhaps the first objective of this section should be
to define the term "bound residue." Like many new terms,
the definition or interpretation of what a bound residue is,
has varied with every individual scientist, and is to a very
large degree dependent upon the extraction techniques used.
To a certain extent, it has been an elusive concept not only
to the pesticide scientist, but to our administrators and
regulatory agencies as well. Concern for bound residues has
varied all the way from total preoccupation with characterization of that elusive few percent of unextractable radiolabeled
chemical, to incredulous disbelief that an unextractable
product should be of any concern whatsoever. Responsible
science, however, dictates that we know at least something
about that unextractable entity. If we can not describe it
precisely as (e.g.) a radiolabeled aspartic acid molecule
linking together other components of a protein, then we should
at least be able to indicate that when passed through another
living organism which is most likely to encounter that treated
product, it is or is not biologically available to that organism,
and if it is available it is eventually eliminated or has
no significant effect on that organism. If it is an unknown
soil residue incorporated into soil organic matter and is
slowly released at a rate representing only a small percent
of what was originally applied, can it be considered safe
and acceptable? These are only a few of the many questions

which we hope to answer in the course of this conference.

I am not aware of any already existing definition of
a bound residue in plants or animals. While it is evident
that individual scientists have developed their own concepts
of bound residues in plants or animals, there does not appear
to have been any coordinated effort to provide a definition.
Last Fall (1974) an American Institute of Biological Sciences-
Environmental Chemistry Task Group attempted to provide such
a definition for soil bound residues. As some of you are
aware, this was a committee organized by AIBS for The Enviro-
mental Protection Agency to develop a series of protocols
designed to provide the information necessary to meet the
Pesticide Registration Guidelines.

The definition which we cast at that time was considered
only an "interim definition" in recognition of the fact that
a more precise or meaningful definition could evolve from
this very conference. The definition: A soil bound residue
is "that unextractable and chemically unidentifiable pesticide
residue remaining in fulvic acid, humic acid, and humin
fractions after exhaustive sequential extraction with nonpolar
organic and polar solvents." In retrospect, perhaps a better
definition would have included reference to plant roots, or
to decomposition by soil microorganisms, or resistance to
release by specific cell free enzymes. It seems entirely
reasonable, however, that analogous definitions could be cast
for unextractable pesticide residues in plants and animals.
Such a definition based on a more nearly universal methodology
has the advantage of providing a standard point of reference
with which to more objectively evaluate individual chemicals
and groups of chemicals. The great difficulty with all fractionation procedures, however, is that the methods employed
either separate out products which are not definite chemical
entities, or form artifacts which do not have the properties
of the original material. Nevertheless, the various fractionation procedures can prove useful for investigation and characterization of bound pesticide residues.

There are many critical questions to be asked concerning
bound residues. Perhaps the three most critical questions
are:
1. What is their nature and/or identity;
2. What is their significance (toxicity, availability,
accumulative nature, etc.); and
3. What is their source?
Within certain limits answering either one of the first
two questions can obviate the need for answers to the other
two. For example, if the radiolabeled bound residue is identified as a natural product its significance can possibly be
assumed on the basis of previous knowledge. On the other
hand if its identity is not determined, but its availability
and/or toxicity is determined to be of no significance, then

its identity becomes academic.

Question three can be important if it is known through hydroponic or sterile rooting medium studies that absorbed radiolabeled pesticide or its plant metabolites are not translocated into edible portions of the plant. Radiolabeled products entering into edible portions of plants grown in pesticide treated nonsterile soil, therefore must come from soil degradation products of the pesticide. Knowledge of the source and the possible soil degradation products available to the plant can provide a clue as to what radiolabeled products might be present in the plant and what extraction procedures are necessary to isolate and characterize them.

Concern for the identity or nature of the bound residue centers around the question of whether or not the bound residue consists of intact pesticide or 1st or 2nd generation degradation products which are adsorbed, incorporated, or entrapped in the plant, animal, or soil matrix and may be released at some future date, or whether it is a common ordinary metabolite which has reached a metabolic state where it can be reincorporated into normal organic building materials. In soil, concern has also been expressed for the long range effects of "polychlorinated or polytrifluoromethylated" soil organic matter on vital soil processes and conditions.

Considerable information exists regarding adsorption mechanisms and sites for pesticides in plants, animals, and soil clay and organic matter. Discussion of this information and its significance will be presented by several of the speakers. There is also a growing body of information which indicates that a number of aliphatic and simple aromatic pesticides are extensively metabolized in one or more systems and subsequently reincorporated as natural products. In the past we have been optimistic in believing that all pesticides would or should be completely mineralized to CO_2, H_2O, N_2, etc., after their function has been fulfilled. In these times of energy crises and shortages we should take comfort in knowing that nature does not wantonly discriminate in the utilization of a simple organic acid, aminoacid, or sugar molecule orginally derived from a pesticide molecule over one derived from its own synthetic efforts. There are a number of pesticides for which complete metabolic pathways are known from the parent pesticide molecule to simple organic acids. For example, the herbicide dalapon is metabolized to pyruvate and alanine (5,6); TCA is metabolized to serine (7); the 2,4-dichlorophenolic portion of 2,4-D goes through a long series of reactions ultimately yielding succinic acid (8). While it is true that most of these metabolic pathways have been worked out in isolated systems free of many competitive adsorptive or metabolic reactions, it is reasonable to expect that at least a small percent of the parent pesticide molecule will be metabolized through to such naturally occurring products in the environment.

We are always relieved in soil metabolism investigations to see large quantities of $^{14}CO_2$ evolved from pesticide treatments. This may be an unreasonable expectation, however, when following metabolism of ^{14}C-aromatic moieties in soil. The formation of humic substances in soil is a dynamic process involving the action of soil microorganisms on plant materials and other organic residues. Macromolecules are formed at the expense of carbohydrates of plant origin. These include bacterial gums, alginic and pestic acids, and other less well-defined polymeric carboxylic acids. Aromatic polyphenols formed by way of quinone oxidation can condense with amino acids to ultimately give humic-like substances. Basidiomycetes as well as other microscopic fungi have been found to degrade lignin (polyphenol) and form appreciable amounts of humic-acid like polymers. Phenolic units from ^{14}C-labeled aromatic compounds have been incorporated into fungus synthesized polymers.

It is generally believed that there is a generic relation between the various humic substances. Fulvic acid is believed to consist of polycondensation material from simpler molecules. Continuation of the polymerization process and chemical modification leads to the less soluble humic acid and eventually to insoluble humin which has the highest molecular weight and structure most resistant to degradation. When radiolabeled pesticide degradation occurs in soil we become concerned when large amounts of the radiolabel remain adsorbed or bound in the soil for long periods of time. The concern for the release of intact pesticide or pesticide metabolites and contamination of subsequent crops is real and justifiable. It is just as real and justifiable, however, to expect that those or other residues will remain in soil for long periods of time in an inocuous manner, or be released so slowly as to be insignificant in comparison to other products produced in the soil system.

Sorenson (9) studied the degradation of ^{14}C-labeled glucose and cellulose in three soils. After a rapid initial breakdown, half-lives of 5 to 9 years were reported for the remaining ^{14}C to be degraded. Other investigators have demonstrated the formation of humin from readily decomposable organic compounds (10,11). These data imply that even readily metabolized compounds are incorporated into humic substances and the extent to which degradation occurs is limited. Although Fuhr (12) has shown that ^{14}C-lignin and -humic acid are not taken up by plants, the significance of plant uptake of pesticides or pesticide products incorporated into soil organic matter is not fully understood. In our own research with chloroaniline residues we have isolated and tentatively identified large polymeric type structures from both soil and isolated microbial cultures. These materials appear analogous in structure to those products of 4-chloroaniline identified by Holland and Saunders (13). The rate of decomposition and release of anilines from such products is not known. The question now arises, given the structure of such a molecule and recognizing that it is made completely of pesticide derived aromatic moieties, "What significance must be attached to this residue?" Is there a general "rule of thumb" which could be put forth indicating that molecules over a given size need not be considered significant in terms of plant uptake?

Summary

Bound and conjugated pesticide residues occur in virtually all biological systems. As indicated at intervals throughout this discussion there are many questions to be asked and perhaps definitions to be cast. Are there also some general protocols which we can recommend regarding isolation, characterization, and evaluating the significance of such residues? Our AIBS-ECTG put forth such recommendations for soil bound residues. Certainly, the recommendations and protocols may change with time as new information and concepts are developed. Can analogous type recommendations be made for plant and animal bound residues? What considerations should be given to conjugates of pesticides and pesticide metabolites? How far must we go in characterizing these various types of residues? Are there extraction schemes which we can outline that will suitably characterize a bound residue from plants, animals or soils and know that in all good conscience this is a reasonable end to that particular requirement, beyond which a more definitive answer is purely academic? Is it reasonable to assume that if a bound residue in plants readily passes through a mono- or polygastric animal, and enters the soil where it is only degraded at an infitesimal rate per year, that this residue can be considered insignificant.

For example, has our knowledge of bound residues progressed to the point that we can recommend a general procedure (or procedures) for the extraction of bound residues from plants, animals and soils (presumably a different procedure is needed for each of the three systems)? The further characterization of the extracts produced by such procedures would be dependent upon the percent of the parent product label present in any specific fraction. In the absence of precise chemical identification, their processing through other biological systems, i.e., soil, animals, etc. would provide an indication of the biological availability of such material. Significance determinations of the availability of the bound residues could be based on several factors:

1. Actual percent of bound material released;
2. Side effects upon the consuming system;
3. Toxicological properties of parent pesticide or most toxic metabolites;
4. Etc.

The advantage of such a system is that it provides every-one with a similar format with which to assess the environmental fate and behavior of their compound and the significance of its bound residues. The disadvantage of such a system is that once established it can become an endpoint beyond which no one, or at least very few people are willing to progress.

The format of this conference was designed with the intent that each of these questions could be faced, discussed, and if adequate information appears to be available, then to resolve the level of need for any further consideration of these questions. In the absence of satisfactory answers it is intended that new and more productive avenues of research will be envisioned which will ultimately enable suitable resolution of this perplexing problem of bound and conjugated pesticide residues.

Literature Cited

1. Williams, R. T., "Detoxication Mechanisms", (1959) 2nd ed., John Wiley & Sons, Inc., New York.
2. Sherwin, C. P., Physiol. Reviews (1922), 2, 264.
3. Berczeller, L., Biochem. Z. (1917), 84, 75.
4. Quick, A. J., J. Biol. Chem. (1932), 97, 403.
5. Kearney, P. C., Kaufman, D. D., and Beall, M. L., Biochem. Biophys. Res. Comm. (1964), 14, 29.
6. Beall, M. L., M. Sc. Thesis. (1964). Univ. Maryland, College Park, Md.
7. Kearney, P. C., Kaufman, D. D., Von Endt, D. W., and Guardia F. S. J. Agr. Food Chem. (1969), 17, 581-584.
8. Kaufman, D. D., In W. D. Guenzi (ed.) "Pesticides in Soil and Water", (1974), Soil Sci. Soc. Am., Madison, Wis. p.164.
9. Sorensen, L. H., Soil Biol. Biochem. (1972), 4, 245.
10. Chekalar, K. I., and Illyuvieva, V. P., Pochvovedenie (1962), 5, 40.
11. Sinha, M. K., Plant and Soil (1972), 36, 283.
12. Fuhr, F., Zeits. Pflanzenernachr. Bodenk. (1969), 121, 43.
13. Holland, V. R., and Saunders, B.C., Tetrahedron (1968), 24, 585.

Biological Activity of Pesticide Conjugates

H. W. DOROUGH

Department of Entomology, University of Kentucky, Lexington, Ky. 40506

Conjugation, by evolutionary design, is a metabolic process whereby endogenous as well as exogenous chemicals are converted to polar components for the purpose of facilitating their removal from the site(s) of continuing metabolic processes. In animals, this is accomplished by the excretion of the more polar metabolites from the body. Elimination may be via the urine and/ or the bile. For example, glucuronide conjugates are excreted mainly in the urine if their molecular weight is approximately 300 or less (1). Those having a molecular weight of 700 or over are excreted largely in the bile, while those of intermediate size are eliminated by both routes. In plants, some elimination of the conjugates may occur but most are simply stored as terminal metabolites in the tissue. Most pesticides, like other exogenous compounds, are subject to a variety of conjugative reaction by living organisms. It is the intent of the current discussion to consider conjugates of these toxicants from the standpoint of their biological activity and potential significance to man and other life.

Conjugated metabolites of pesticides are but one of the many types of residues which may result from those chemicals used to control various pests. True, they are not among those metabolites such as DDE, paraoxon, etc. which are immediately recognized by all pesticide chemists and most laymen, but, nevertheless, they are pesticide residues and must be treated as such. This is stated simply to infer that the significance of pesticide conjugates should not be over estimated, under estimated, or ignored. But, rather, they must be evaluated in much the same fashion as any other pesticide residues and judgements of their significance based on sound scientific data.

While this view appears highly virtuous and is congruent in respect to one's scientific inclinations, pesticide conjugates possess some rather unique characteristics which severely challenge its practicality. Of uppermost importance is the fact that conjugation increases the polarity of the pesticide or its metabolite. Consequently, the resulting metabolites become very much akin to each other and to a multitude of endogenous chemicals found in all biological systems. Thus, the problems involved in isolation, purification and identification of conjugates are magnified tremendously as compared to apolar metabolites, and indeed have proven to be insurmountable in far too many cases.

Much could be said to document the adverse effects increasing polarity has on attempts to identify and subsequently determine the biological activity of pesticide metabolites, especially those formed by conjugation. However, it is necessary only to peruse past progress in this regard to see the real importance of metabolite polarity. With few exceptions, one will find that the ease and rapidity with which metabolites of a particular pesticide have been identified and toxicologically evaluated is inversely proportional to the polarity of the individual metabolites under consideration.

Concomitant with the water soluble nature of pesticide conjugates is a general lack of knowledge relative to the synthesis of these compounds. Even in those instances where the pesticide-contributing moiety has been identified and the endogenous chemical fairly well defined, there has been little success in chemically synthesizing the suspect intact conjugate. As a result, sufficient quantities of the metabolites are not available for determining their biological activity as is routinely done with non-conjugate metabolites of pesticides. While there is no doubt that the chemical synthesis of many pesticide conjugates is extremely difficult, the primary reason that synthetic conjugates are not commonly available is that few concentrated efforts have been placed on their preparation. This will likely remain the case until such time biological activity data relative to pesticide conjugates are included in the numerous requirements for commercial utilization of pesticidal chemicals. There is ample evidence to suggest that pesticide conjugate synthesis is feasible (2-4) and that pesticide conjugates can, and will, be prepared if ever deemed essential for evaluating the safety of pesticides to man and the environment.

Having attempted to place pesticide conjugates in their proper prospective as pesticide residues with certain properties quite unlike the non-conjugate materials, it is now desirous to address generally the significance of pesticide metabolites whether they be free or conjugated.

Metabolite Significance.

The term "metabolite" is used here to denote any derivative of the parent pesticide molecule formed subsequent to its preparation. This includes products formed spontaneously during storage, and those formed chemically and biochemically after their application. It would exclude impurities remaining in the product following synthesis and normal purification procedures.

Generally, a metabolite should be judged significant or

potentially significant unless adequate information is available
which reasonably assures that estimated maximum levels of
exposure (a) to man will in no way adversely effect his well-
being or that of future generations, and (b) to other animal
species will not endanger the survival or integrity of the exposed species.

The reason for considering man separate from other animal
species is simple. With man, there can be no compromise, but
with other animal species some harm may have to be tolerated for
the benefit of man. However, the degree of harm must be minimal
and of short duration.

For a pesticide metabolite to become significant in animals,
it must be available. While the manner of exposure to animals
may have a bearing on the ultimate significance of the metabolite,
the most important thing is exposure, per se. The basic sources
of exposure are shown in Table I.

TABLE I
Common Sources of Pesticide Metabolites.

External	Internal
1. Food	1. Enzymatic Formation
2. Environmental Contaminates	2. Non-Enzymatic Formation

Of the external sources, the food must be considered as a
major means by which man is exposed to pesticide metabolites,
especially the conjugates. Successful production of most crops
require pesticide treatments of some type, and it is generally
in this way that metabolites find their way into the diet.

"Environmental contaminates" is a catch-all phrase to cover
all external sources of pesticide metabolites other than food.
Pesticide metabolites in the air, on dust particles, and on
various surfaces are examples of this type of exposure. This
usually is not a major source of conjugate residues. Internal
sources of metabolites refer to those generated in the animal
body. They may be formed in a number of ways, chemically and
biochemically, in both man and in animals which constitute a
portion of his diet.

In the broadest sense, there are 3 different categories of
pesticide metabolites, free, conjugated and bound. The free
metabolites are those which are derived from the parent molecule
and have not reacted further with natural components of a
biological system. With apolar pesticides, the free metabolites
are usually considered as those extractable from the substrate
and which partitioned from water into an organic solvent such as
chloroform or ether.

Conjugated metabolites are derivatives of the pesticide which
have reacted chemically with a natural component of the organism

to form a new material. Generally, this involves a free metabolite, usually hydroxylated, conjugated as a glucuronide, glucoside, sulfate, etc. These metabolites are usually extractable
from the substrate with polar solvents but do not partition from
water into solvents such as chloroform or ether.

Bound or unextracted metabolites also are conjugates but
are derivatives of the pesticide which can not be removed from
the substrate by thorough extraction. Little is known of their
chemical nature, but it is suspected that they represent derivatives of the pesticide which have reacted with components of
the organism such as proteins, cell membrane and/or various
other cellular inclusions.

Assuming that the metabolic pathway of a pesticide has been
completely defined, how does one go about determining the significance of these metabolites? Ask a toxicologist and he will
stress the need for acute and chronic toxicity studies. An
ecologist will emphasize the need for determining their impact
on the environment, while an enzymologist would question their
effect on animal enzyme systems. This type of response can go on
indefinitely. Actually, it is very difficult to determine just
what information should be obtained for evaluating the significance of a metabolite.

Asking the right question is just part of the problem.
There remains the task of obtaining appropriate information to
answer the question. All experimental approaches and procedures
are not the same and, thus, some must be better than others.
The best approach should be determined to the best of our
ability before engaging in research designed to evaluate metabolite significance.

To gather data is one thing; to properly utilize it is
another. Data collected in studies designed to determine
metabolite significance will be meaningful only if correctly
interpreted. It is imperative that we know before conducting
the experiments that the results are subject to interpretation
and that the qualifications of the interpretors are as good as
technology will allow.

In summary, then, there are 3 basic questions to be asked,
and answered, before initiating research in the area of
metabolite significance: 1. What do we need to know? 2. How do
we go about obtaining this information? 3. How can this
information best be utilized?

It is readily apparent that an interdisciplinary approach is
essential if these questions are to be answered satisfactorily.
Some insight into the scientific diversity required in determining
metabolite significance may be gained by considering just a few
areas of concern (Table II) which must be considered with every
metabolite which is encountered.

Potential adverse effects of pesticide metabolites are not
confined to this list. Nor should any list made at any time,
even by the most qualified scientists, be considered as final.

Like the requirements for determining metabolite significance,
the factors to be considered will continue to change as our
knowledge increases.

TABLE II
Some Possible Detrimental Expressions of Biologically
Active Pesticide Metabolites

Acute Toxicity	Teratogenesis
Carcinogenesis	Reduced Fecundity
Mutagenesis	Altered Behavior

        Regardless of the area of concern selected for evaluation,
there is certain information about any metabolite which would
aid in that evaluation. Among the more basic requirements are:
1. Levels attainable in the body. 2. Fate in the animal. 3.
Site(s) of concentration and/or storage of the metabolites and
its derivatives.
        By knowing the maximum levels likely to be encountered by
an animal, the duration of these levels, and the sites of concentrations of the metabolite or its derivatives, an expert in
any area of concern could better determine the potential
significance of the chemical. More important, this information
would aid in designing research to more clearly define the
significance of a metabolite.
        Those who have worked in the area of pesticide metabolism
readily acknowledge the difficult task involved in evaluating
the significance of pesticide metabolites. It is known, for
example, that the majority of the terminal residues of many
pesticides exist as conjugates and bound metabolites. Almost no
direct information exists which might be used to determine their
significance in animals. Therefore, their potential significance
must be estimated using indirect evidence until pesticide
conjugates are identified and evaluated individually.
        While not all pesticide metabolites which possess biological
activity are necessarily significant, none is significant
without possessing some type of biological activity. As a
starting point, then, the potential significance of pesticide
conjugates can be estimated to some degree by gaining an insight into the biological activity of any conjugate, pesticide
or non-pesticide. The critical point initially is not the type
of activity, but only if conjugates might be expected to be
biologically active. Naturally, activity falling within those
areas of concern mentioned earlier (toxicity, carcinogenesis,
etc.) would be of special interest, particularly if associated
with pesticide conjugates.
        If biological activity of conjugates is indicated, the next
step would be to consider the aforementioned information basic

to evaluating the significance of metabolites, i.e., levels attainable in the body, fate in animals, and storage in the body. Currently, the only information available on pesticide conjugates, and this is meager, relates to the fate of the metabolites in the animal. The potential for accumulation and storage, however, may be estimated from the metabolism data.

Finally, one must consider the possible use of simple screening techniques for determining the chronic effects of pesticide conjugates on biological systems. Such assays are used extensively in estimating the potential carcinogenic, mutagenic and teratogenic characteristics of drugs, chemicals, and various environmental pollutants, and may hold promise for similar evaluations of the pesticide conjugates.

The following discussion is predicated on the approach outlined above. No attempt was made to cover all known conjugates, or even all pesticide conjugates. Rather, data were selected that demonstrated the points being made and which would serve as a nucleus for continued discussions of the biological activity and significance of pesticide conjugates.

Conjugate Nomenclature.

A discussion of pesticide conjugates is made exceedingly awkward because of the absence of a simple, consistent system of nomenclature. The terms aglycone and glycone apply only to the glycoside conjugates where the former denotes the non-sugar moiety and the latter the sugar moiety of the conjugate. Pesticides are conjugated with a number of different endogenous materials other than sugar, and often times their identity is not known. In these cases, there is no simple terminology which readily differentiates the exogenous moiety from the endogenous portions of the conjugate.

In this paper, a simple self-explanatory system of nomenclature applicable to all conjugates formed from the reaction of an exogenous compound with an endogenous compound is used.

Definitions:

Exocon - That portion of a conjugate derived from an exogenous compound. Used to denote this portion when existing as a precusor to conjugation, a part of the conjugate complex, or after cleavage of the conjugate linkage.

Endocon - That portion of a conjugate derived from an endogenous compound.

This system is particularly useful when one of the components of the conjugate is unknown. From this standpoint, it is very appropriate that the terms exocon and endocon be used in a paper dealing with pesticide conjugates.

Another area which needs clarification terminology is the differentiation of unbound and bound conjugates. Usually, the term conjugate is used to define those exocon-endocon complexes which can be extracted from the biological substrates. This, then, indicates that those not extracted, the bound residues, are not conjugates. Of course, these residues are conjugates and should be designated as such.

From the biological activity standpoint, the important thing is the availability of conjugates to living organisms, particularly to animals and those plants consumed in the animal diet. The extraction characteristics of the residues really have little meaning unless related to their bioavailability. With this in mind, pesticide conjugates should be categorized as follows:

Bioavailable Conjugates - If from animals and plants, those pesticide conjugates which, when administered orally to animals, are absorbed from the gastrointestinal tract. If from soils, those conjugates which are taken up by plants and/or soil-inhabiting animals.

Bound, or Bio-unavailable, Conjugates - If from animals and plants, those pesticide conjugates which, when administered orally to animals, are not absorbed from the gastrointestinal tract and are excreted in the feces. If from soils, those conjugates which are not taken up by plants and/or soil-inhabiting animals.

With the animal- and plant-derived pesticide conjugates, it is rather simple to determine bioavailability as defined above. For example, combined extractable and unextractable ^{14}C-conjugates could be given orally to rats and the urine and feces radioassayed. That material excreted via the urine would be bioavailable; usually, that eliminated in the feces would be unavailable and classified as bound. That the fecal ^{14}C-residues indeed were not absorbed from the gut could be confirmed by canulating the bile duct and radioassaying the bile.

The availability of pesticide conjugates in soils could be determined by growing various plants in soils that had the free metabolites removed. Availability to soil-inhabiting animals could be evaluated by assaying earthworms, insects, etc. at designated times after being placed in the conjugate-containing soil. Certain criteria, such as test species, exposure times, etc., would have to be established and standardized, but these should be worked out easily by pesticide chemists and biologists who have experience in pesticide uptake studies.

It is important to note that the bioavailability studies do not necessarily have to be qualitative in nature. Once conjugation has been established by conventional means, it is essential only to quantitate the fate of the residues in plants and animals. Naturally, the bioavailable conjugates should be identified and/or their toxicological significance determined.

Conjugation and Biological Activity.

It is quite likely that the vast majority of conjugate reactions taking place in living organisms has little direct influence on the biological activity of the exocon involved. The
reason for this is two-fold. First, the precusors themselves may
be void of any biological activity and the process of conjugation
serves only to enhance elimination. Secondly, an exogenous
compound, even if biologically active in its initial form, is
subject to a number of biochemical reactions (hydrolysis, oxidation, etc.) which may yield an inactive exocon prior to conjugation. Again, the role of conjugation would be related to
elimination rather than directly to deactivation. This type of an
effect on the biological activity of toxic exogenous compounds
may be referred to as "indirect deactivation".

"Direct deactivation" by conjugation may be defined as the
direct reaction of an active exocon with an endocon to yield an
inactive derivative. With pesticides, this usually involves
conjugation of a toxic free metabolite formed by hydroxylation of
the parent molecule. Certain pesticides, drugs, etc. containing
nitrogen may react directly with endogenous chemicals to form N-
conjugates. In any event, the immediate exogenous precursor,
the exocon, must be active and the resulting conjugate inactive
if the reaction is to be considered a direct deactivation.

If the indirect deactivation reactions were combined with
those involving direct deactivation, then there is no question
but that conjugation may be classified generally as a detoxication mechanism. This is an important concept in considering
the biological activity of pesticide conjugates. Since so
little is actually known concerning the latter, it is somewhat
comforting to know that the possibility of their being biologically active, or toxicologically significant, is not very great.
This line of thinking is probably the major factor contributing
to our current lack of knowledge about the pesticide conjugates.
However, it is not without substantial validity and should be
kept in mind in making any predictions relative to the potential
significance of conjugate metabolites.

Contrary to the situation with pesticides, conjugative
deactivation of drugs has been demonstrated for a number of
compounds. As early as the 1880's, it was reported that the
hypnotic activity of chloral hydrate was lost when converted to
trichloroethyl glucuronide (5,6). Only when the glucuronide
was administered at very high doses was there any sign of
activity and this probably resulted from the free alcohol formed
upon hydrolysis of the compound in the gut. Over the years, many
other conjugate metabolites of drugs have been shown to be void
of therapeutic activity. These include drugs exhibiting anti-
bacterial, hypnotic, and various other types of biological
activity (7-11). Recently, it has been suggested that the
antihypotensive agent dopamine (3,4-dihydroxyphenethylamine)

2. DOROUGH *Biological Activity of Pesticide Conjugates* 19

generated from L-dopa [3-(3,4-dihydroxyphenyl)-L-alanine] in dogs
was inactivated by conjugation (12). While the endocon was not
characterized, strong acid hydrolysis of the plasma released
high levels of dopamine. Dopamine infusions confirmed that such
levels would produce excessive cardiac stimulation or hyperten-
sion if present in the free form.

Some of the conjugates just discussed represent indirect
deactivation while others are clearly direct deactivations. Unless the exocon is isolated, identified and assayed, it is
impossible to differentiate between indirect and direct deactivation reactions. To establish that a conjugate is inactive is
very meaningful, but this information does not indicate whether
cleavage of the material would yield an active or inactive
exocon. The importance of knowing the type of conjugative
deactivation should become obvious in the following two sections
where the toxicities of specific conjugates and their exocons
are considered separately.

Indirect Deactivation. Examples where conjugation has no
direct influence on the biological activity of exogenous compounds
are shown in Fig. 1. Meprobamate is a tranquilizer with an
LD_{50} to mice of 800 mg/kg. In animals, the drug is metabolized
to hydroxymeprobamate (Fig. 1), a product virtually non-toxic to
mice (13). Once this action has occurred, the toxicological
consequences of glucuronide conjugation are nil. The meprobamate
glucuronide is non-toxic and is devoid of pharmacological
activity (14).

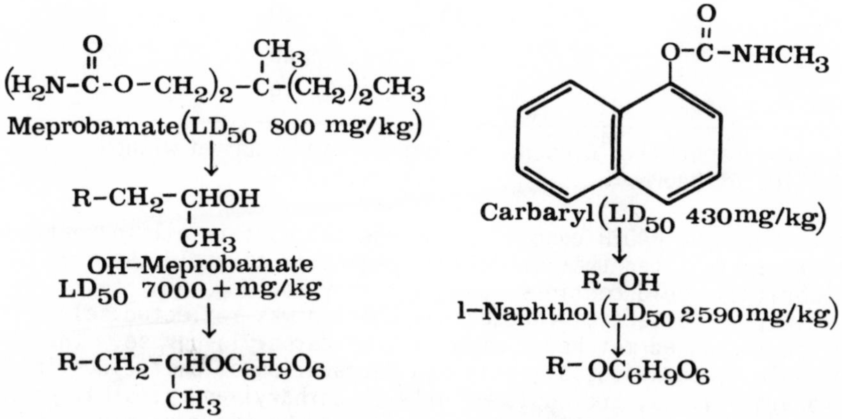

Figure 1. *Indirect conjugative deactivation of the drug meprobamate (13) and the insecticide carbaryl (15) by glucuronidation*

Carbaryl, an insecticide, has an LD_{50} to rats of 430 mg/kg. One of the major metabolites of carbaryl formed in most animal systems is 1-naphthol (15). This compound is only one-sixth as toxic to rats as carbaryl and further detoxication resulting from conjugation would not be of great importance.

Even though conjugation is not the initial step in the deactivation of some exogenous compounds, its role in the overall detoxication process of such compounds should not be minimized. Compounds like hydroxymeprobamate and 1-naphthol may be relatively non-toxic, but without efficient conjugation mechanisms the excretion rates would not be sufficient to prevent their accumulation to toxic levels.

A lack of an efficient conjugative system may also result in the accumulation of toxic exogenous materials which otherwise would be rapidly degraded. Carbaryl, for example, is metabolized extremely efficiently by rat liver microsomes fortified with 2 micromoles of NADPH and 5 micromoles of UDPGA (16). Under these conditions almost 60% of the carbamate was converted to glucuronide conjugates (Table III). As glucuronidation was supressed by limiting the quantity of UDPGA, the conjugates were reduced as expected. The unexpected results occurred with the parent compound where its total metabolism was reduced in proportion to the reduction in conjugation.

TABLE III

Effect of Reduced Glucuronide Conjugation on Total Carbaryl Metabolism by Rat Liver Microsomes + NADPH (16).

Umoles	% Distribution of Metabolites		
UDPGA added	Carbaryl	Free	Glucuronides
5	26	17	57
3	32	20	48
0.3	55	20	25
0	71	19	10

Carbaryl-naphthyl-^{14}C used. No metabolism occurred without addition of NADPH.

Since the NADPH concentration was the same in all incubations it was anticipated that oxidative metabolism to yield the free metabolites would continue, and that they would accumulate. However, the effect of reduced conjugation was reflected solely in the increased metabolic stability of carbaryl, per se. The possible consequences of decreased degradation of pesticides in vivo are obvious, and could occur with carbaryl and possibly other toxicants if conjugation was reduced by the monoamine oxidase inhibitors or other drugs (17).

Direct Deactivation and Reactivation. Although direct deactivation by conjugation occurs less frequently than indirect, its toxicological significance is much greater. As mentioned earlier, the exocon is an active product and, if not conjugated, may not be deactivated and/or eliminated. Moreover, the exocon may be released if the conjugate is consumed by animals and thereby be in a form to express its biological activity. From this standpoint, direct deactivation of toxicants by conjugation, particularly in plants, could provide a source of biologically active materials, the exocons, which otherwise would not be available. There is also a very real possibility that the conjugate would provide protection of the toxic exocon against metabolic degradation and/or facilitate transportation of the exocon to its site of action.

While there are numerous examples of direct deactivation by conjugation, the data in Table IV clearly illustrates the potential toxicological significance of this phenomenon. Cyclohexylamine(1) and the 4- and 5-hydroxy derivatives of carbaryl (4) are toxic exocons of glycoside conjugation (Fig. 2).

TABLE IV
Toxicity of Exocon and Conjugate to Mice When
Administered IP

Compounds	24 Hr LD_{50}, mg/kg
Cyclohexylamine[a]	100
Glucuronide form	600
4-Hydroxycarbaryl[b]	55
Glucoside form	1550
5-Hydroxycarbaryl[b]	50
Glucoside form	950

a (1), b (4)

The cyclohexylamine has an LD_{50} of 600 mg/kg when conjugated as a glucuronide. A more pronounced deactivation is noted with the carbaryl metabolites. In the free form the LD_{50} to mice is approximately 50 mg/kg; their glucoside conjugates have LD_{50} values of 1550 and 950 mg/kg for the 4- and 5-hydroxy carbaryl derivatives, respectively.

Obviously, the conjugation of cyclohexylamine and the hydroxycarbaryl compounds is an effective detoxication mechanism. With these particular toxicants, there are no indications that the conjugate forms protected the exocons from metabolic degradation or served as a "carrier" of the exocons

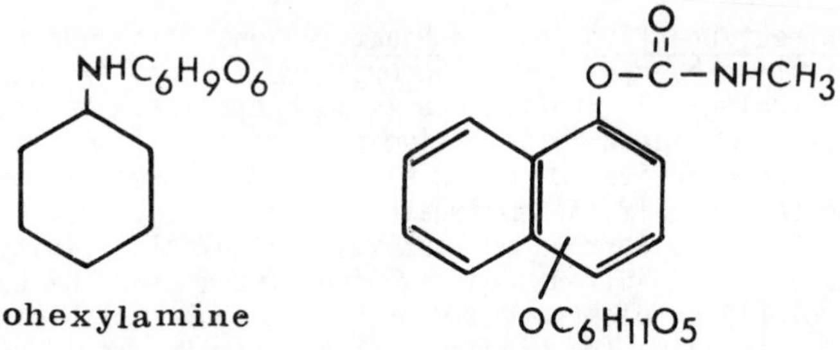

Cyclohexylamine
Glucuronide

4-and 5-Hydroxycarbaryl
Glucoside

Figure 2. Direct detoxication of exocons by glycoside conjugation
(see Table IV)

to the site of action.

Results of similar studies with various chemotherapeutic
agents do suggest that active exocons are released when the
conjugated drugs are administered (18-20). Generally, the conjugates are less active than the free compounds although the
estrogen, estriol, was no more active than its glucuronide when
both forms were administered orally (21). When injected, the
estriol retained its activity while the glucuronide was only
weakly active. These data suggest that the primary site of
conjugate cleavage to yield the exocon was in the gut. They also
demonstrate that the route of administration can affect the
biological activity of conjugates to a greater degree than that
of nonconjugated compounds.

Activation.

Although there are many conjugates of drugs, and of some
pesticides, which exhibit varying degrees of biological activity,
there is little evidence suggesting that the desired effects,
therapeutic or pesticidal, are dependent upon conjugative
activation. In fact, there is no indication of such an
occurrence relative to pesticides and their toxic action. The
activity reported for the conjugates usually is not as great as
for the parent compound, or liberated exocon, and the observed
activity is probably due, at least partially, to reactivation.
This is not too surprising with most pesticides since a relatively high lipid solubility is required for them to penetrate to
the site of action. It is difficult to imagine an inactive,
apolar chemical being converted to a potent nerve poison by a
process such as conjugation which increases its polarity.

Carcinogenicity. Acute toxicity is not always the most
obvious or most significant type of detrimental biological
activity exhibited by conjugates or other chemicals. Carcino-

genesis, mutagenesis and teratogenesis may be induced by certain chemicals and these dastardly consequences must be accepted as a possibility from exposure to pesticide conjugates. Thus far, pesticide conjugates have not been strongly implicated as causative agents of the above abnormalities. However, a brief discussion of the carcinogenic activity of certain non-pesticidal conjugates justifiably increases one's concern about the potential significance of pesticide conjugates.

Perhaps the most striking form of biological activity attributed to conjugate compounds is the ability of certain ones to induce cancer. Most of the studies relative to carcinogenic conjugates have centered around the aromatic amines, especially 2-acetylaminofluorene (AAF). While the parent compound is an active carcinogen, it has been established beyond doubt that metabolic activation is required for its carcinogenic properties to be expressed (22-24).

The first step in the biochemical activation of AAF (Fig. 3) is the formation of N-hydroxy-2-acetylaminofluorene (N-OH-AAF). Many studies have demonstrated that this metabolite is a greater carcinogen than AAF and that it, or a subsequent metabolite, binds more efficiently with hepatic protein and nucleic acids (22,23,25-27).

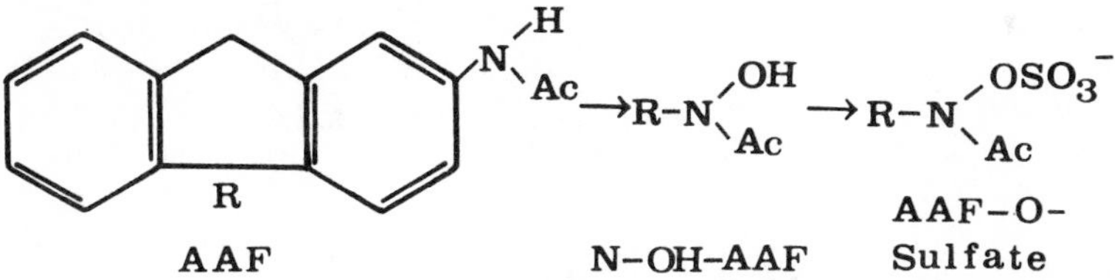

Figure 3.   Metabolic activation of the carcinogen 2-acetylaminofluorene. The sulfate form has been identified as the active carcinogen.

More recently, several investigators have shown that the real active carcinogen involved is the sulfate ester of N-OH-AAF (28-31). Glucuronide and phosphate esters are also formed from the N-OH-AAF (30,32), but the evidence overwhelmingly supports the sulfate ester as the active carcinogen.

There is evidence, however, that conjugates other than sulfates are potent carcinogens. 2-Naphthylamine, for example, induces bladder cancer when implanted therein (33). A more active carcinogen is produced when the compound is conjugated as an O-glucuronide. There are many other examples where conjugates are implicated as active carcinogens and this type of biological activity cannot be ignored when considering pesticide conjugates.

Chemotherapeutic Conjugates.

It is quite natural for one to think in negative terms when considering the possible biological activity of pesticide conjugates. Nevertheless, there is a substantial number of conjugates which exhibit very desirable pharmacologic action, and further demonstrate that certain conjugates are definitely biologically active.

Glycosides constitute the bulk of the chemotherapeutic conjugates. Most are naturally occurring and are derived from plants even for commercial purposes. Their source, chemistry and therapeutic uses have been previously covered in detail (34), and only a few examples will be presented here.

The cardioactive glycosides are commonly used drugs which upon hydrolysis yield one or more sugars and a host of rather complex alcohols. Digitoxin, a cardiotonic drug, is obtained from Digitalis sp. and is the 3-O-glycoside of digitoxigenin (Fig. 4). Other glycosides occur in the plant but digitoxin is the most active. The many other cardioactive glycosides (ouabain, lantoside C, etc.) are similar in structure to digitoxin but vary markedly in biological activity.

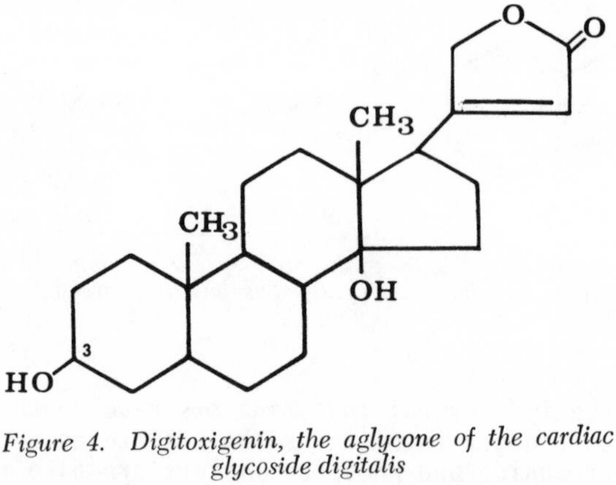

Figure 4. Digitoxigenin, the aglycone of the cardiac glycoside digitalis

In addition to the cardioactive glucosides, many other plant conjugates have therapeutic value. The plant source and chemical structures of the aglycones vary widely, as do the prescribed uses of these drugs. Their biological activity is such that they are used as cathartics, emetics, diuretics, vasoconstrictors, and as antirheumatic agents.

Although not of current commercial use, certain plasma protein-nitrogen mustard conjugates show promise as anti-tumor agents (35-36). As with most other conjugation, the toxicity of mustards to animals was reduced by protein conjugation but the

inhibitory potency of the chemical was not drastically altered.
This is demonstrated in Table V by the data obtained with
aniline mustard and its protein conjugates. It was hypothesized
that the proteins enveloped the cytotoxin groups and protected
them from hydrolytic degradation before they reached their
active site.

TABLE V
Toxicity and Tumor Inhibitory Potency of Aniline Mustard
and its Protein Conjugates in Mice (36).

Compound	Mg/kg		$LD_{50}/$
	LD_{50}	ID_{90}	LD_{90}
Aniline Mustard	112	1.4	80
Globulin Conjugate	254	1.9	134
Fibrinogen Conjugate	450	2.3	193

ID_{90} - 90% inhibitory dose.

Protein conjugation of the carcinogen, 2-anthrylamine,
served as an immunizing agent when administered to rats prior to
a single oral dose of the carcinogen (37). Up to 50% tumor
inhibition was achieved, leading the authors to conclude that
they had produced animals resistant to the action of 2-
anthrylamine in producing neoplasia.

The in vivo conjugation of some drugs may yield compounds
which contribute significantly to the intended therapeutic
action of the non-conjugated material administered. One such
case was noted in a study designed to explain why patients with
renal failure showed an increased sensitivity to the hypotensive
effect of methyldopa (38,39). The evidence obtained showed that
methyldopa-O-sulfate, formed metabolically, and normally eliminated rapidly, accumulated in the plasma of patients with impair-
ed elimination. It was this conjugate, the author believed,
that was acting in the same manner as methyldopa and, consequently,
gave an additive effect when the patients were again dosed with
the antihypertensive drug. Possibly, conjugates of other drugs
contribute to the biological activity of the free therapeutic
agent in the same way. However, it is not likely to be recognized unless some unique physiological responses occur which
alter the expected action of the drug.

Effects on Plants.

It is unlikely that pesticide conjugates other than those
formed from herbicides would have any appreciable effect on
plant growth. With the herbicides, however, conjugates could be
formed which retained herbicidal activity directly or which

could be cleaved to yield an active exocon. Evidence that such
conjugates are formed from 2,4-D has recently been reported (40).
    Using cultured callus tissues of soybean cotyledon, 2,4-D
was shown to be rapdily conjugated with a variety of amino acids
which, in turn, were synthesized and their biological properties
evaluated. Of the 20 amino acid conjugates of 2,4-D tested for
their ability to stimulate cell division and elongation, all
were active to some degree. Selected data from these studies are
shown in Table VI. In many cases, the growth stimulation of the
conjugates exceeded that produced by 2,4-D.

TABLE VI
Biologically Active 2,4-D Amino Acid Conjugates (40).

	Tissue Response	
Conjugates	Elongation, %[a]	Cell Division, Mg[b]
2,4-D, Free	39	224
Glutamic Acid	64	288
Phenylalanine	49	433
Arginine	71	274

[a] Elongation of Avena coleoptile sections at 10^{-5} M
concentration.

[b] Mg soybean callus tissue at 10^{-6} M concentration.

It was pointed out that 2,4-D could be formed metabolically
from the conjugates and, thus, might be the active component.
The point is, however, that the conjugates did elicit a growth
response and were active either directly or served as a mechanism
for obtaining greater concentrations of 2,4-D at the site of
action. The authors concluded that amino acid conjugation of
2,4-D could not always be considered as a detoxication mechanism.

Fate of Pesticide Conjugates in Animals.

    Only rarely is it possible to obtain pure pesticide conju-
gates in quantities suitable for thorough acute and chronic
toxicological evaluations. Moreover, the data currently
available do not demonstrate that such evaluations are essential
for establishing the safety of pesticidal chemicals. The fact
remains, however, that "proof of safety" of the pesticide conjugates has yet to be documented. This situation requires that we
continue to evaluate the significance of these metabolites using
whatever approaches likely to yield useful information.
    In our laboratory, we have taken an indirect approach in
attempting to establish parameters for estimating conjugate
significance. The studies have centered around the carbamate

insecticides and are designed to determine the fate of certain conjugate metabolites in mammals. Just as fate studies are useful in predicting the safety of parent pesticides, they should be of value in estimating the potential significance of their conjugates. Since plants would probably provide the major source of the conjugates to man, we have chosen to use plants to generate conjugate and bound residues from radioactive carbamates, and to use these in our fate studies with rats. Also, a limited number of radioactive conjugates of the carbamate insecticide carbaryl (1-naphthyl N-methylcarbamate) has been chemically synthesized and similarly studied in the rat. Some of the data obtained in these studies are presented below.

Carbaryl, when injected into bean plants, is metabolized sequentially into water soluble compounds, or conjugates, and then to unextractable residues or bound metabolites (41). After 20 days, there are sufficient amounts of ^{14}C-conjugates and ^{14}C-bound materials resulting from a carbaryl-naphthyl-^{14}C treatment to administer to rats (Fig. 5) (42). Most of these metabolites contain the intact carbamate ester linkage (41,43).

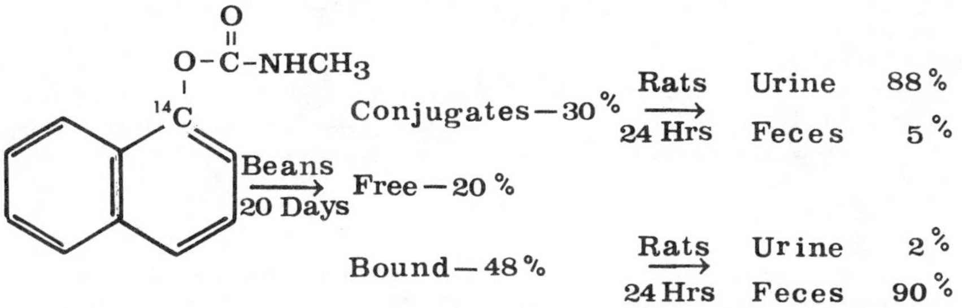

$$
\begin{array}{l}
\text{O} \\
\text{\textbardbl} \\
\text{O-C-NHCH}_3
\end{array}
$$

	Rats 24 Hrs	Urine	88 %
Conjugates—30 %		Feces	5 %
Free — 20 %			
Bound — 48 %	Rats 24 Hrs	Urine	2 %
		Feces	90 %

Beans
20 Days

^{14}C

Figure 5. Nature of carbaryl naphthyl-^{14}C in bean plants 20 days after injection and fate of conjugated and bound materials when administered orally to rats (42)

When the conjugate metabolites were given orally to rats, over 90% of the radiocarbon was excreted within 24 hours (Fig.5), mostly via the urine. Although the bound residues were excreted equally as efficiently, almost all of the elimination occurred in the feces. These data prove that the conjugate metabolites were absorbed from the gut and were available to the animal. The bound residues did not appear to be available to the animal since elimination was rapid and via the feces. Using these data alone, one would have to predict that the conjugate metabolites were potentially more significant than the bound ones.

The animal data obtained with the carbaryl plant metabolites are similar to those using several other carbamate insecticides, except ethiofencarb. The latter compound [(2-ethylthiomethyl)-phenyl N-methylcarbamate] is a Bayer product under development in the United States by Chemagro. Unlike most carbamates, this material is hydrolyzed within the plants and, even with the ^{14}C-ring-labeled compound, there is no appreciable buildup of the bound residues (Fig. 6). Hydrolysis of the ester to yield $^{14}CO_2$ was indicated by the fact that only 18% of the injected radio-carbon remained in plants 20 days after injection with ethiofen-carb-carbonyl-^{14}C. After the same period, 98% of the ring-^{14}C material could be recovered from the plants.

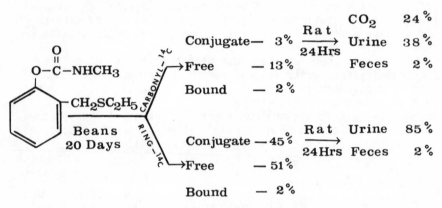

Figure 6.   Nature of ethiofencarb ^{14}C-metabolites in bean plants 20 days after injection and fate of conjugated materials when administered orally to rats (42)

The ^{14}C-carbonyl conjugates were metabolized by rats to respiratory $^{14}CO_2$ (24%) and to products eliminated in the urine (38%). Only 2% of the dose was eliminated in the feces. Fecal elimination of the ^{14}C-ring conjugate was approximately the same, but no $^{14}CO_2$ was produced and most of the dose (85%) was voided in the urine. The bound ^{14}C-materials were not administered to rats because of the excessive bulk needed to obtain the required radiocarbon. As for the usefulness of these data in estimating metabolite significance, two things are obvious. First, the terminal residues of ethiofencarb in bean plants are predomin-ately non-carbamate in nature which should lessen their potential significance. Second, the conjugative metabolism does not pro-duce the bound metabolites usually encountered in plants and, thus, need no evaluation. Add to this the fact that the conju-gates are almost all ethiofencarb hydrolysis products which are rapidly eliminated from the body, and the significance of the conjugates become even less apparent.

In addition to working with the total ^{14}C-conjugate and bound carbaryl metabolites formed in plants, the glucosides of certain free metabolites have been synthesized and fed to rats. The primary purpose of these studies was to test the metabolic stability of the exocon-O-glucoside linkage and to compare the

fate of conjugates in animals with that of the free metabolite.
1-Naphthyl-^{14}C glucoside was excreted intact, about 20% of
the dose, in the urine of rats treated orally (Table VII) (43).
The major metabolite was the glucuronide form (24%) while naphthyl
sulfate and 1-naphthol in the urine each constituted 10% of the
dose. Identical studies with the ^{14}C-glucose-labeled conjugate
confirmed that the naphthyl glucoside in the urine contained the
same sugar moiety as administered. It also showed that the glucuronide was not formed by the oxidation of 1-naphthyl glucoside
but resulted from the cleavage of the glucoside linkage to form
1-naphthol, which was then conjugated with endogenous glucuronic
acid. The free hydrolytic metabolite of carbaryl, 1-naphthol,
was excreted more rapidly than its glucoside, apparently because
the free hydroxyl group allowed direct glucuronidation.

TABLE VII
Fate of 1-Naphthol-^{14}C and its Glycoside in Rats (44).

| Compound | Percent of Dose in 0-24 Hour Urine as | | | |
	1-Naphthol	Naphthyl Glucoside	Naphthyl Glucuronide	Naphthyl Sulfate
1-Naphthol-^{14}C	1	0	73	15
Naphthyl-^{14}C Glucoside	10	19	24	10
Naphthyl Glucoside-^{14}C	0	16	.1	0

Unlike 1-naphthyl glucoside, the glucoside linkage of 4- and
5-hydroxycarbaryl glucoside was almost completely cleaved by
the rats (45). While 30% of the dose was excreted, only a small
amount, 2 to 3% of the dose, contained the intact glucoside bond;
over 90% of this was the parent conjugate while the remainder
was the 4- or 5-glucoside of 1-naphthol. Other radiocarbon eliminated by the rats appeared to lack the 1-naphthyl moiety. The
naphthyl-^{14}C-labeled forms of the 4- and 5-hydroxycarbaryl
glucosides have not been synthesized and, consequently, the fate
of the free and conjugated metabolites have not been compared.

Screening for Carcinogenic/Mutagenic Potential.

That certain conjugate compounds, formed in vivo, can induce
cancer in animals is an established fact. This does not mean
that the pesticide conjugates will exhibit carcinogenic and/or
mutagenic properties, but it does add considerably to one's concern about just what type of biological activity they do, or do
not possess. Because of the problems relating to the isolation,

identification, synthesis, etc., of pesticide conjugates, it is currently impossible to fully evaluate the carcinogenic and mutagenic potential of these metabolites. However, there are techniques available which have proven useful in estimating this potential for a variety of chemicals, including pesticides, that may be applicable to pesticide conjugates (46-49).

The most promising method for screening pesticide conjugates for mutagenic and carcinogenic activity appears to be one utilizing bacterial strains designed specifically for this purpose (48,49). Four strains of Samonella typhimurim were developed which could not synthesize histidine but were reverted back to the wild type by particular mutagens and carcinogenes. While the genetics and biochemistry of the system are quite complex, the assays are very simple. Basically, the procedure calls for counting the number of colonies which develop from the mutant strains in the presence of the test substance and sufficient histidine for a few cell divisions. If no mutation occurs, there is no colony formation because the mutant strains cannot biosynthesize the compound. However, if mutation occurs, the colonies thrive and their growth rate can be correlated to the potency of the mutagen.

Preliminary tests have been conducted in our laboratory using the screening technique with several pesticides (42). The experiments showed that the system was very sensitive to nitrosocarbaryl and captan (Table VIII). Both of these materials previously have been shown to be potent mutagens (47,50). No effect was observed with the other insecticides at the concentrations tested. Certain of these are shown in Table VIII.

TABLE VIII
Bacterial Mutagenesis of Pesticides and Related Compounds (42).

Compound	Minimum concentration, ug/plate, for	
	Mutagenesis	Growth Retardation
Heptachlor epoxide	100 +	2500 +
Diazinon	100 +	1000 +
Carbaryl	100 +	2500 +
Nitrosocarbaryl	1	200
Captan	2.5	50

Salmonella typhimurium strain TA 1535; method of Ames et al., 1973 (48). + indicates that these concentrations were inactive.

Some of those pesticides which did not show any activity are among those pesticides banned because of their carcinogenic properties. Thus, this test, like all others, will not provide all the answers. However, all means available must be used when

dealing with such a vital topic and future studies using the
bacterial carcinogenic/mutagenic screening test will include
isolated pesticide conjugates.

Significance of Pesticide Conjugates.

Sufficient numbers of conjugates have demonstrated varying
forms of biological activity to establish that the pesticide conjugates are potentially biologically active. It seems necessary,
therefore, to determine if, and which, conjugates are active and
the significance of this activity. Taking into account current
technical problems discussed earlier, one possible approach to
accomplishing this is outlined in Table IX.

TABLE IX
Sequential Approach to Determining the
Significance of Pesticide Conjugates

Factors to be Determined	Results & Priority		
1. Bioavailability --------------	a. Available	- high	→ 2
	b. Unavailable	- low	
2. Degradation/excretion rates --	a. Slow	- high	→ 3,4
	b. Rapid	- low	→ 3
3. Carcinogenic/mutagenic potential --------------------	a. Positive	- high	→ 4
	b. Negative	- low	
4. Ninty-day feeding studies ----	a. Effect	- high	→ 5
	b. No effect	- low	
5. Tests 1-4 --------------------	a. Positive	- high	→ 6
	b. Negative		
6. Identity and synthesis of active component(s) ---------	a. Successful	- high	→ 7
	b. Unsuccessful	- higher (halt production)	
7. Full toxicological significance ---------------	a. Hazardous (halt production)		
	b. Safe		

Those conjugates found to be bioavailable should be isolated,
their fate in animals determined, and assayed for carcinogenic/
mutagenic potential. A concentrated effort is needed immediately
to select the best screening method for the latter. If the bioavailable conjugates accumulate in the animal and/or show

carcinogenic/mutagenic potential, a 90-day feeding study of the crude ^{14}C-conjugate preparation should be performed. Levels fed should be the maximum which could occur in a normal diet, and the criteria used for determining detrimental effects the same as those used in similar studies of major free pesticide metabolites. In addition, total accumulation of ^{14}C-residues in the animals over the 90-day period should be determined.

After the feeding study has been completed, all data collected to that point should be reviewed. Those conjugates which accumulated extensively, showed strong carcinogenic potential and/or produced ill effects in the 90-day feeding study should be further evaluated. Components of the crude conjugate preparation should be isolated and identified. If this proved unsuccessful, the discontinued use, or development, of the parent pesticide would have to be seriously considered. The synthetic conjugates should be re-evaluated as before and those found potentially harmful subjected to a full-scale toxicological study. Results of these studies would determine the commercial fate of the parent compound.

It is not important whether one accepts or rejects the approach suggested above. The important thing is that this approach, and others, be considered, and that a well designed and well organized program be set in motion to determine the significance of pesticide conjugates.

Literature Cited.

1. Smith, R.L. and R.T. Williams, "Implication of the conjugation of drugs and other exogenous compounds" in Glucuronic Acid: Free and Combined, pp 457-491, G.J. Dutton (ed). Academic Press, New York (1966).
2. Steller, W.A. and W.W. Brand, J. Agr. Food Chem. (1974) 22, 445.
3. Ecke, G., J. Agr. Food Chem. (1973) 21, 792.
4. Cardona, R.A. and H.W. Dorough, J. Agr. Food Chem. (1973) 21, 1065.
5. Kulz, E., Arch. Ges. Physiol. (1882) 28, 506.
6. Kulz, E., Z. Biol. (1884) 20, 157.
7. McChesney, E.W., J. Pharmacol. Exptl. Therap. (1947) 98, 368.
8. Woods, L.A., J. Pharmacol. Exptl. Therap. (1954) 112, 158.
9. Koechlin, B.A., W. Kern and R. Engleberg, Antibiot. Med. Clin. Therapy (1959) 6, Suppl. 1, 22.
10. Keberle, H., K. Hoffmann and K. Bernhardt, Experientia (1962) 18, 105.
11. Glazko, A.J., L.M. Wolf, W.A. Dill and A.C. Bratton, J. Pharmacol. Exptl. Therap. (1949) 96, 445.
12. Tjandramaga, T.B., L.I. Goldberg and A.H. Anton, Proc. Soc. Exp. Biol. Med. (1973) 142, 424.
13. Ludwig, B.J., J.F. Douglas, L.S. Powell, M. Meyer and F.M. Berger, J. Med. Pharm. Chem. (1961) 3, 53.
14. Tsukamoto, H., H. Yoshimura and K. Tatsumi, Chem. Pharm.

Bull. (1963) 11, 1134.
15. Dorough, H.W., J. Agr. Food Chem. (1970) 18, 1015.
16. Mehendale, H.M. and H.W. Dorough, Pest.Phy. (1971) 1, 307.
17. Culver, D.J., T. Lin and H.W. Dorough, J. Econ. Entomol. (1970) 63, 1369.
18. Schuller, J., Z. Biol. (1911) 56, 274.
19. Dodgson, K.S., G.A. Garton, A.L. Stubbs and R.T. Williams, Biochem. J. (1948) 42, 357.
20. Wilder Smith, A.E. and P.C. Williams, Biochem. J. (1948) 42, 253.
21. Odell, A.D., D.I. Skill and F.G. Marrian, J. Pharmacol. Exptl. Therap. (1937) 60, 420.
22. Miller, E.C., J.A. Miller and H.A. Hartmann, Cancer Res. (1961) 21, 815.
23. Miller, E.C., J.A. Miller and M. Enomoto, Cancer Res., (1964) 24, 2018.
24. Irving, C.C. Conjugates of N-hydroxy Compounds In: W.H. Fishman (ed.). Metabolic Conjugation and Metabolic Hydrolysis, pp 53-119. New York: Academic Press, Inc. (1970).
25. Miller, J.A. and E.C. Miller, Progr. Exptl. Tumor Res., (1969) 11, 273.
26. Marroquin, F. and E. Farber, Cancer Res. (1965) 25, 1262.
27. Lotlikar, P.D., J.D. Scribner, J.A. Miller and E.C. Miller, Life Sci. (1966) 5, 1263.
28. DeBaun, J.R., J.Y. Rowley, E.C. Miller and J.A. Miller, Proc. Soc. Exp. Biol. Med. (1966) 129, 268.
29. DeBaun, J.R., E.C. Miller and J.A. Miller, Cancer Res. (1970) 30, 577.
30. Irving, C.C., Xenobiotica (1971) 1, 387.
31. Weisburger, J.H., R.S. Yamamoto, G.M. Williams, P.H. Matsushima and E.K. Weisburger, Cancer Res. (1972) 32, 491.
32. Lotlikar, P.D. and M.B. Wasserman, Biochem. J. (1970) 120, 661.
33. Allen, M.J., E. Boyland, C.E. Dukes, E.S. Horning and J.G. Watson, Brit. J. Cancer (1957) 11, 212.
34. Claus, E., V. Tyler and L. Brady, "Pharmacognosy" 6th ed., pp 79-130, Lea and Febiger, Philadelphia (1970).
35. Szekerke, M., R. Wade and M.E. Whisson, Neoplasma (1972) 19, 199.
36. Szekerke, M., R. Wade and M.E. Whisson, Neoplasma (1972) 19, 211.
37. Peck, R.M. and E.B. Peck, Cancer Res. (1971) 31, 1550.
38. Myhre, E., O. Stenback, E.K. Brodwall and T. Hansen, Scand. J. Clin. Lab. Invest. (1972) 29, 195.
39. Myhre, E., E.K. Brodwall, O. Stenback and T. Hansen, Scand. J. Clin. Lab. Invest. (1972) 29, 195.
40. Feung, C., R.O. Mumma and R.H. Hamilton, J. Agr. Food Chem. (1974) 22, 307.
41. Dorough, H.W. and O.G. Wiggins, J. Econ. Entomol. (1969) 62, 49.

42. Marshall, T.C. and H.W. Dorough, Unpublished data (1975).
43. Kuhr, R.J. and J.E. Casida, J. Agr. Food Chem. (1967) 15, 814.
44. Dorough, H.W., J.P. McManus, S.S. Kumar and R.A. Cardona, J. Agr. Food Chem. (1974) 22, 642.
45. Cardona, R.A. and H.W. Dorough, Unpublished data (1973).
46. Epstein, S.S. and H. Shafner, Nature (1968) 219, 385.
47. Legator, M.S., F.J. Kelly, S. Green and E.J. Oswald, Ann. N.Y. Acad. Sci. (1968) 160, 344.
48. Ames, B.N., F.D. Lee and W.E. Durston, Proc. Nat. Acad. Sci. (1973) 70, 782.
49. Ames, B.N., W.E. Durston, E. Yamasaki and F.D. Lee, Proc. Nat. Acad. Sci. (1973) 70, 2281.
50. Elespuru, R.K. and W. Lijinsky, Fd. Cosmet. Toxicol. (1973) 11, 807.

3

Pesticide Conjugates—Glycosides

D. S. FREAR

Metabolism and Radiation Research Laboratory, Agricultural Research Service,
U.S. Department of Agriculture, Fargo, N. Dak. 58102

Abstract

Selected examples of O-glucoside, N-glucoside, S-glucoside, glucose ester, acylated glucoside and gentiobioside metabolites of pesticides and plant growth regulators are discussed from the standpoint of the isolation, identification, metabolism and significance in plants and insects.

Introduction

In the past, pesticide metabolism studies have emphasized the isolation, identification, and toxicity of primary reaction products. Recently, however, increased interest has been focused on the nature and significance of a variety of conjugated pesticide metabolites, including a number of glycosides. Increasing evidence suggests that the rate and extent of glycoside formation is a significant factor in regulating the biological activity and the selectivity of pesticides and their toxic metabolites (Figure 1). However, relatively few studies have reported the isolation and identification of pesticide metabolites as glycosides. Many workers simply note that unknown polar metabolites are present. Other reports only show that unknown polar metabolites are hydrolyzed by acids, bases or various glycosidic enzymes. Also, little information is available concerning the distribution, specificity, activity and regulation of the enzyme systems responsible for the biosynthesis, hydrolysis and further metabolism of glycoside metabolites. The continuing search for more selective and less persistent pesticides, together with increased concern about the nature of "terminal" pesticide residues, suggest that

Mention of a trademark or proprietary product does not constitute a guarantee or warranty of the product by the U. S. Department of Agriculture and does not imply its approval to the exclusion of other products that may also be suitable.

additional basic studies of glycoside metabolism should provide
valuable and needed insights into the behavior and fate of
pesticides.

An examination of the natural products literature reveals
that a vast number and variety of glycosides are present in
nature. Comparative metabolism studies, however, have shown
that pesticides and their metabolites are normally conjugated as
β-glucuronides in vertebrates and as β-glucosides in plants,
insects and other invertebrates (1). Since the following paper
will discuss glucuronides, the present discussion will be
limited to glucosides and their metabolites in plants and
insects.

Several different types of simple and complex glucosides
have been isolated and identified including; a variety of
O-glucosides, several N-glucosides as well as a few S-glucosides,
glucose esters, acylated glucosides and gentiobiosides. However,
no attempt will be made to catalog or discuss all of the
pertinent references in the literature. Instead, selected
examples of each type of glucoside will be used to illustrate
some of the methodology that has been used and the problems that
have been encountered in the isolation, identification and
metabolism of this important class of conjugated pesticide
metabolites.

O-Glucosides

Many investigators have suggested that O-glucosides
represent a major class of conjugated pesticide metabolites in
plants and insects. A number of pesticides are substituted
phenols, and many pesticides are metabolized to phenols or
alcohols by oxidation or hydrolysis. It is generally assumed
that these phenols or alcohols are conjugated as O-glucosides in
plants and insects. However, in very few studies has a
metabolite actually been isolated and identified as a glucoside.
Tentative identification is generally provided by enzymic
hydrolysis and/or analysis of the aglycones after hydrolysis.
Frequently, the identification and analysis of the carbohydrate
moiety is omitted because of difficulties in removing naturally
occurring carbohydrate impurities.

Naturally occurring phenolic compounds exhibit a variety of
metabolic activities in plants and animals, and their role as
mediators of metabolism is well established (2-5). Phenolic
pesticides or pesticide metabolites also affect a number of
metabolic processes and biological functions. A possible means
for regulating cellular concentrations of biologically active
phenolic intermediates in pesticide metabolism is outlined in
Figure 2. Support for such a hypothesis has been provided by a
number of studies. Examples of such studies will be mentioned
briefly. A review of carbamate insecticide metabolism in plants
and insects by Kuhr (6) suggests that differences in the toxicity

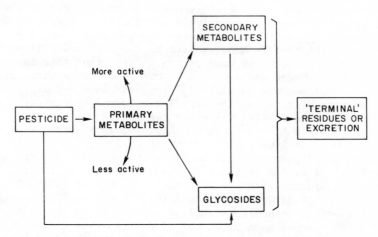

Figure 1.   *Role of glycosides in pesticide metabolism and the bioregulation of pesticide toxicity*

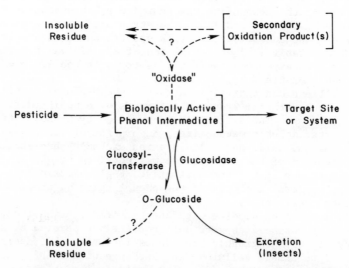

Figure 2.   *Proposed scheme for the bioregulation of phenol intermediates in the metabolism of pesticides*

of these chemicals to various insects may be due, in part, to
differences in the rate of conjugation of primary metabolites as
glucosides and other water-soluble conjugates. Studies by Bull
and Whitten (7) indicate that enzymic O-glucosylation of p-
methylsulfonyl phenol, a metabolite of O,O-dimethyl O-(4-methyl-
thiophenyl)phosphate, is more active in resistant than in
susceptible tobacco budworms. A recent review of several studies
by Still, Rusness and Mansager (8) suggests that O-glucosylation
of 2-hydroxychlorpropham (isopropyl-5-chloro-2-hydroxycarbanilate)
may provide an effective means for regulating the phytotoxicity
of this phenolic intermediate of chlorpropham (isopropyl-3-
chlorocarbanilate) metabolism in plants. Also, the direct
involvement of phenolic intermediates in the formation of
"terminal" pesticide residues has been suggested in recent plant
metabolism studies with the herbicide, cisanilide (cis-2,5-di-
methyl-1-pyrrolidinecarboxanilide) (9). These studies indicate
that a reactive phenolic metabolite of cisanilide is oxidized
further and serves as a precursor of an insoluble residue
fraction. Both plants and insects have active phenol oxidase
systems. The significance of these enzymes in the formation of
"terminal" pesticide residues should be determined.

The primary mechanism of O-glucosylation in plants (1,
10-11) and insects (1, 11-13) appears to involve UDPG as the
most effective glucosyl donor, a UDP-glucosyltransferase, and
various phenol and alcohol acceptor groups.

Examples of several O-glucoside metabolites isolated from
pesticide treated plants are shown in Figure 3. Metcalf et al.
(14) extracted the O-glucoside of 2,3-dihydro-2,2-dimethyl-3-
keto-7-hydroxybenzofuran from cotton leaves with aqueous ethanol.
Isolation of the metabolite was achieved by chromatography on
silicic acid columns and TLC. The structure of the O-glucoside
was determined unequivocally by IR spectra and mass spectra of
the TMS derivative. A molecular ion at m/e 628 was reported
together with expected fragment ions for both the glucose and
the phenol moieties.

Still and Mansager (15) extracted the O-glucoside of
2-hydroxychlorpropham from soybean roots by a modified Bligh-Dyer
procedure (16). The glucoside was purified by n-BuOH extraction
of the water soluble metabolites, adsorption of impurities on
basic aluminum oxide and cellulose ion exchange chromatography.
Identification of the O-glucoside was obtained by GLC-MS
analysis of the acetylated derivative, β-glucosidase hydrolysis,
and mass spectral analysis of the phenol aglycone and its
methylated derivative.

In metabolism studies with the herbicide, cisanilide, two
O-glucoside metabolites were not completely separated by a
variety of chromatographic procedures including TLC, adsorption
on Amberlite XAD-2, cellulose ion exchange with DE-52 and gel
filtration on Biogel P-2 (9). The analytical problems
associated with a mixture of O-glucoside metabolites and a

probable degradation of TMS or acetylated derivatives during
attempted GLC separation were circumvented by enzyme hydrolysis
of the isolated glucoside mixture, TLC separation of the
aglycones, IR, MS and FT-PMR analysis of aglycone structures,
and quantitative analysis of glucose with glucose oxidase. The
separation of the mixed O-glucoside metabolites by a more
effective and less destructive chromatographic technique, such
as HPLC, would permit a direct analysis of each glucoside.

N-Glucosides

Numerous pesticides are substituted anilines. Others may be
metabolized to anilines by oxidation, reduction or hydrolysis.
In plants, several herbicides are either directly or indirectly
metabolized to N-glucosides (17-24). Proposed pathways for the
formation of chloramben (3-amino-2,5-dichlorobenzoic acid),
dinoben (3-nitro-2,5-dichlorobenzoic acid), propanil (3',4'-
dichloropropionanilide) and pyrazon (5-amino-4-chloro-2-phenyl-
3(2H)-pyridazinone) N-glucoside metabolites are shown in
Figure 4.

Studies with chloramben (17, 19, 21, 25, 26) and pyrazon
(23, 24, 27, 28, 29) have shown that the rate and extent of
N-glucoside formation is an important factor in the movement,
phytotoxicity and selectivity of these herbicides in plants.
The phytotoxicity of substituted aniline metabolites, such as
3,4-dichloroaniline, formed by the hydrolysis of propanil (22)
may also be affected by N-glucoside formation.

The in vitro biosynthesis of N-glucosyl chloramben and
other N-glucosyl arylamines has been reported in studies with
plants (30, 31). A UDP-glucosyl transferase from soybean was
specific for the nucleotide glucosyl donors UDPG and TDPG, but
exhibited a relatively broad specificity toward acceptor
arylamines (Figure 5).

The N-glucosides of chloramben and pyrazon are stable in
vivo and appear to persist in soybean (26) and sugarbeet (24) as
"terminal" metabolites. Propanil metabolism studies in rice
(22, 32), however, suggest that the N-glucoside of 3,4-dichloro-
aniline may undergo further metabolic reactions to yield other
glycosides and a methanol-insoluble "lignin" complex.

The N-glucosides of arylamines are hydrolyzed by dilute
acids, but do not appear to be hydrolyzed by β-D-glucosidase.
At the present time, the configuration of the N-glucosidic
linkage has not been established. It is postulated, however,
that inversion of configuration occurs during the UDP-glucosyltransferase catalyzed reaction and that the β-D-glucose anomer
is formed.

Heterocyclic N-glucosides have also been reported.
Kamimura et al. (33) identified the N-glucoside and the O-
glucoside of 3-hydroxy-5-methylisoxazole, a soil fungicide, as
major metabolites in cucumber, tomato and rice plants, and in

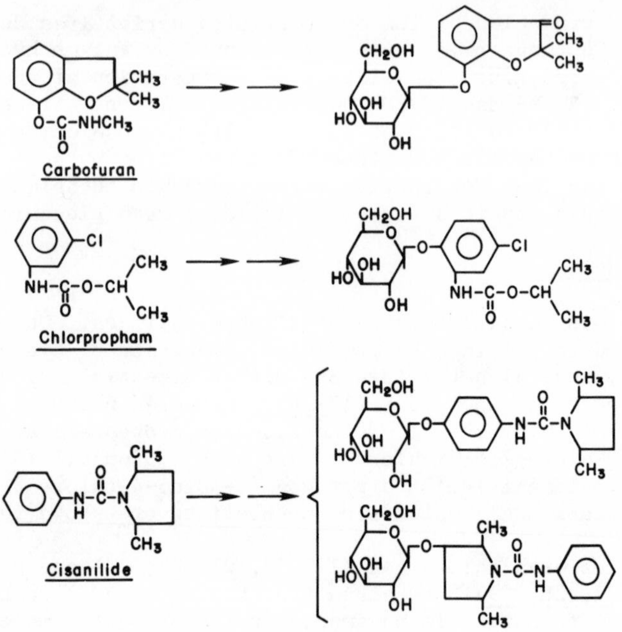

Figure 3.   Pesticide metabolites characterized as O-glucosides

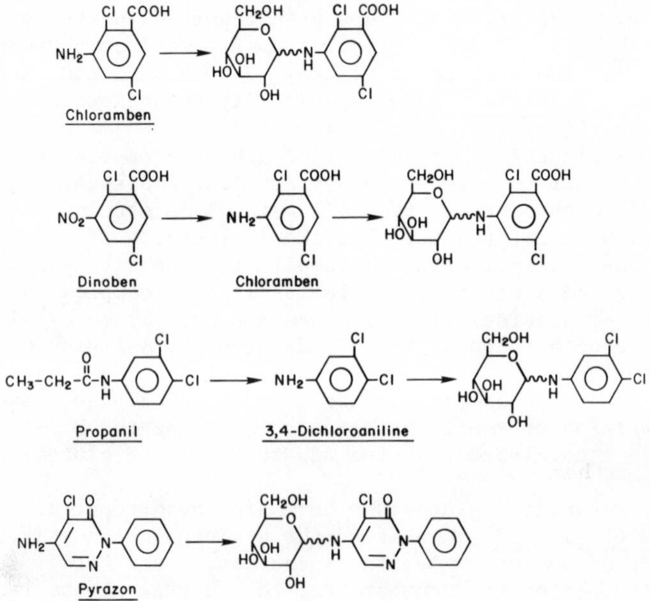

Figure 4.   Pesticide metabolites characterized as N-glucosides

tobacco callus. The N-glucoside was quite stable to acid
hydrolysis and was not hydrolyzed by β-glucosidase. The
structure of the isolated N-glucoside metabolite was determined
by analysis of MS, PMR, IR and UV spectra and by identification
of the acid hydrolysis products. Linkage of the aglycone to the
C-1 of glucose was established by a failure to detect a free
reducing group and by the isolation of methyl penta-O-methyl-β-
glucopyranoside after permethylation and methanolysis of the
glucoside. Optical rotary dispersion studies suggested that the
glucose linkage was the β-configuration.

Studies on the fate of both the O- and the N-glucoside
metabolites in cucumber seedlings showed that the O-glucoside
was hydrolyzed and converted to the N-glucoside while the N-
glucoside remained unchanged (33). Recent in vitro enzyme
studies (34) support these findings and show a requirement for
UDPG as the glucosyl donor (Figure 6).

Zeatin[6-(4-hydroxy-3-methyl-trans-2-butenylamino)purine], a
plant hormone with cytokinin activity, and 6-benzylaminopurine,
a related synthetic cytokinin, are also metabolized to N-glucosides in plants (35). In studies with 6-benzylaminopurine and de-
rooted radish seedlings, N-glucosides accounted for 90% of the
extractable metabolites after 24 hours (35, 36). N-glucosylation
occurs at either the 7- or 9-position of the purine ring as shown
in Figure 7. Both glucosides appear to be stable metabolites.
The 7-glucosides of cytokinins are quite stable and appear to
accumulate, possibly as storage forms (35, 37, 38). In radish
seedlings, 7-glucosylzeatin was not translocated (39). Guern et
al. (40), however, reported that synthetic 6-benzylamino-9-β-D-
glucosylpurine was readily translocated without appreciable
enzymic modification in chick pea. The physiological significance
of cytokinin N-glucosides is obscure.

In studies by Parker et al. (35), N-glucoside metabolites of
cytokinins were isolated by cellulose ion exchange chromatography,
TLC and paper chromatography. Structures were determined by
ultraviolet and mass spectroscopy. Glucose was determined with
glucose oxidase after acid hydrolysis with a polystyrene
sulphonic acid resin (H+ form) at 120° for 1 hour.

Glucose Esters

Several pesticide and plant growth regulators are acids or
readily hydrolyzable esters. Studies with a number of these
pesticides have shown that they are rapidly complexed as water-
soluble metabolites and easily hydrolyzed to the free acid by
treatment with a dilute base or an acid. In some of these
studies, it has been speculated that glucose esters were formed.
Unfortunately, very few of these water-soluble complexes have
been isolated or identified. However, a number of reports (41-
49) have shown that auxins and plant growth regulators are
metabolized to glucose esters in higher plants (Figure 8).

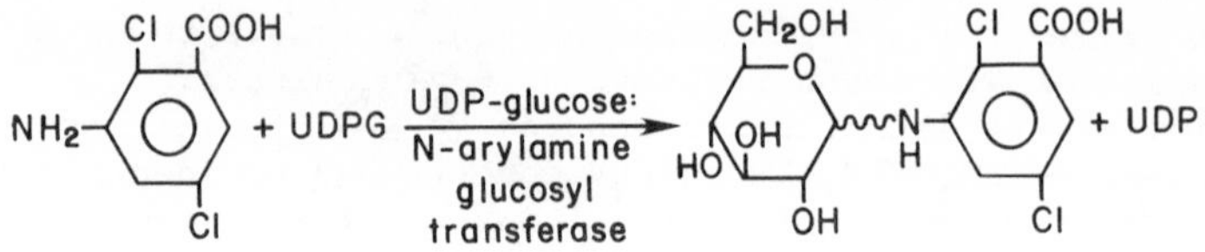

Figure 5. *Biosynthesis of* N-*glucosyl chloramben*

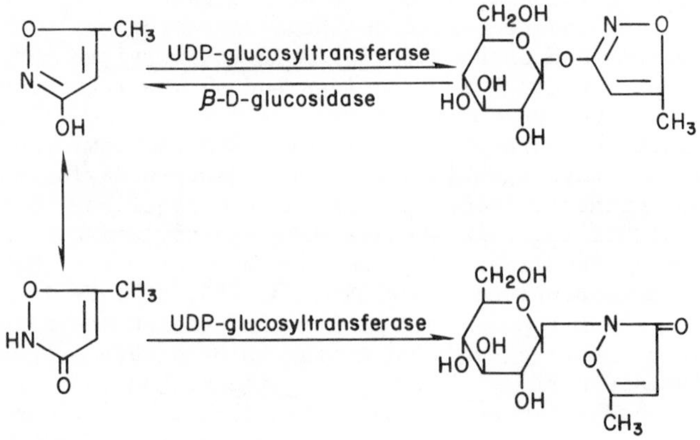

Figure 6. *Proposed scheme for the metabolism of 3-hydroxy-5-methylisoxazole in plants*

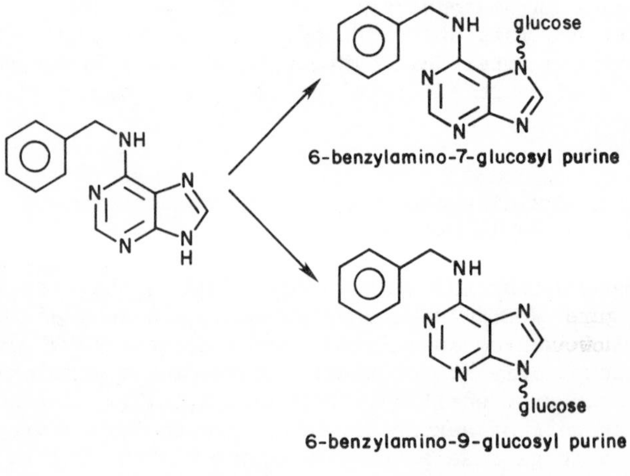

6-benzylamino-7-glucosyl purine

6-benzylamino-9-glucosyl purine

Figure 7. *Cytokinin metabolism in plants—N-glucoside formation*

Enzymatic synthesis of the 2-0, 4-0 and 6-0 esters of IAA
(indole-3-acetic acid) and glucose have been reported by
Kopcewicz et al. (47). A crude enzyme from mature sweet corn
kernels required ATP and CoA as cofactors and suggested that
IAA-CoA thiol ester formation was required for acylation of
glucose. A proposed reaction sequence is shown in Figure 9.
The presence of three isomeric forms was attributed to acyl
migration. Zenk (45) has suggested that the biosynthesis of
the 1-0-ester of IAA and glucose proceeds by a different
mechanism and is catalyzed by a UDP-glucosyltransferase, as shown
in Equation 1. Additional support for a UDP-glucosyltransferase
mechanism of glucose ester biosynthesis has been provided by
Jacobelli et al. (50) and Corner and Swain (51). In studies with
enzymes isolated from germinating lentils and geranium leaves,
the formation of several hydroxybenzoic acid and hydroxycinnamic
acid glucose esters required UDPG. Obviously, additional in
vitro studies are needed to understand the mechanism(s) of
glucose ester biosynthesis.

In addition to glucose esters, the biosynthesis and
isolation of several IAA-myo-inositol esters have also been
reported (47, 48).

Glucose esters are sensitive to mild alkaline hydrolysis,
and are also hydrolyzed by acids. Hydrolysis with β-glucosidase
has been reported for glucose esters of IAA, NAA (1-naphthalene-
acetic acid) and 2,4-D (2,4-dichlorophenoxy acetic acid).
However, the glucose ester of abscisic acid was not hydrolyzed
by β-glucosidase (42). It has been suggested by several authors
that glucose ester biosynthesis and hydrolysis may be an
important factor in the bioregulation of plant hormone levels
(47, 52). It is interesting to speculate that the selective
phytotoxicity of a number of herbicides may also be affected by
differential rates of glucose ester formation and hydrolysis.

S-Glucosides

Various glucosinolates have been isolated as natural
products from several plant species. A general structure for
these S-glucosides is shown in Figure 10. The R group may be
aliphatic or partly aromatic. Glucosinolates do not appear to
be hydrolyzed by emulsin, but are hydrolyzed in certain plant
species by a mixture of enzymes called myrosinase.

Recent investigations (53-55) of benzylglucosinolate biosynthesis in plants have shown that an intermediate, phenyl-
acetothiohydroximate, is glucosylated by a UDP-glucosyltransferase (Figure 11a). Examples of pesticide S-glucosides are
limited. However, studies by Kaslander et al. (56) have shown
that the fungicide, dimethyldithiocarbamate, is metabolized in
potato slices and cucumber seedlings to the S-glucoside
(Figure 11b). In insects, Gessner and Acara (57) have shown

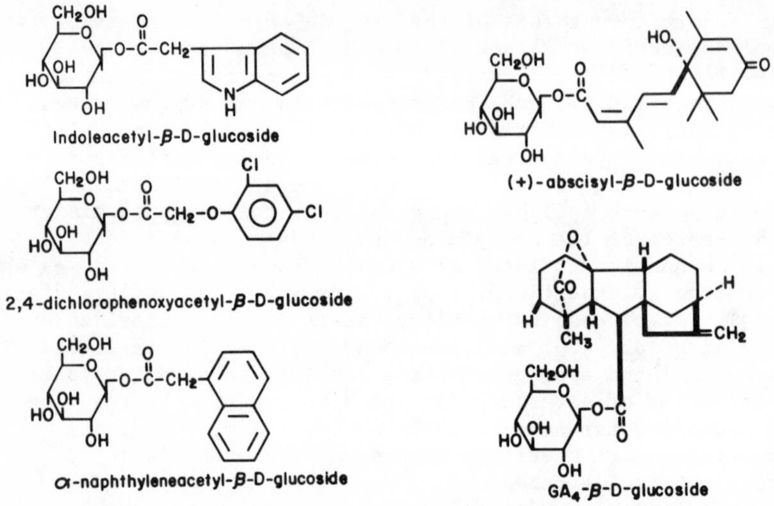

Indoleacetyl-β-D-glucoside

2,4-dichlorophenoxyacetyl-β-D-glucoside

α-naphthyleneacetyl-β-D-glucoside

(+)-abscisyl-β-D-glucoside

GA₄-β-D-glucoside

Figure 8. Glucose esters of auxins and plant growth regulators

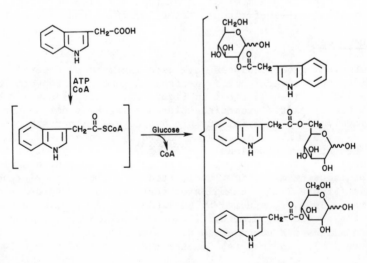

ATP
CoA

Glucose

CoA

Figure 9. Proposed biosynthesis of 2-0, 4-0, and 6-0 glucose esters of
IAA

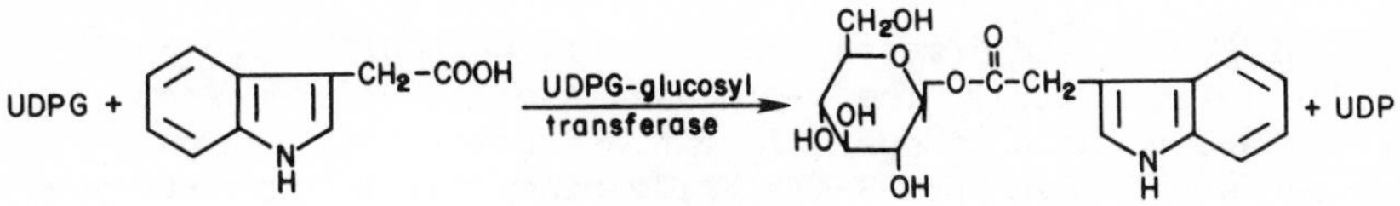

Equation 1.

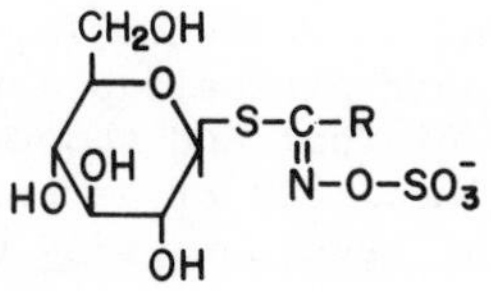

Figure 10.   *General structure of glucosinolates*

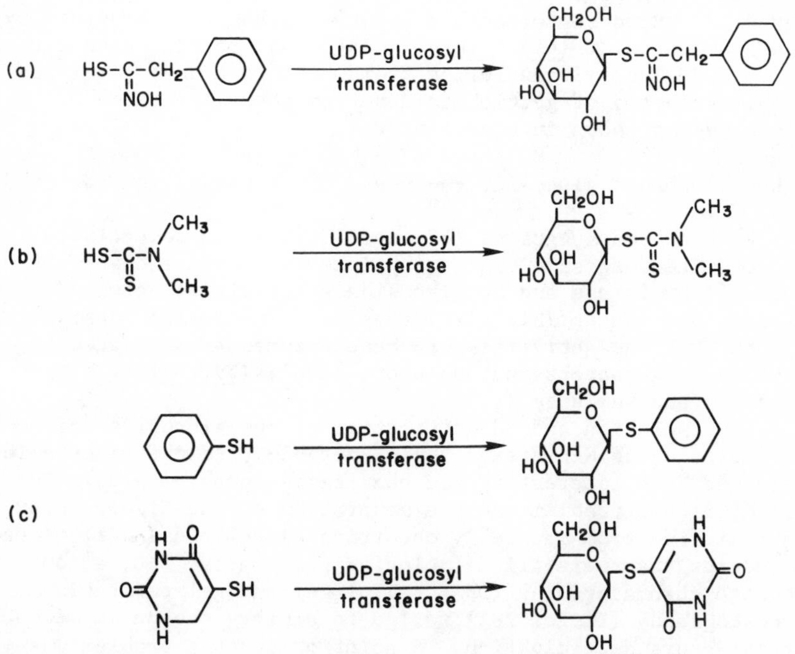

(a)

(b)

(c)

Figure 11.   Biosynthesis of S-glucosides in plants and insects

that thiophenol and 5-mercaptouracil are metabolized to S-glucosides (Figure 11c). The S-glucosides were isolated from
excreta and treated tissue. In vitro studies with fat body
tissues established that S-glucoside biosynthesis was catalyzed
by a UDP-glucosyltransferase.

Complex Glycosides

Recently, a few studies have demonstrated that glucosides
of pesticide metabolites may be subject to further metabolism
in plants.

Extensive early studies by Miller (58) showed that several
xenobiotic alcohols and phenols were conjugated as gentiobiosides by higher plants. Apparent differences in enzyme
specificity toward aglycones and their glucosides were noted,
and sharp contrasts in the ability of root and shoot tissues of
different species to form either glucosides or gentiobiosides
were reported. These glucosides and gentiobiosides were not
translocated from tissues in which they were formed.

Studies with the herbicide, diphenamid (N,N-dimethyl-2,2-
diphenylacetamide), have shown that a primary oxidation product,
N-hydroxymethyl-N-methyl-2,2-diphenylacetamide, is conjugated as
a β-glucoside and a β-gentiobioside (59). Structures were
determined by MS of acetylated derivatives and analysis of
hydrolysis products. Time course studies (60) suggest that the
glucoside is a precursor of the gentiobioside as shown in
Figure 12. Support for such a hypothesis has been provided by
Yamaha and Cardini (61). They isolated and partially characterized a UDP-glucosyltransferase from wheat germ that catalyzes
the biosynthesis of gentiobiosides from phenolic-mono-β-D-
glucosides as shown in Equation 2.

   UDPG + phenol glucoside ⟶ UDP + phenol gentiobioside

It has been suggested (62) that diphenamid selectivity may
be determined, to some extent, by inherent differences in UDP-
glucosyltransferase and/or glycosidase activities between
tolerant and susceptible plant species. It is also interesting
to note that the activities of these enzymes are affected by
environmental factors such as ozone levels (59, 60) or light
intensity and humidity (63).

The isolation and identification of the glucoside and
gentiobioside of N-hydroxymethyl-N-methyl-2,2-diphenylacetamide
presented some interesting and challenging problems (59). The
most difficult problem was the separation of the highly polar
gentiobioside from naturally occurring glycolipids. All of the
procedures that were tried including preparative TLC, anion
exchange chromatography (DEAE cellulose) and gel permeation
chromatography (Biogel P-2) failed to achieve the purity needed
for structure determination. A solution to this problem was

achieved by reaction of the partially purified gentiobioside
with Tri-Sil Z, TLC separation of the TMS derivative, hydrolysis
of the isolated TMS derivative (90% MeOH reflux at 70°C for 2
hours) and final TLC of the original gentiobioside. In our
experience, TMS derivatives of glucosides are usually stable
enough to be handled in this manner and provide a simple method
for changing the polarity of a glycoside metabolite for
separation purposes. The mild conditions needed for the
recovery of the unchanged glycoside are also helpful in
situations where the glycoside is easily hydrolyzed.

   Another interesting lesson learned in diphenamid metabolism
studies (59) was that failure to demonstrate metabolite
hydrolysis with emulsin does not preclude the presence of a
β-D-glucoside. The glucoside metabolite of N-hydroxymethyl-N-
methyl-2,2-diphenylacetamide was not hydrolyzed by emulsin, and
the gentiobioside was hydrolyzed partially at the β-1,6 linkage
to yield the glucoside. Optical rotation studies with the
glucoside and its tetraacetate clearly indicated, however, that
the metabolite was a β-glucoside. In retrospect, PMR analysis
may have provided an unequivocal assignment of anomeric
configuration. Even though emulsin is the enzyme generally used
to hydrolyze β-D-glucosides, other β-glucosidases with different
substrate specificities are available from a variety of sources.
One of these enzymes, a crude hesperidinase from Aspergillus
niger, contained a β-glucosidase that catalyzed the complete
hydrolysis of both the glucoside and the gentiobioside
metabolites. Enzyme hydrolysis with hesperidinase at pH 5.25
provided an excellent means of isolating and identifying the
acid labile N-hydroxymethyl aglycone. Even mild acid hydrolysis
of the glucoside resulted in the loss of formaldehyde and the
formation of N-methyl-2,2-diphenylacetamide.

   Several plant pigments have been identified as malonate
hemi-ester derivatives of β-glucosides (64-66). Spectral and
chemical studies with acylated betacyanins (64) and isoflavones
(65) have established that the O-malonyl group is located at the
C-6 of the glucose moiety. Hahlbrock (67) isolated a malonyl
CoA transferase from parsley cell cultures that catalyzed the
transfer of malonate from malonyl CoA to flavone glycosides.
Moore and Wilson (68) reported the enzymatic hydrolysis of
acylated flavone glycosides and showed that partially purified
enzymes from parsley and chick pea leaves catalyzed the
hydrolysis of the malonate ester linkage.

   Recently, Shimabukuro et al. (69) reported the isolation
and tentative identification of 6-O-malonyl-β-D-glucoside of p-
nitrophenol as a major metabolite of fluorodifen (p-nitrophenyl-
α,α,α-trifluoro-2-nitro-p-tolyl ether) in peanut. In these
studies, glucoside metabolites of fluorodifen were extracted
with 80% methanol and the concentrated aqueous extract was
partitioned with hexane and then with isopropyl ether to remove
unreacted fluorodifen and free p-nitrophenol. Two major

glucoside metabolites, a neutral glucoside and an anionic
malonyl glucoside of p-nitrophenol were separated by gel
filtration on Sephadex G-10 and cellulose ion exchange chromatography on DE-52. The isolated malonyl glucoside was stable to
weak acids. Only limited hydrolysis to the glucoside was
reported during its purification. Mild alkaline hydrolysis,
however, rapidly liberated the glucoside. Enzymatic hydrolysis
was achieved with hesperidinase, but not with emulsin. Similar
resistance to emulsin hydrolysis has been reported by Minale
et al. (64) for the 6-O-malonyl-β-D-glucoside of betanidin. The
structure of the isolated O-malonyl-β-D-glucoside of p-nitrophenol was established by MS after methylation with diazomethane
and acetylation with acetic acid anhydride and $ZnCl_2$. Deuterium
exchange of one or both of the methylene protons of the malonyl
group was demonstrated and the malonyl moiety was tentatively
located at the C-6 of the glucose. Efforts are now underway to
make an unequivocable structure assignment based on PMR and
possibly ^{13}CMR studies. A proposed pathway for the biosynthesis
of this acylated glucoside is shown in Figure 13.

        Studies on the isolation and identification of complex
glucoside metabolites have just begun. The extent, variety and
significance of these metabolites remain largely unknown. In
plants, complex glycoside metabolites may be intermediates in
the formation of "terminal" pesticide residues. Recent studies
by Still and Mansager (70) indicate that phenolic intermediates
in the metabolism of chlorpropham by alfalfa may be glycosylated
as a homologous series of methanol-water soluble oligosaccharide
derivatives. Partial hydrolysis of these highly polar complex
glycoside metabolites was achieved by repeated treatment with
cellulase. Additional studies on the isolation, identification,
and biosynthesis of complex glycoside metabolites are needed.

Methodology

        A variety of techniques and procedures are available for
the isolation and identification of glycoside metabolites.
Unfortunately, an adequate discussion or consideration of the
many methods that have been used is not possible in the time
available this morning. Besides, each glycoside metabolite must
be considered, to a large extent, as a unique isolation and
identification problem. What works in one situation may not work
in another, depending on the nature of the particular glycoside,
the aglycone and the endogenous materials in the tissue extract.
It is fortunate, therefore, that a variety of techniques and
approaches are available to the investigator.

        Several points should be stressed, however, in the
isolation and identification of glycoside metabolites: (a) the
importance of protecting labile glycosidic linkages from
hydrolysis during extraction and chromatographic separation; (b)
the necessity for freeing the isolated glycoside from

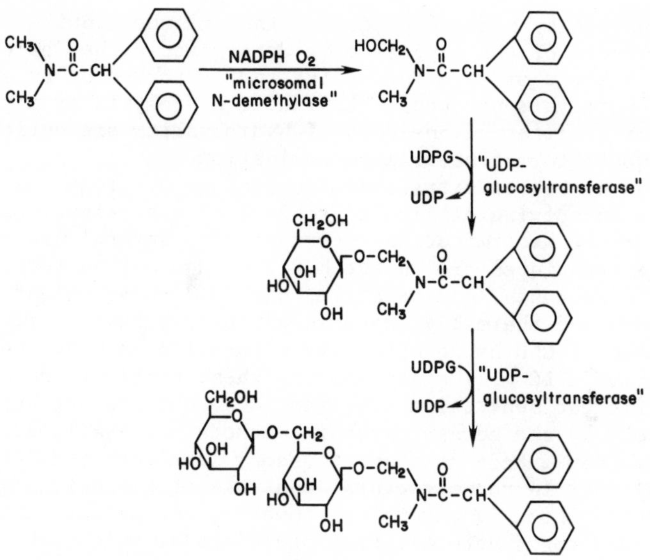

Figure 12.   Diphenamid metabolism in tomato—proposed bio-synthesis of gentiobioside and metabolite

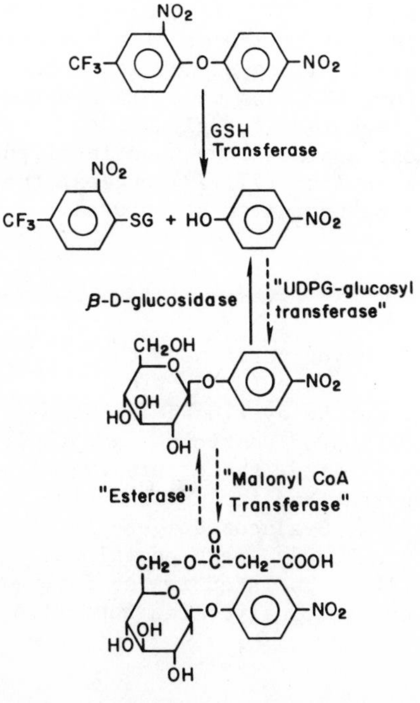

Figure 13.   Proposed biosynthesis of p-nitrophenyl-6-0-malonyl-β-D-glucoside in peanut

contamination by natural products; and (c) the requirement that
the structure of the glycoside be determined either as the
intact molecule or as a derivative and supported by an analysis
of the hydrolysis products. Enzymatic hydrolysis by a glycosidase or qualitative analysis of hydrolysates are obviously
not adequate for structural determination.

In a number of situations, current methodology has not been
adequate in solving difficult problems of glycoside isolation
and structure determination. Fortunately, several new and
improved techniques are on the horizon, and may be very helpful.

In some cases, non-volatile, high molecular weight
glycosides and thermally unstable glycosides have limited the
usefulness of GLC as an effective separation method. Hopefully,
preparative HPLC will solve some of these problems, and provide
more efficient separation of glycosides that are not separated
adequately by the chromatographic methods now available.

Electron impact mass spectroscopy is a very sensitive and
powerful tool in the structural analysis of glycosides (71-74).
Ion fragmentation is extensive, however, and primary molecular
ions needed for a determination of molecular weight and
elemental composition are often absent. Recent studies have
shown that chemical-ionization (CI) and field desorption (FD)
mass spectra of glycosides exhibit strong quasi molecular ion
peaks and limited ion fragmentation (74-76). These techniques
should provide a sensitive and more direct means for
determining the structure of intact glycosides.

Nuclear magnetic resonance (NMR) spectroscopy is another
important method for structure elucidation in studies with
carbohydrates and their derivatives (77, 78). In the past, the
use of NMR spectroscopy in pesticide metabolism studies has been
limited. Recently, however, the availability of new instrumentation, particularly Fourier transform NMR, has made NMR spectroscopy a practical and very useful technique for glycoside
structural analysis. Thus far, most applications have involved
PMR spectroscopy. However, recent reviews (77, 79) suggest that
nuclei other than protons may also be useful in structural
studies of glycoside metabolites.

Summary

Glucosides often account for a major portion of the
pesticide metabolites in plants and invertebrates. Their
importance and significance should not be overlooked. Even
though the isolation and identification of pesticides and their
metabolites as glycosides has been limited, the diversity of
isolated glucosides is already apparent and includes; 0-glu-
cosides, N-glucosides, glucose esters, S-glucosides, acylated
glucosides and gentiobiosides. Undoubtedly, many additional
glucosides and other glycosides will be isolated in the future.
Information about the nature and extent of glycoside formation

is needed before their role in pesticide metabolism can be determined.

At the present time, glucosides have been implicated as significant factors in the bioregulation of pesticide toxicity and selectivity in plants and insects. Unfortunately, the enzyme systems reponsible for glucoside formation and hydrolysis have not been studied to any great extent.

Also, the further metabolism of glucoside metabolites has received little attention. In plants, the possible role and significance of these complex glycoside metabolites in the formation of "terminal" pesticide residues has been suggested.

Literature Cited

1. Smith, J. N., Advan. Comparative Physiol. Biochem. (1968), 3, 173.
2. Finkle, B. J., In: "Phenolic Compounds and Metabolic Regulation," pps. 3-24 (1967), Eds. Finkle, B. J. and Runeckles, V. C., Appleton-Century-Crofts, N.Y.
3. Hanson, K. R., Zucker, M., and Sondheimer, E., In: "Phenolic Compounds and Metabolic Regulation," pp. 69-93 (1967), Eds. Finkle, B. J., and Runeckles, V. C., Appleton-Century-Crofts, N.Y.
4. Ramwell, P. W., and Sherratt, H. S. A., In: "Biochemistry of Phenolic Compounds," pp. 457-510 (1964), Ed. Harborne, J. B., Academic Press, N.Y.
5. Daly, J. W., In: "Phenolic Compounds and Metabolic Regulation," pp. 27-66 (1967), Eds. Finkle, B. J., and Runeckles, V. C., Appleton-Century-Crofts, N.Y.
6. Kuhr, R. J., J. Agr. Food Chem. (1970), 18, 1023.
7. Bull, D. L., and Whitten, C. J., J. Agr. Food Chem. (1972), 20, 561.
8. Still, G. G., Rusness, D. G., and Mansager, E. R., ACS Symposium Series No. 2 (1974), "Mechanism of Pesticide Action," pp. 117-129.
9. Frear, D. S., and Swanson, H. R., Pest. Biochem. Physiol. (1975), 5, 73.
10. Pridham, J. B., Phytochemistry (1964), 3, 493.
11. Towers, G. H. N., In: "Biochemistry of Phenolic Compounds," pp. 249-294 (1964), Ed. Harborne, J. B., Academic Press, N.Y.
12. Trivelloni, J. C., Enzymologia (1964), 26, 6.
13. Mehendale, H. M., and Dorough, H. W., J. Insect Physiol. (1972), 18, 981.
14. Metcalf, R. L., Fukuto, T. R., Collins, C., Borck, K., Abd El-Aziz, S., Munoz, R., and Cassil, C. C., J. Agr. Food Chem. (1968), 16, 300.
15. Still, G. G., and Mansager, E. R., Phytochemistry (1972), 11, 515.

16. Bligh, E. G., and Dyer, W. J., Can. J. Biochem. Physiol. (1959), 37, 911.
17. Colby, S. R., Science (1965), 150, 619.
18. Swanson, C. R., Kadunce, R. E., Hodgson, R. H., and Frear, D. S., Weeds (1966), 14, 319.
19. Colby, S. R., (1966), Weeds 14, 197.
20. Colby, S. R., 152nd Amer. Chem. Soc. Meeting, (1966), Abstr. No. A-33, New York.
21. Frear, D. S., In: "Degradation of Herbicides," 2nd Edition, Eds. Kearney, P. C., and Kaufman, D. D., Marcel Dekker, Inc., N.Y. (In press).
22. Still, G. G., Science (1968), 159, 992.
23. Ries, S. K., Zabnik, P. J., Stephenson, G. R., and Chen, T. M., Weed Sci. (1968), 16, 40.
24. Stephenson, G. R., and Ries, S. K., Weed Sci. (1969), 17, 327.
25. Stoller, E. W., Plant Physiol. (1966), 44, 854.
26. Swanson, C. R., Hodgson, R. H., Kadunce, R. E., and Swanson, H. R., Weeds (1966), 14, 323.
27. Frank, R., and Switzer, C. M., Weed Sci. (1969), 17, 365.
28. Stephenson, G. R., and Ries, S. K., Weed Res. (1967), 7, 51.
29. Frear, D. S., Hodgson, R. H., Shimabukuro, R. H., and Still, G. G., Advan. Agron. (1972), 24, 327.
30. Frear, D. S., Swanson, C. R., and Kadunce, R. E., Weeds (1967), 15, 101.
31. Frear, D. S., Phytochemistry (1968), 7, 381.
32. Yih, R. Y., McRae, D. H., and Wilson, H. F., Science (1968), 161, 376.
33. Kamimura, S., Hishikawa, M., Saeki, H., and Takahi, Y., Phytopathology (1974), 64, 1273.
34. Murakoshi, I., Ikegama, F., Kato, F., Tomita, K., Kamimura, S., and Haginiwa, J., Chem. Pharm. Bull. (1974), 22, 2048.
35. Parker, C. W., Wilson, M. M., Letham, D. S., Cowley, D. E., and MacLeod, J. K., Biochem. Biophys. Res. Commun. (1973), 55, 1370.
36. Wilson, M. M., Gordon, M. E., Letham, D. S., and Parker, C. W., Jour. Exptl. Bot. (1974), 25, 725.
37. Deleuze, G. G., McChesney, J. D., and Fox, J. E., Biochem. Biophys. Res. Commun. (1972), 48, 1426.
38. Parker, C. W., and Letham, D. S., Planta (1973), 114, 199.
39. Gordon, M. E., Letham, D. S., and Parker, C. W., Annals of Bot. (1974), 38, 809.
40. Guern, J., Doree, M., and Sadorge, P., In: "Biochemistry and Physiology of Plant Growth Substances," pp. 1155-1167 (1968), Eds. Wightman, F., and Setterfield, G., Runge Press, Ottawa.
41. Koshimizer, H. Fukui, and Mitsui, T., Agr. Biol. Chem. (1968), 32, 789.
42. Milborrow, B. V., J. Exptl. Bot. (1970), 21, 17.
43. Hiraga, K., Yokota, T., Murofushi, N., and Takahashi, N., Agr. Biol. Chem. (1972), 36, 345.

44. Shindy, W. W., Jordon, L. S., Jolliffe, V. A., Coggins, C. W., and Kumamoto, J., J. Agr. Food Chem. (1973), 21, 629.
45. Zenk, M. H., Nature (1961), 191, 493.
46. Ehmann, A., Carbohydrate Res. (1974), 34, 99.
47. Kopcewicz, J., Ehmann, A., and Bandurski, R. A., Plant Physiol. (1974), 54, 846.
48. Ehmann, A., and Bandurski, R. S., Carbohydrate Res. (1974), 36, 1.
49. Thomas, E. W., Loughman, B. C., and Powell, R. G., Nature (1964), 204, 286.
50. Jacobelli, G., Tabone, M. J., and Tabone, D., Bull. soc. chim. biol. (1958), 40, 955.
51. Corner, J. J., and Swain, T., Nature (1965), 207, 634.
52. Hiraga, K., Yamane, H., and Takahaski, N., Phytochemistry (1974), 13, 2371.
53. Matsuo, M., and Underhill, E. W., Biochem. Biophys. Res. Commun. (1969), 36, 18.
54. Matsuo, M., and Underhill, E. W., Phytochemistry (1971), 10, 2279.
55. Matsuo, M., Kirkland, D. F., and Underhill, E. W., Phytochemistry (1972), 11, 697.
56. Kaslander, J., Sijpesteinj, A. K., and VanderKerk, G. J. M., Biochem. Biophys. Acta (1961), 52, 396.
57. Geasner, T., and Acara, M., Jour. Biol. Chem. (1968), 243, 3142.
58. Miller, L. P., In: "Phytochemistry," p. 311, Vol. I (1973), Ed. Miller, L. P., Van Nostrand Reinhold Co., N.Y.
59. Hodgson, R. H., Frear, D. S., Swanson, H. R., and Regan, L. A., Weed Sci. (1973), 21, 542.
60. Hodgson, R. H., Dusbabek, K. E., and Hoffer, B. L., Weed Sci. (1974), 22, 205.
61. Yamaha, T., and Cardino, C. E., Arch. Biochem. Biophys. (1960), 86, 133.
62. Schultz, D. P., and Tweedy, B. G., J. Agr. Food Chem. (1971), 19, 36.
63. Schultz, D. P., and Tweedy, B. G., J. Agr. Food Chem. (1972), 20, 10.
64. Minale, L., Piatelli, M., DeStefano, S., and Nicolaus, R. A., Phytochemistry (1966), 5, 1037.
65. Beck, A. B., and Knox, J. R., Aust. J. Chem. (1971), 24, 1509.
66. Kruezaler, F., and Hahlbrock, K., Phytochemistry (1973), 12, 1149.
67. Hahlbrock, K., FEBS Letters (1972), 28, 65.
68. Moore, A. L., and Wilson, S. B., Biochem. J. (1972), 129, 18P.
69. Shimabukuro, R. H., Walsh, W. C., Stolzenberg, G. E., and Olson, P. A., Abs. Weed Sci. Meetings (1975), p. 65, No. 171 (Washington, D. C.).

70. Still, G. G., and Mansager, E. R., Chromatographia (1975), 8, 129.
71. Kochetkov, N. K., and Chizhov, O. S., Advan. Carbohydrate Chem. (1966), 21, 39.
72. Pearl, I. A., and Darling, S. F., Phytochemistry (1968), 7, 831.
73. Kochetkov, N. N., and Chizhov, O. S., Methods Carbohyd. Chem. (1972), 6, 540.
74. Lönngren, J., and Svensson, S., Advan. Carbohydrate Chem. Biochem. (1974), 29, 41.
75. Lehman, W. D., Schulten, H. R., and Beckey, H. D., Org. Mass Spectrom. (1973), 7, 1103.
76. Foltz, R. L., Chemtech (1975), January, 39.
77. Angyal, S. J., In: "The Carbohydrates" IA 2nd Ed., (1972), pp. 195-215, Eds. Pigman, W., and Horton, D., Academic Press, N.Y.
78. Coxon, B., Methods Carbohydrate Chem. (1972), 6, 513.
79. Wilson, N. K., and Stothers, J. B., In: "Topics in Stereochemistry," (1974), 8, Eds. Eliel, E. L., and Allinger, N. L., 1, John Wiley and Sons, N.Y.

4

Recent Advances in the Isolation and Identification of Glucuronide Conjugates

JEROME E. BAKKE

Metabolism and Radiation Research Laboratory, Agricultural Research Service,
U.S. Department of Agriculture, Fargo, N. Dak. 58102

The chemistry and biochemistry of glucuronic acid and glucuronide conjugates has been adequately reviewed in a book edited by
G. J. Dutton (1). The chapter by Jayle and Pasqualini reviews
methods for extraction, fractionation, and identification of
steroid glucuronides. The present discussion will cover methods
that have recently been applied to the extraction, separation,
and identification of glucuronide conjugates from mammalian
plasma, urine, and bile. These methods will be discussed in the
sequence in which they are usually applied in practice and with
bias toward the methods that the author has used.

Extraction of Glucuronides

Four materials have recently been successfully applied to the
extraction or concentration of glucuronides present in aqueous
solutions. These are: Amberlite XAD-2$^{(R)}$ (Rohm & Haas Co.), a
synthetic polystyrene polymer, Porapak Q$^{(R)}$ (Waters Assoc.), a gas
chromatography column packing, liquid anion exchangers (tetraheptylammonium chloride, Eastman Organic Chemicals; methyl tri-
caprylyl ammonium chloride, General Mills), and Sephadex LH-20
(Pharmica Fine Chemicals).

Amerlite XAD-2$^{(R)}$. This bead form polymer has been used for
the extraction of glucuronide conjugates from aqueous solutions
(urine and salt solutions; 2,3). The XAD-2 column is first washed
with methanol to remove contaminants and then washed with water
prior to adding the aqueous solution containing the glucuronides.
Materials in the sample that do not bind to the polymer are washed
through the column with water and the glucuronides and other bound
materials are recovered from the column by elution with organic
solvents (usually methanol or acetone).

Porapak Q$^{(R)}$. Columns of Porapak Q have been used in our
laboratory for the extraction of glucuronides and other metabolites from urine, plasma, and various aqueous solutions or

extracts. Porapak Q is used in the same manner as XAD-2 and is
assumed to function in the same manner, i.e., by essentially a
reversed phase chromatographic process.

It is possible to effect some separation of the materials
bound to Porapak Q by elution with different organic solvents or
stepwise gradients of methanol or acetone in water. Aschbacher
(4) utilized a stepwise gradient of methanol in water to separate
urinary diethylstilbesterol (DES) from DES-glucuronide. The
glucuronide eluted from the Porapak Q with 80% methanol. The
DES was eluted from the column with methanol. In another case,
the separation of the urinary metabolites from crufomate (5),
using a stepwise elution series of hexane, diethyl ether, methanol, it was possible to effect a separation of urinary metabolites into two solubility classes. Hexane displaced the water
from the Porapak Q column, diethyl ether eluted two nonpolar
metabolites, and methanol eluted the remaining twenty-one metabolites which included seven glucuronides.

The plasma metabolites from crufomate, which contained the
same seven glucuronides, also bound to Porapak Q when the plasma
was applied directly to the column. Again, the metabolites were
quantitatively recovered by subsequent elution of the column with
methanol.

Some compounds that do not absorb to Porapak Q from aqueous
solution can be retarded on the column by an apparent salting out
reversed phase process. For example, the major urinary metabolite from cyclophosphamide (6) was not bound to Porapak Q from a
simple aqueous solution; however, in urine it remained on the
column until the bulk of the urinary solids had been eluted.

Liquid Anion Exchangers. Mattox et al. (7) have demonstrated the applicability of liquid anion exchangers dissolved in
organic solvents for the extraction of glucuronides from aqueous
solutions. This extraction is assumed to involve mainly an ion
exchange process. The more polar steroid glucuronides were extracted less efficiently; however, the completeness of the extraction could be increased if the aqueous phase was made 4M with
ammonium sulfate. The glucuronides were recovered from the organic phase by extraction with ammonium hydroxide.

All three of the above procedures have applicability to the
extraction of glucuronides from aqueous media; however, none of
these processes is specific for glucuronides. The XAD-2 and
Porapak Q will extract many nonpolar materials and the liquid
anion exchangers will extract any anion that can compete with
the counter ion that is present.

Sephadex LH-20 (H_2O). A 120 X 2 cm column of water-equili-
brated LH-20 has been used in our laboratory for the preliminary
fractionation of urinary and plasma metabolites from xenobiotics
(6). Using water as the eluent, this column usually separates
these metabolites by a typical reversed phase chromatographic

process into polarity or solubility classes. In three separate
studies, all the glucuronide conjugates of xenobiotic metabolites
appeared in one fraction that had an elution volume range of 0.4
to 0.9 of the total column volume. The xenobiotics studied were
crufomate (seven glucuronides), 0,0-dimethyl-0-(3,5,6-trichlor-
pyridyl)phosphorothionate (one glucuronide) and propachlor (sheep
2 and rat 5 glucuronides).

Since many urinary solids elute from this column with this
glucuronide containing-fraction, it was advantageous to use the
Porapak Q column before or after use of the LH-20 column to remove many very polar materials and inorganic salts. The general
applicability of this procedure for the concentration of glucuronides will have to await its application in many more cases.

Counter Current Distribution. Counter current distribution
has been applied to the separation of test mixtures of glucuronides and sulfate esters (8). The application of this technique
to the separation of these two groups of conjugates from biological fluids was not reported; however, the authors relate that
such application would require a prepurification to remove materials that would interfere with the partition systems used (salts).
This technique may have application to the fractionation or concentration of glucuronides if the salts can be removed using
XAD-2 or Porapak Q.

Fractionation of Glucuronides

Several methods have been applied to the separation of
glucuronide conjugates. These have been anion ion exchange cellulose columns, counter current distribution, the amino acid
analyzer colums, paper chromatography, LH-20 columns, and gas
liquid chromatography of glucuronides rendered volatile by derivatization. A number of these methods have been well studied with
test mixtures but have not, as yet, been applied in actual practice.

Anion Exchange Cellulose. Knaak et al. (9) separated four
urinary glucuronide conjugates of carbaryl metabolites using a
column of DEAE cellulose. The metabolites were eluted from the
column using a tris-HCl buffer gradient. Knaak et al. (10) later
pointed out with other compounds, that this technique did not
adequately separate closely related glucuronides and demonstrated
that acetyl and trimethylsilyl derivatives of methyl esters of a
test mixture of glucuronides could be separated by gas liquid
chromatography. This procedure was suggested as a method for
identifying glucuronides if standards were available.

Van Der Wal and Huber (11) have studied the separation of a
test mixture of steroid glucuronides by high-pressure liquid
chromatography (HPLC) using anion exchange celluloses. ECTEOLA-
celluloses were found to be best suited for HPLC using acetate

buffers. The application of this procedure to glucuronides from
biological fluids has not been reported.

DEAE-Sephadex$^{(R)}$. Paulson and Jacobsen (12) used a DEAE
Sephadex column eluted with a gradient of water to M KBr for the
partial separation of glucuronides from sulfate esters. KBr was
used for the gradient to facilitate the preparation of micro pellets for infrared analysis.

This DEAE Sephadex-KBr gradient technique was applied to the
separation of the glucuronides from crufomate (5). The metabolites in the urinary glucuronide fraction from the LH-20 column
[see Extraction of Glucuronides: LH-20 (H_2O)] separated into five
fractions on the DEAE column. The last four fractions to elute
from the column contained three, one, two, and one glucuronides,
respectively. This was determined by derivatization, glc, and
interpretation of the mass spectral data. The fractions were
sufficiently free from other urinary constituents after elution
from the column and removal of the KBr using Porapak Q for derivatization and glc.

Sephadex LH-20 (HCO_3^-). A 115 X 0.9-cm column of LH-20 equil-
ibrated and eluted with 0.065 M ammonium bicarbonate was used to
partially fractionate the glucuronides from crufomate metabolism
(5). The metabolites from the glucuronide fraction from the LH-20
(H_2O) column [see Extraction of Glucuronides: LH-20(H_2O)] separated into three glucuronide-containing fractions. These fractions
contained 4, 2, and 1 glucuronides, respectively, as determined by
derivatization, glc, and mass spectral interpretation. The fraction containing four glucuronides separated into two glucuronide
containing fractions on glc. One contained three glucuronides
that have not been separated. These were the same glucuronides
that were inseparable using the DEAE column.

The LH-20 (HCO_3^-) and DEAE-KBr columns have not given ideal
separations of glucuronide mixtures. Closely related glucuronides
do not separate. If the glucuronide derivatives are stable to glc,
some separations are possible. Mass spectrometry has been the
only method used in this laboratory for the detection of mixtures
of glucuronide derivatives in such samples.

Cation-Exchange Resin Chromatography. Urinary glucuronides
of metabolites from terbutryne (2-methylthio-4-tert-butylamino-6-
ethylamino-s-triazine) were separated using the amino acid analysis column eluted with a citrate buffer gradient. The column
technique has been reported (13) and a preliminary report on the
structures of the glucuronides has been presented (14). This
column separated the five glucuronides into four fractions. The
two glucuronides that did not separate were separated by glc of
the perTMS derivatives. Three of the glucuronides were characterized as S-glucuronides and two as alkyl-0-glucuronides (mass
spectrometry).

The general methods used to isolate these glucuronides consisted of the following steps. The metabolites from terbutryne
were extracted from the urine on either an XAD-2 or Porapak Q
column and eluted from the column with methanol. The residue in
the methanol eluate was dissolved in water, adjusted to pH 3 and
chromatographed on the amino acid analyzer. The buffer salts were
removed from the separated fractions using the Porapak Q column
technique. The fractions were further separated from contaminat-
ing materials by paper and thin-layer chromatography. For final
purification prior to mass spectral analysis, the metabolites were
silylated with excess N,N-bis(trimethylsilyl)trifluroacetamide
containing 1% trimethylchlorosilane and gas chromatographed
(SE-30). Glucuronidase and hydrolysis studies were also used to
characterize the aglycones.

The cation exchange technique is probably applicable only in
cases where the aglycone contains centers of basicity which would
give the glucuronide the zwitterionic character of amino acids.
It is of interest that these glucuronides eluted from the resin
without the use of organic solvents when they would bind to XAD-2
or Porapak Q from aqueous solution.

Liquid Anion-Exchange Paper Chromatography. Mattox et al.
(15) have described a paper chromatographic system for the separation of steroid glucuronides using the liquid anion exchanger,
tetraheptylammonium chloride (THAC), in the mobile phase. Although this system has not been applied (to the author's knowledge) in xenobiotic metabolism studies, it would appear to be of
value if the glucuronides can be recovered from the chromatograms
for structural characterization. The technique has only been
applied to a test mixture of steroid glucuronides and it is unknown how much preliminary cleanup of biological fluids would be
required before the system would become functional.

Gas-Liquid Chromatography. Utilization of gas-liquid chromatography (glc) for the isolation of glucuronides requires the
conversion of these conjugates to volatile derivatives. The
volatilization of glucuronides by derivatization not only makes
it possible, in many cases, to purify and possibly separate glucuronides by glc, but also makes it possible to obtain mass spectral
data for structural determination.

The derivatives that have been used for the volatilization
of glucuronides are the methyl (aglyconyl-2,3,4-tri-0-acetyl-
glucopyranosid)uronates (acetyl-methyl; 10, 16); the totally
methylated glucuronides (permethyl; 16, 17, 18, 19, 20, 21);
the methyl (aglyconyl-2,3,4-tri-0-trimethylsilyl-glucopyranosid)-
uronates (TMS-methyl; 10, 16, 22, 23); and the totally trimethyl
silylated glucuronides (perTMS; 23, 24). In all cases, the
aglycone will also be derivatized if functional groups are present
that react with the reagent used.

In general, the permethyl glucuronides have shorter retention times on the glc column than the other derivatives (16). They also exhibit good thermal stability; however, the permethyl deriv- ative of chloramphenicol glucuronide was not stable to glc (21). The acetyl-methyl glucuronides exhibit much less stability during glc separation (16, 24, 25); however, they are stable to mass spectral analysis (24, 25).

The perTMS and TMS-methyl derivatives of glucuronides exhibit good stability to glc and their retention times tend to be between those of the permethyl and acetyl-methyl glucuronides (16). The perTMS derivatives offer an added advantage in that the glucur- onide can be trapped from the glc and the TMS groups easily re- moved with aqueous methanol to recover the intact glucuronide. The glucuronide can then be subjected to other derivatization techniques and/or glucuronidase hydrolysis.

Further investigation of the use of glc as a method for separation and purification of glucuronide derivatives will be needed before any general rules as to the expected stability of various glucuronides and glucuronide-derivative types can be made. However, the perTMS and TMS-methyl derivatives of approximately twenty glucuronide conjugates of xenobiotic metabolites have been subjected to glc at this laboratory and all exhibited good thermal stability. These derivatives ranged in mass from 600 to 860 and included glucuronide conjugates of metabolites from diethyl stil- besterol (one glucuronide), terbutryne (five), crufomate (seven), propachlor (five), and 3,5,6-trichloropyridin-2-ol (one). These were all chromatographed using temperature programming (10° per min from $100^{\circ}C$) on 6' X 1/8" 3% SE-30 columns.

Of these twenty perTMS glucuronides (characterized by mass spectrometry), five eluted from the glc between 225 and $234^{\circ}C$, seven eluted between 240 and $250^{\circ}C$, and eight between 270 and $280^{\circ}C$. It is, therefore, apparent that, under the conditions used, glc cannot be relied upon to separate closely related TMS-methyl or perTMS glucuronides. However, if mass spectrometry is used, especially high resolution, for the identification (characteriza- tion), structures can be assigned to the glucuronides in a mixture.

Identification Methods--Mass Spectrometry

The classical methods involving glucuronidase hydrolysis and identification of the aglycone will not be covered in this dis- cussion. The properties of glucuronidase have been covered (1), and methodology for identification of aglycones involves the skills required to identify any metabolite. This discussion will involve mainly electron impact mass spectrometry as a tool to characterize the structure of derivatized glucuronide conjugates.

Mass spectrometry is rapidly becoming the method of choice for the initial characterization of glucuronide conjugates. In con- junction with hydrolysis (enzymatic or chemical) and identification of the aglycone using infrared, proton magnetic resonance, and mass

spectra, most glucuronides can be identified. If the glucuronide
can be rendered volatile by derivatization, its mass spectrum can
be obtained by conventional means (i.e., electron impact). If the
derivatized glucuronide is unstable or does not give sufficient
aglycone containing fragment ions by electron impact, chemical ionization can be attempted (24) or possibly field desorption (27,
28). These latter two methods require special ion sources not
generally available on existing spectrometers.

If the derivatized glucuronide is stable to glc, the gas
chromatograph can be used as a means of introduction of the derivative into the mass spectrometer (21, 23).

The presence of a derivatized glucuronide in the sample introduced into the mass spectrometer is readily determined by fragment
ions in the mass spectrum that are characteristic for the derivatized glucuronic acid moiety. These fragment ions have been
determined for the permethyl derivatives (18, 21), the acetyl-
methyl derivatives (29), and the TMS-methyl and perTMS derivatives
(23). These fragment ions are listed in Table I.

Table I.   Fragment Ions Diagnostic for the Derivatized Glucuronic
           Acid Moiety.

Permethyl (m/e)	Acetyl-Methyl (m/e)	TMS-Methyl (m/e)	PerTMS (m/e)
233 or 232[1]/	317	407	465
201*[2]/	257	406	464
141	215	317*	375*
116	197	217	217
101	173	204	204
88	155*	423[3]/	481[3]/
75	127		
	43		

1/   m/e 232, when present, indicates a phenolic glucuronide;
     m/e 233, when present, indicates an aliphatic glucuronide.
2/   An asterisk indicates an intense ion.
3/   These ions, when present, indicate an aliphatic O-glucuronide.

Some glucuronides will form the 4-5 dehydro analogue of the
derivatized glucuronic acid moiety during either derivatization,
glc, or thermally in the mass spectrometer (16, 21, 30). This
chemical or thermal degradation process cannot be distinguished
from the electron impact fragmentation mode which gives glucuronic
acid moiety fragment ions of the same mass. This degradation

product, though undesirable, still gives interpretable mass spectral data since usually only the highest mass ion for each derivative listed in Table I is missing. This degradation reaction
(elimination of methanol, acetic acid or trimethylsilanol) will
also be apparent in the molecular ion region.

In interpretation of electron impact mass spectra for the
presence of acetyl-methyl glucuronides, one must be aware that
the mass spectrum from methyl (1,2,3,4-tetra-0-acetyl-gluco-
pyranosid)uronate contains all the fragment ions listed in
Table I and no molecular ion. Also the mass spectrum from permethyl glucuronic acid contains all ions in Table I except for
the m/e 232 and/or 233 ions and also no molecular ion (21). The
ions at m/e 232 and/or 233 are not always present or can be of
very weak intensity in the spectra from permethyl glucuronides.
Therefore, unless the sample has been subjected to separation
techniques which would assure that derivatized glucuronic acid
is not present in the sample, other ions in the mass spectra must
be present to confirm the presence of a glucuronide. These are
ions that contain both the glucuronic acid and aglycone moieties
(Table II). These ions are important for determining the mass of
the glucuronide (if the molecular ion is not present) and the mass
of the aglycone moiety which will be discussed later.

Of the fragment ions listed in Table I for the TMS-methyl
and perTMS derivatives of glucuronides, only the ions at m/e 217
and 204 are present in mass spectra from methyl(1,2,3,4-tetra-0-
trimethylsilyl-glucopyranosid) uronate and perTMS-glucuronic acid
(23). The remaining fragment ions listed for each derivative are
confirmation that a glucuronide is present, i.e., something other
than TMS is the aglycone.

Once the presence of a derivatized glucuronide is established, the mass of the aglycone (which will be appropriately derivatized) can easily be established if the molecular ion is present.
The number of methyl, acetyl, or TMS moieties added to the aglycone during derivatization can be determined using the appropriate deuterium labeled reagents since the number of derivatizable
functional groups on the glucuronic acid moiety remains constant.

If the molecular ion is not present, its mass can usually be
deduced from the fragment ions listed in Table II. The ions listed result from obvious fragmentations and eliminations from the
glucuronic acid moiety.

The deduced molecular ion can usually be confirmed by the
presence in the mass spectra of the aglycone containing fragment
ions or rearrangement ions listed in Table III. It is interesting to note a difference between these aglycone ions from the
various derivatives of aromatic glucuronides. The TMS derivatives
give aglycone containing ions that result from the elimination
of the glucuronic acid moiety with rearrangement of a TMS to the
aglycone. The permethyl and acetyl-methyl derivatives give
aglycone-containing ions that result from the same elimination
except that a proton rearranges to the aglycone.

Table II. Fragment Ions used to Predict the Mass of the Molecular
Ion.

Permethyl (m/e)	Acetyl–Methyl (m/e)	TMS–methyl & PerTMS (m/e)
M −30 (CH_3O)	M −59 (CH_3−COO)	M −15 (CH_3)
M −31 (CH_3OH)	M −60 (CH_3−COOH)	M −90 [$(CH_3)_3SiOH$]
		M −105. [CH_3+ $(CH_3)_3SiOH$]

Table III. Aglycone Containing Fragment Ions from Various
Aliphatic and Aromatic Glucuronides[1]

Derivative	Glycosidic linkage	m/e	Fragment ion
Permethyl	Gl–O(aro)–Agl	(M −232)	[HO–Agl]$^{+\cdot}$
	Gl–O(aliph)Agl	(M −249)	[Agl]$^{+\cdot}$
Acetyl–methyl	Gl–O(aro)–Agl	(M −316)	[HO–Agl]$^{+\cdot}$
TMS–methyl	Gl–O(aro)–Agl	(M −334)	[TMS–OAgl]$^{+}$
	Gl–S(aro)–Agl	(M −334)	[TMS–S–Agl]$^{+}$
	Gl–O(aliph)–Agl	(M −423)	[Agl]$^{+}$
PerTMS	Gl–O(aro)–Agl	(M −392)	[TMS–OAgl]$^{+\cdot}$
	Gl–S(aro)–Agl	(M −392)	[TMS–S–Agl]$^{+\cdot}$
	Gl–O(aliph)–Agl	(M −481)	[Agl]$^{+}$

[1] Gl   = The derivatized glucuronic acid moiety.
     Agl   = The derivatized aglycone moiety.
     aro   = Aromatic acetal linkage.
     aliph = Aliphatic acetal linkage.

The fragment ions in Tables I and III can also be used to
predict the presence of a derivatized glucuronide and its molecular ion if it can be established that a glucuronide was present
and not a mixture of glucuronic acid and some metabolite.
These ions are especially important if the molecular ion and the
ions in Table II are not present or are of very weak intensity.
This method could be applied to the TMS derivatives with more
confidence since the ions in Table I would rule out the presence
of glucuronic acid.   The method could be applied to permethyl
derivatives with equal confidence if either the m/e 232 or 233
ion were present.
      In summary, mass spectrometry of glucuronide derivatives can
yield the following information: a) The presence of a glucuronide
from the fragment ions listed in Table I.   b) The molecular weight
of the derivatized aglycone. c) The number of derivatizable functional groups on the aglycone by use of dueterated derivatization
reagents.   d) The elemental composition of the derivatized aglycone
from high resolution data.   e) The type of glycosidic linkage,
whether aliphatic or aromatic from the designated ions listed in
Tables I and III.   This information, along with data from the free
aglycone, usually yields enough information to direct synthesis
efforts toward the right compound for identification of the aglycone.
      The author prefers to use the perTMS derivatives for the fol-
lowing reasons: a) A one-step derivatization.   b) Relatively good
stability of these derivatives to glc.   c) The presence of a perTMS
glucuronide can be determined with confidence.   d) Most glucuronides have given molecular ions and all have given M -15 fragment
ions.   4) The TMS moieties can be easily removed after glc purification for subsequent glucuronidase studies.   5) The presence of
functional groups on the aglycone that react with diazomethane can
be determined by comparing the mass spectrum from the perTMS derivative with that from the methyl TMS derivative.
      The major disadvantage to the TMS derivatives is their labil-
ity to hydrolysis.   The mass spectrum should be obtained as quickly
as possible after preparation or trapping from the glc.   Also,
liquid chromatographic techniques are not applicable to these derivatives due to their ease of hydrolysis.
      Two examples will point out the applicability of mass spectrom-
etry and exact mass determinations to the identification and differentiation of isomeric glucuronides.
      Two of the seven perTMS glucuronides of crufomate metabolites
gave the same molecular ion and fragment ions in Table II.   Both
mass spectra contained the fragment ions diagnostic for perTMS
glucuronides (Table I) except that one contained the m/e 481 ion.
The elemental composition of the m/e 481 ion was determined by
precise mass measurement.   Its composition was consistent with the
ion resulting from the homolytic cleavage of the bond between the
exocyclic acetyl oxygen and the aglycone with the charge remaining
on the glucuronic acid moiety.   The mass spectrum from the other

perTMS glucuronide contained an intense chlorine containing M -392 (Table III) fragment ion (m/e 344) which indicated a phenolic glucuronide.

The two perTMS glucuronides were trapped from the gas chromatograph, the TMS groups removed with aqueous methanol, and the free glucuronides subjected to glucuronidase hydrolysis. The same aglycone was obtained from both glucuronides. The structure of the aglycone was confirmed by synthesis to be 4-(1',1'-dimethyl-2'-hydroxyethyl)-2-chlorophenol.

The TMS-methyl derivatives of both glucuronides were prepared. The mass spectrum from the TMS-methyl derivative of the perTMS glucuronide that gave the m/e 481 ion showed that it had reacted with two moles of diazomethane and the m/e 481 ion now appeared at m/e 423 (Table I). Therefore, the glycoside linkage in this glucuronide was through the aliphatic hydroxyl of the aglycone.

The mass spectrum from the TMS-methyl derivative of the other glucuronide showed that it had reacted with one mole of diazomethane and still contained the intense chlorine-containing ion at m/e 344 (M -334, Table III).

The above mass spectral data from the two derivatives along with identification of the aglycone established the structures of these two isomeric glucuronides. In this example, it was essential that the two glucuronides be separated from one another before mass spectral analysis, for glucuronidase studies would have given the same aglycone. This separation was effected using the LH-20(HCO$_3$$^-$) column.

Two isomeric glucuronides were isolated (as perTMS derivatives separated by glc) from the urine of sheep dosed with propachlor (N-isopropyl-α-chloroacetanilide). The perTMS derivatives of both isomers gave molecular ions of weak intensity and the fragment ions listed in Table II. Fragment ions were also present which were diagnostic for the presence of perTMS glucuronides (Table I). One perTMS derivative gave a weak intensity m/e 481 ion (Table I) and an intense M-481 ion (Table III). The other gave an intense M -392 ion (Table III). From these data, it was assumed that the former was an aliphatic glucuronide and the latter an aromatic glucuronide.

The aglycones were obtained by glucuronidase hydrolysis. Mass spectra were obtained from the TMS derivatives of the aglycones and exact masses of the molecular ions and major fragment ions were obtained.

The molecular ions from both TMS-aglycones had the same elemental composition. Both mass spectra contained intense fragment ions at M -79. The 79 amu leaving group was calculated to have a mass of 78.98389, i.e., a negative mass defect, which indicated the presence of oxygen, sulfur, or silicon or a combination of any two or all of these elements. A computer search for elemental compositions for this mass gave SO$_2$CH$_3$ as the best fit. This SO$_2$CH$_3$ group was assumed to have replaced the chlorine in the original propachlor molecule.

The most informative differences in the mass spectra from the two TMS–aglycones were that the aliphatic TMS–aglycone gave an intense ion at M -103 [103 = $(CH_3)_3SiOCH_2$] usually indicating the TMS ether of a primary alcohol, and the aromatic TMS–aglycone gave an intense ion at M -42 which resulted from the elimination of propene from the molecular ion. Both of these were confirmed by exact mass determinations. The elimination of propene from the aromatic TMS–aglycone supported the presence of the N–isopropyl group and, therefore, the placement of the SO_2CH_3 moiety on the acetyl group. The structure of the aromatic aglycone was confirmed by synthesis and was shown to be ring–hydroxylated in the para position.

The aliphatic aglycone has not been synthesized however; the absence of the M -propene ion and the presence of the M -$(CH_3)_3$-$SiOCH_2$ ion place the hydroxyl group on a methyl group of the isopropyl moiety with the assumption that the SO_2CH_3 is on the acetate.

From these data, the structures were assigned to the two isomeric glucuronides. This example demonstrates the value of high resolution mass spectrometry for determining the elemental compositions of fragment ions in the assignment of structures because of the unexpected metabolic transformations of the xenobiotic which led to glucuronide formation. In this case, separation of the two glucuronides would not have been essential, for glucuronidase studies would have demonstrated the presence of two aglycones.

Literature Cited

1. Dutton, G. J., "Glucuronic Acid," Academic Press, New York, N.Y., 1966.
2. Mattox, Vernon R., Goodrich, June E. Vrieze, Wiley D., Biochemistry (1969) 8, 188.
3. Chang, T.T.L., Kuhlman, Ch. F., Schillings, R. T., Sisenwine, S.F., Tio, C. O., Ruelius, H. W., Experientia (1973) 29, 653.
4. Aschbacher, P. W. (personal communication).
5. Bakke, J. E. (unpublished data).
6. Bakke, J. E., Feil, V. J. Fjelstul, C. E., Thacker, E. J., J. Agr. Food Chem. (1972) 20, 384.
7. Mattox, Vernon R., Litwiller, Robert D., Goodrich, June E., Biochem. J. (1972) 126, 533.
8. Assandri, Alessandro, Perazzi, Antonio, J. Chromatog. (1974) 95, 213.
9. Knaak, J. B., Tallant, N. J., Bartley, W. J., Sullivan, L. J., J. Agr. Food Chem. (1965) 13, 537.
10. Knaak, J. B., Eldridge, J. M., Sullivan, L. J., J. Agr. Food Chem. (1967) 15 605.
11. Van Der Wal, Sj., Huber, J.F.K., J. Chromatog. (1974) 102, 353.
12. Paulson, Gaylord D., Jacobsen, Angela M., J. Agr. Food Chem. (1974) 22, 629.

13. Bakke, J. E., Robbins, J. D., Feil, V. J., J. Agr. Food Chem., (1967) 15, 628.
14. Larsen, G. L., Bakke, J. E.   Abstr. 168th ACS National Meeting, Atlantic City (1974) Division of Pesticide Chemistry Abstr. #4.
15. Mattox, Vernon R., Goodrich, June E., Litwiller, Robert D., J. Chromatogr. (1972) 66, 337.
16. Imanari, Toshio, and Tamura, Zenzo. Chem. Pharm. Bull. (1967) 15, 1677.
17. Thompson, R. M., Desiderio, D. M., Biochem. Biophys. Res. Comm. (1972) 48, 1303.
18. Thompson, R. M., Gerber, N., Seibert, R. A., Desiderio, D. M., Res. Commun. Chem. Pathol. Pharmacol. (1972) 4, 543.
19. Gerber, N., Seibert, R. A., Thompson, R. M., Res. Commun. Chem. Pathol. Pharmacol. (1973) 6, 499.
20. Gerber, N., Olsen, G. D., Smith, R., Griffin, D., Deinzer, M. L., Pharmacologist (1974) 16, 225.
21. Thompson, R. M., Gerber, N., Seibert, R. A., Desiderio, D. M., Drug Metabolism and Disposition (1973) 1, 489.
22. Horning, E. C., Horning, M. G., Ikekawa, N., Chambaz, E. M., Jaakonmaki, P. I., Brooks, C.J.W., J. Gas Chromatog. (1967) 5, 283.
23. Billets, Stephen, Lietman, Paul S., Fenselau, Catherine, J. Med. Chem. (1973 16, 30.
24. Chang, T.T.L., Kuhlman, Ch. F., Schillings, R. T., Sisenwine, S. F., Tio, C. O., Ruelius, H. W., Experientia (1973) 29, 653.
25. Paulson, G. D.   (private communication).
26. Games, D. E., Games, M. P., Jackson, A. H., Olavesen, A. H., Rossiter, M., Winterburn, P. J., Tetrahedron Letters (1974) 2377.
27. Schulten, H. R., Games, D. E., Biomed. Mass Spec., in press.
28. Paulson, G. D., Zaylskie, R. G., Dockter, M. M., Anal. Chem. (1973) 45, 21.
29. Björndahl, H., Hellerqvist, C. G., Lindberg, B., Svensson, S., Angew. Chem. Int. Ed. Engl. (1970) 9, 610.

5

Amino Acid Conjugates

RALPH O. MUMMA and ROBERT H. HAMILTON

Departments of Entomology and Biology, Pesticide Research Laboratory and Graduate Study Center, The Pennsylvania State University, University Park, Pa. 16802

Amino acid conjugates of pesticides have been reported in both plants and animals. Perhaps the most well known conjugates are the glycine conjugates of acidic materials which are commonly found in animals (1-3). The glycine conjugate of benzoic acid, hippuric acid, was first isolated by Liebig in horse urine in 1829 (4). Knoop found in 1904 that long chain phenylalkyl acids fed to dogs were excreted as glycine conjugates of either phenyl-acetic acid or benzoic acid suggesting the existence of the β-oxidation pathway (5). Nicotinic acid also is excreted in urine of man as the glycine conjugate (nicotinuric) while phenylacetic acid is excreted as the glutamine conjugate (5). Birds excrete both these substances as the diornithine conjugates (5).

Most of the pesticides that are recognized to form amino acid conjugates in plants are acidic insecticides, fungicides and herbicides but primarily the latter. The aspartic acid conjugate of indole-3-acetic acid, phenoxy herbicides and auxin-like plant growth regulators has been reported in plants by numerous laboratories (6-25). To further complicate the picture Feung et al. (26, 27) identified six additional amino acid conjugates (glutamic acid, alanine, valine, leucine, phenylalanine and tryptophan) of 2,4-D in soybean callus tissue. Figure 1 shows some examples of simple amino acid conjugates all involving conjugation through an α-amide bond. More complex amino acid conjugates have been reported such as the glutathione conjugate of triazines and diphenylether herbicides (28-33). However, in this case the glutathione is conjugated by means of a sulfur-carbon bond and its biochemical origin is different from amide linked amino acid conjugates. The glutathione conjugates are covered in more detail in another chapter (see Chapter by D. H. Hutson).

Some other amino acid conjugates not linked through amide linkages with the α-amino group have been reported. For example 3-amino-1,2,3-triazole has been reported conjugated with alanine, glycine or serine (34) and alanine conjugates of N,N-dialkyldithio carbamates have also been reported (35).

Although amino acid conjugates of auxin type herbicides would now appear to be commonly found in plants; their levels, mechanism of formation and biological significance remain to be evaluated. With regard to levels of specific conjugates, Feung et al. (36) have shown that while 2,4-D amino acid conjugation was common to five plant callus tissues, the kind and percentage of each conjugate was species specific. In addition the levels and amounts of conjugates also varied with time, thus when soybean callus was incubated with 2,4-D the ether-soluble metabolites (amino acid conjugates) are very rapidly formed (Figure 2) but degrade with time (8). This was particularly significant in the case of the glutamic and aspartic acid conjugates (Figure 3). Other metabolites accumulated in longer exposure times at the expense of the glutamic and aspartic conjugates. Although no enzymatic conjugating system has yet been demonstrated there is evidence that the aspartic conjugation system is inducible (37). The biological significance of these conjugates also demands greater attention since the 2,4-D amino acid conjugates are biologically active, they stimulate plant cell division and cell elongation at concentrations typical of auxins (38).

The chemical, physical and biological properties, the isolation, identification and analytical methods for amino acid conjugates will be discussed. Since this report is to reflect the state of the art of work with amino acid conjugates, most of the examples will be taken from our own investigations of amino acid conjugates of 2,4-D. These data represent the most extensive investigation of amino acid conjugates.

Chemical and Physical Properties

Most amino acid conjugates behave as weak acids. They are soluble in water under basic conditions and insoluble under acidic conditions. At pH 7 they are usually soluble in polar organic solvents such as methanol, ethanol, 1-butanol and acetone. At pH 3 or lower most amino acid conjugates are nonionized and extractable into ethyl ether. Also 1-butanol extracts the conjugates out of water at all pH's. Some amino acid conjugates complicate the extraction procedure because of being difunctional such as the dicarboxylic amino acids, glutamic and aspartic acids, and the basic amino acids such as arginine, lysine, and histidine. The latter three are ionized or zwitter ions at all pH's and do not extract with ethyl ether. At pH 3 four extractions are usually necessary to effectively extract the amino acid conjugates out of water with ethyl ether.

The amino acid conjugates of 2,4-D can be easily crystallized from aqueous-alcoholic solvents under acidic conditions. All possess low volatility and high melting points. Most of these conjugates are relatively stable in acid and basic solutions at room temperatures. The amino acid conjugates are readily hydrolyzed in

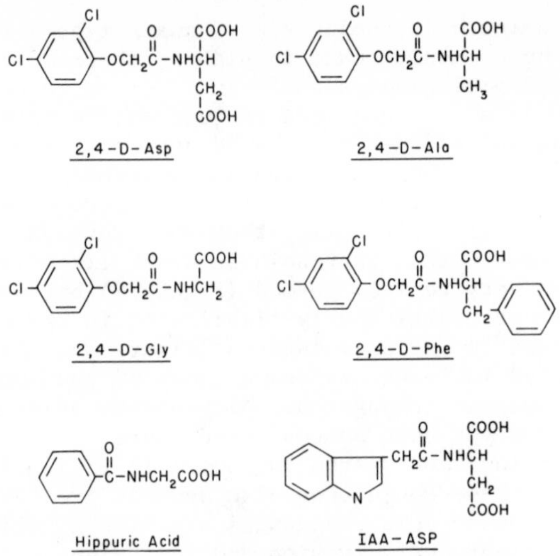

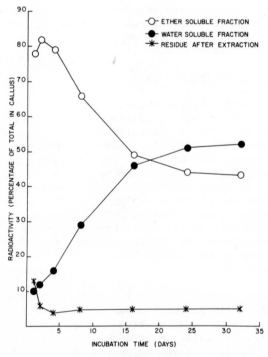

Figure 1. Typical amino acid conjugates

Figure 2. Distribution of the radioactivity taken up
by the soybean callus in water-soluble, ether-soluble
(amino acid conjugates), and residue fraction

6 N HCl at 70°C in 24 hours to free amino acids and the parent
acidic pesticide.

Amino acid conjugates readily form esters with alcohols and
this sometimes causes problems in the purification of the isolated
metabolites. Since methanol, ethanol, and 1-butanol are used in
our isolation procedure and in the chromatographic solvents,
significant amounts of these esters were isolated and readily
detected by mass spectroscopy.

All the amino acid conjugates of 2,4-D that were tested were
partially hydrolyzed by Emulsin (27, 39) which is a crude enzyme
preparation used to hydrolyze β-glucosides. With these results
one could assume the isolated amino acid conjugates actually were
β-glucosides, however, Emulsin hydrolyzed the synthetic amino
acid conjugates also. Evidently the enzymatic preparation contains
sufficient peptidase to effect the hydrolysis.

Nineteen amino acid conjugates of 2,4-D were prepared by the
reaction of 2,4-dichlorophenoxyacetylchloride with the corresponding L-amino acid in aqueous sodium hydroxide (40) as is
shown in Figure 4.

The N^{α}-lysine conjugate was prepared in a slightly different
manner. The basic ε-amino group of lysine was derivatized to a
carbobenzoxy group which was eventually removed by hydrogenation
(42). A slightly modified reaction has been used to prepare the
amino acid conjugates of indole-3-acetic acid (43).

Since both paper and thin-layer chromatography are so important in the isolation, purification, and identification of amino
acid conjugates, special emphasis must be placed on the proper
selection of good solvent systems. Since theoretically nearly
twenty amino acid conjugates are possible, the chromatographic
solvent system must be able to separate most of the conjugates.
Table I shows the mobility of ten selected amino acid conjugates
of 2,4-D in thin-layer (TLC) and paper chromatographic (PC) solvent systems (42). Valine, leucine and isoleucine conjugates have
similar chromatographic properties and are difficult to separate.

Chromatographic techniques must be used for identification
purposes when only trace amounts (<0.01 µg) of radiolabeled
materials are available. Thus multiple chromatographic solvent
systems are required to produce sufficient data to permit identification of the unknown with a reasonable degree of confidence.
Obviously, previously synthesized standards must always be chromatographed along with the unknowns since chromatographic metabolites are not always reproducible due to variations in temperature, humidity, etc. However, when sufficient sample is available
(>0.01 µg) mass spectrometric analysis can be extremely useful.
For example, nearly all of the amino acid conjugates of 2,4-D or
of indole-3-acetic acid gave molecular ions and characteristic
fragmentation patterns typical of both the amino acid and of the
parent pesticide (42, 43). The upper region of the spectra (>m/e
219) is characteristic of the specific conjugate and particularly
useful for identification purposes. Figure 5 illustrates the

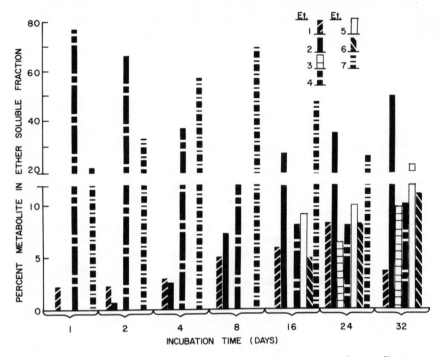

Figure 3.   *Relative amounts of the ether-solubles isolated from soybean callus tissues grown for different times in 2,4-D-1-^{14}C. E_{t}^{4} = glutamic acid conjugate, E_{t}^{7} = aspartic acid conjugate.*

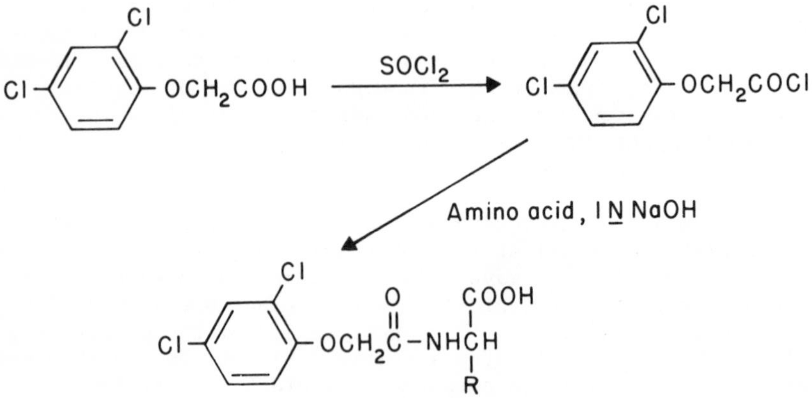

Figure 4.   *General scheme for the synthesis of amino acid conjugates of 2,4-D*

Table I. R_f Values of Amino Acid Conjugates of 2,4-D.

Com-pound	Solvent System[a]						
	TLC						PC
	I	II	III	IV	V	VI	VII
Gly	0.00	0.27	0.00	0.17	0.00	0.80	0.68
Ala	0.35	0.33	0.48	0.20	0.54	0.77	0.74
Ser	0.14	0.17	0.12	0.11	0.35	0.73	0.65
Val	0.36	0.39	0.52	0.27	0.50	0.74	0.79
Leu	0.40	0.42	0.56	0.29	0.58	0.75	0.82
Ile	0.42	0.43	0.56	0.27	0.58	0.74	0.82
Asp	0.17	0.01	0.26	0.03	0.31	0.71	0.36
Glu	0.13	0.02	0.21	0.03	0.30	0.71	0.43
Phe	0.33	0.37	0.49	0.25	0.49	0.74	0.80
Trp	0.27	0.28	0.32	0.18	0.42	0.80	0.80

[a]I, benzene–dioxane–formic acid (90:25:2, v/v/v); II, chloro-form–methanol–concentrated ammonium hydroxide (70:35:2, v/v/v); III, diethyl ether–petroleum ether (60–70°)–formic acid (70:30:2, v/v/v); IV, benzene–triethylamine–methanol–concentrated ammonium hydroxide (85:15:20:2, by vol); V, benzene–methanol–cyclohexane–formic acid (80:10:20:2, by vol); VI, 1-butanol–acetic acid–water (90:20:10, v/v/v); and VII, 1-butanol–95% ethanol–3 $\underline{N}$ ammonium hydroxide (4:1:5, v/v/v).

mass fragmentation (>m/e 219) of 2,4-D-Ile. It gives a strong
molecular ion (53%, m/e 333) and fragments typical of the amino
acid. The main fragments >m/e 219 can be grouped into four types
as follows: (a) parent-Cl (P-35); (b) P-COOH (P-45); (c) P-H$_2$O
(P-18); and (d) P-side chain fragmentation. The side chain
fragmentation is similar to the side chain fragmentation previously reported for peptides and derivatives and are characteristic
of the amino acid (44).

Figure 6 shows the prominent mass spectral ions arising from
the fragmentation of the 2,4-D portion of the molecule and are
typical for all the amino acid conjugates as well as 2,4-D (42).
The presence of the characteristic chlorine isotope peaks permits
identification of metabolites even when significant impurities
are present.

Isolation, Purification and Identification

In our hands the procedure of the extraction of the plant
tissue and the time involved in this procedure depends upon the
tissue being examined. Pesticide metabolism studies with plant
callus tissues offers many advantages over using the whole plant.
Callus tissue does not require a lighted growth chamber. It is
sterile, uses inexpensive equipment, requires little space,
usually does not contain many interfering substances and offers
versatility in comparing metabolism in different plants at the
same time by using different plant callus tissue. Whole plants,
on the other hand, do require controlled environmental growth
chambers or greenhouse space and contain significant interfering
phenolic substances and pigments. Obtaining sterile intact plants
is also not usually feasible and restricts the method of treatment
if microbial metabolism is to be avoided.
Usually we can isolate and identify a metabolite from plant
callus tissue in 1/5 to 1/15 the time it takes to work with the
whole plant. The large amount of plant pigments and sugars often
causes streaking of chromatograms and thus does not give good
separations as is typical of callus tissue. Therefore, most of
our identification of metabolites have been performed with plant
callus tissues. However, once the metabolites have been identified the relative amount and types of metabolites must be

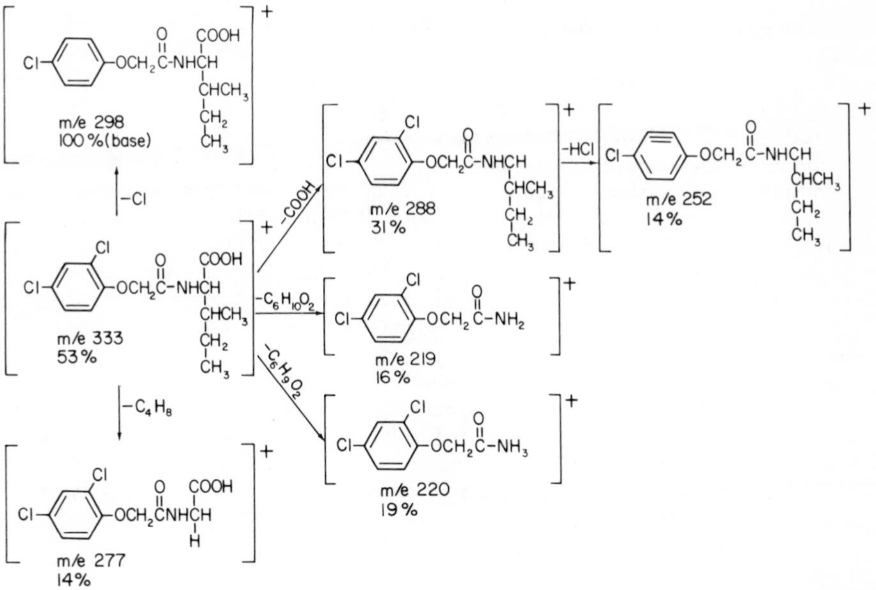

Figure 5. *Prominent mass spectral ions arising from fragmentation of 2,4-D-Ile (m/e >219)*

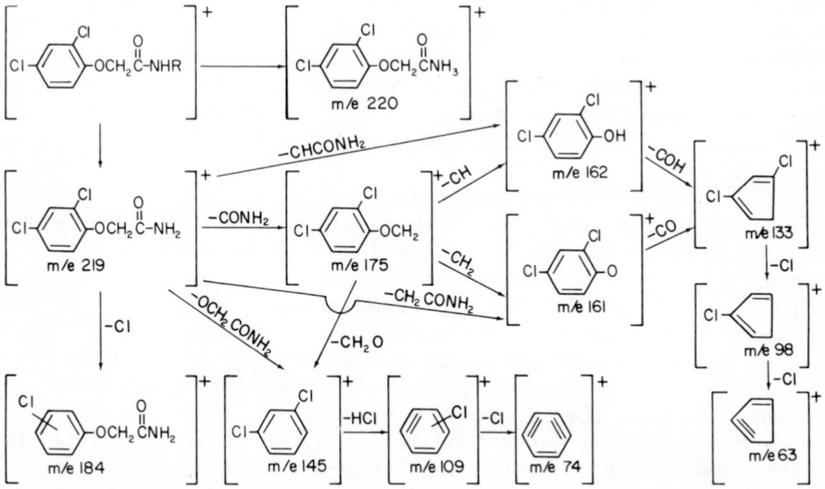

Figure 6. *Prominent mass spectral ions arising from the fragmentation of the 2,4-D portion of the molecule of amino acid conjugates of 2,4-D*

determined in the whole plant in order to determine the significance of the metabolites.

Figure 7 shows the isolation scheme of amino acid conjugates
of 2,4-D from plant callus tissue. The frozen tissue was homogenized in 95% ethanol in a Waring Blendor. The homogenate was
filtered with suction, and the residue rinsed thoroughly with 80%
ethanol. The pooled filtrate was evaporated and the aqueous
concentrate (adjusted to pH 3.0) was extracted four times with
ethyl ether. The aqueous layer contains the glucosidic conjugates
including the hydroxylated metabolites. The water fraction was
then usually extracted four times with equal volumes of 1-butanol
(water saturated). The ethyl ether fraction contains the amino
acid conjugates of 2,4-D as well as any free 2,4-D. The ethyl
ether extract was concentrated and the components separated on
paper chromatography. The radioactive compounds were eluted from
the paper chromatograms and purified by thin-layer chromatography.
The solvent systems that we found best suited for our work are
given in Table I. The thin-layer tanks always contained a paper
liner which does affect the mobility of the solvent system and the
silica gel layer was activated (135°C for 4-8 hr). Figure 8 shows
a typical separation on paper chromatograms of whole plant and
callus tissue extracts. As indicated the callus tissue extracts
give much better resolution. Usually only one additional thin-
layer chromatographic separation of each eluted band is necessary
to obtain sufficient purity for mass spectral analysis. All compounds eluted from radioactive bands are subjected to acid hydrolysis (6 $\underline{N}$ HCl, 70°C, 24 hr), enzymatic hydrolysis (Emulsin,
Nutritional Biochemical Company) and chromatographic characterization in all the listed thin-layer and paper chromatographic
solvents, including comparison with standard synthetic compounds.
Following this procedure the sample is analyzed in a mass spectro-
meter·via a solid probe inlet. Unfortunately mass spectrometric
analysis destroys the sample and requires a relatively pure finite
(>0.01 μg) amount of compound. This amount of material is sometimes hard to obtain when metabolites are present in small quantities. When insufficient amounts of sample are available for mass
spectrometric analysis, identification must be based purely upon
chromatographic data.

Since the concentration of amino acid conjugates in the
callus tissue varies greatly with time of exposure ($\underline{8}$) it is desirable to determine the concentration of metabolites after
several time intervals as is illustrated in Figure 2 for soybean
callus treated with 2,4-D. The ether-soluble fraction (conjugates)
decreases with time while the water-soluble metabolites increase.
In fact the major conjugates (glutamate and aspartate) both increase and later decrease over definite time intervals, first the
glutamic and then the aspartic (Figure 3).

A comparison of the metabolism of 2,4-D (Table II) shows the
relative amount of 2,4-D-Asp and 2,4-D-Glu in the ethyl ether
extract of six different plant callus tissues (carrot, jackbean,

Callus Tissue (2,4-D)

1. Ethanol homogenization

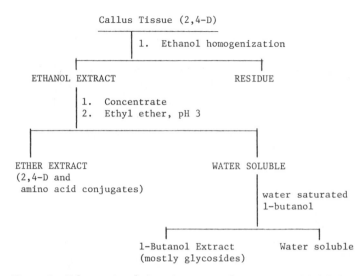

ETHANOL EXTRACT   RESIDUE

1. Concentrate
2. Ethyl ether, pH 3

ETHER EXTRACT   WATER SOLUBLE
(2,4-D and
amino acid conjugates)
                                 water saturated
                                 1-butanol

1-Butanol Extract   Water soluble
(mostly glycosides)

Figure 7. *Scheme of isolation of amino acid conjugates of 2,4-D from plant callus tissue*

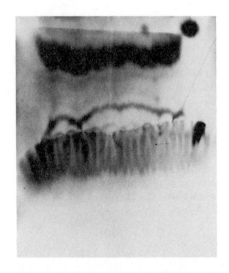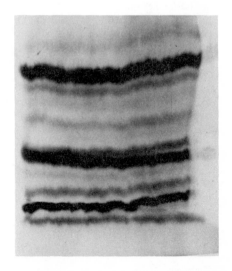

Figure 8. *Radioautography of decending paper chromatograms of ether-soluble (pH 3.0) metabolites of 2,4-D-1-^{14}C isolated from soybean plant (A) and soybean callus tissue (B). Solvent system: 1-butanol–ethanol (95%)–ammonium hydroxide (3N) (4:1.25:1, v/v/v); Whatman No. 1 paper.*

soybean, sunflower, tobacco and corn) (36). Once the metabolites
have been thoroughly characterized in the callus tissue, it is
easier to recognize and quantify them in the whole plant extracts.
All whole plants and plant callus tissue examined in our laboratories to date, contained some amino acid conjugates but in
varying amounts.

Table II. Relative Percentage of 2,4-D-Asp and 2,4-D-Glu in Six
           Callus Tissues Incubated with 2,4-D for 8 Days

Callus Tissue	2,4-D-Asp	2,4-D-Glu
Carrot	0	23.8
Jackbean	1.2	32.5
Soybean	3.7	12.9
Sunflower	0	5.0
Tobacco	0.8	6.7
Corn	1.6	1.3

Biological Properties and Metabolism

     Unfortunately the literature does not contain many examples
where the biological activity of amino acid conjugates has been
determined. Twenty amino acid conjugates of 2,4-D and several
amino acid conjugates of indole-3-acetic acid have been reported
to possess biological activity (38, 39). They stimulate plant
cell division and cell elongation (38, 39). Table III indicates
the elongation of Avena coleoptile sections and Table IV the
stimulation of soybean cotyledon callus tissue induced by selected
amino acid conjugates of 2,4-D. As indicated in these Tables the
amino acid conjugates of 2,4-D are biologically active at physiological concentrations (10^{-6}-10^{-7} M) and in some cases considerably more active than 2,4-D. Their physiological effect is therefore typical of the effect of the parent herbicide. At higher
than physiological concentration these amino acid conjugates
possess herbicidal properties and the D-amino acid conjugates of
2,4-D have been observed to stimulate fruit growth (41). In
addition we have determined the toxicology of a number of L-amino
acid conjugates of 2,4-D in rats and showed their LD_{50} to be
similar to that of 2,4-D.

Table III. Growth of Soybean Cotyledon Callus Tissue Induced
by Amino Acid Conjugates of 2,4-D

2,4-D or Conjugate	Relative Percent Greater or Less than 2,4-D			
	10^{-5}M	10^{-6}M	10^{-7}M	10^{-8}M
Control (no additive)		All Died		
2,4-D-Gly	+13	- 4	-14	- 35
2,4-D-Glu	+56	+29	+53	+ 85
2,4-D-Leu	+26	+28	+35	+155
2,4-D-Phe	+92	+93	+ 9	+110

Table IV. Elongation of Avena coleoptile Sections Induced by
Amino Acid Conjugates of 2,4-D

2,4-D or Conjugate	% Elongation			
	10^{-5}M	10^{-6}M	10^{-7}M	10^{-8}M
2,4-D	39	74	45	39
2,4-D-Asp	57	35	26	22
2,4-D-Ile	45	55	47	24
2,4-D-Phe	49	59	32	26
2,4-D-Try	66	41	24	22

The amino acid conjugates are capable of being metabolized to
other biologically active compounds (8). 2,4-D-Glu is metabolized
by soybean callus tissue to 2,4-D, 2,4-D-Asp and to the hydroxylated metabolites; 4-hydroxy-2,5-dichlorophenoxyacetic acid and
4-hydroxy-2,3-dichlorophenoxyacetic acid. Interestingly 2,4-D-Glu
is more rapidly metabolized by soybean callus tissue than is
2,4-D, (Table V) especially to the hydroxylated metabolites. Of
special note is that 2,4-D-Glu is metabolized to other amino acid
conjugates, particularly the aspartic conjugate.

Table V.   Relative Percentage Metabolites of 2,4-D and 2,4-D-Glu
           Incubated with Soybean Callus Tissue

Ether Soluble			Water Soluble		
Metabolite	% In Tissue*		Metabolite	% In Tissue*	
	2,4-D	2,4-D-Glu		2,4-D	2,4-D-Glu
2,4-D-Asp	3.7	11.7	(4-OH-2,5-D,	26.3	54.9
2,4-D-Glu	12.9	6.7	4-OH-2,3-D)		
2,4-D	33.7	6.0	2,4-D	0.8	4.2
Others	11.9	1.9	Others	6.7	11.7
Total	62.2	26.1	Total	33.8	70.8

*2,4-D Incubated 12 days, 2,4-D-Glu Incubated 8 days.

Analytical Methods

      Although amino acid conjugates of pesticides have been iso-
lated for many years no comprehensive investigation has been
reported concerning the development of analytical methods for
these compounds.
      Recently, in this laboratory Arjmand (45) developed an
analytical method for the analysis of nineteen metabolites of
2,4-D including the amino acid conjugates. This technique involved the gas chromatographic analysis of the trimethyl silyl
(TMS) derivatives. He showed that sixteen amino acid conjugates
could be separated and quantified when analyses were performed in
two separate columns (OV-1 and OV-17 stationary phases) with
temperature programming conditions. A typical separation is shown
in Figure 9. The proper derivatization reagent and conditions
were found to be important. Hexamethyldisilazane gave monosily-
lated amino acid conjugates while more stronger silylating
reaction conditions always resulted in a mixture of mono- and
disilylated products.
      All the TMS derivatives of the amino acid conjugates of 2,4-D
were stable and gave a linear response with a flame ionization
detector in the range of 1-10 μg. Unfortunately electron capture
detectors were not applicable since temperature program conditions
were employed and the TMS derivatives do not work well with this
detector. Figure 10 shows a typical GLC separation of the ether
extract of soybean callus tissue fortified with 30 ppm amino acid
conjugates and hydroxylated 2,4-D metabolites. Unfortunately the
percentage recovery of the amino acid conjugates from the fortified callus tissue varied greatly as evidenced in Table VI.

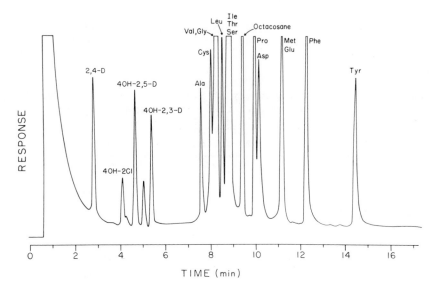

Figure 9. GLC separation of TMS derivatives of 2,4-D metabolites and amino acid conjugates. Column: 1% OV-17 on 80/100 mesh Supelcoport, 6′ × 4 mm i.d. glass. Temperature programmed at 5°/min up to 280°C, initial temperature 180°C.

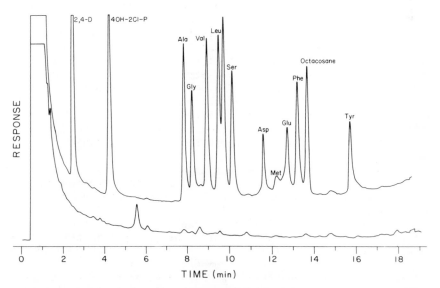

Figure 10. GLC of ether extract of soybean callus tissue fortified with 2,4-D metabolites and amino acid conjugates. Lower tracing is control tissue extract without fortification. Column: 2% OV-1 on 100/120 mesh Supelcoport, 6′ × 4 mm i.d. glass. Temperature programmed at 5°/min up to 280°C, initial temperature 180°C.

Recovery ranged from 18.5 to 91.4%. No investigation was conducted on ways to improve the recovery which obviously needs
further study.

Table VI. Percentage Recovery of 2,4-D-Conjugates from Soybean
           Callus Tissue Fortified with 30 ppm each

Compound	ppm Recovered	% Recovery
2,4-D	25.6	85.49
2,4-D-Ala	21.7	72.33
2,4-D-Val	27.4	91.40
2,4-D-Leu	24.5	81.77
2,4-D-Asp	5.6	18.53
2,4-D-Phe	21.6	71.88

Discussion

     Amino acid conjugates are obviously more wide spread in
plant tissue than once envisioned. Aspartic acid conjugates have
been found to be most abundant, however, conjugates with glutamic
acid, alanine, valine, leucine, phenylalanine and tryptophan have
been identified. Probably as more tissues and plants are examined
conjugates with additional amino acids will be found. Since
different plant tissues contain different concentrations of amino
acid conjugates, perhaps the concentration of the conjugate in the
tissue reflects a free amino acid pool size and should be examined
further. Since it is now clear that the glutamic acid conjugate
is a major metabolite, a number of reports of the aspartic acid
conjugates must be examined critically especially since it is
difficult to separate the glutamic and aspartic conjugates by
chromatography.
     Although it seems clear that amino acid conjugates are impor-
tant in plant callus tissue, just how significant these compounds
are in the whole plant remains to be proved. Unfortunately,
almost all of the 2,4-D metabolism studies with callus tissue has
been performed at physiological concentrations and metabolism of
2,4-D at herbicidal concentrations might be significantly different. The concentration of amino acid conjugates found in soybean callus tissue exhibited a temporal relationship. The highest
concentration was found the first day of exposure of the plant to
2,4-D and the amino acid conjugates steadily decreased with a
concomitment increase in the concentration of the nonbiologically

active hydroxylated metabolites. Whether a similar temporal
relationship exists in the whole plant remains to be determined.
     Although all the amino acid conjugates of 2,4-D possess
auxin-like properties, whether they express these properties as
the amino acid conjugate or some derivative or hydrolyzed product
is not known. The evidence would suggest that perhaps the conjugates may have in vivo biological activity. The fact that they
are so rapidly formed and stimulate plant cell division (at
physiological concentrations) in excess of that of 2,4-D is very
suggestive. Venis (37) has shown the amino acid conjugation of
indole-3-acetic acid and other aromatic acids is catalyzed by an
auxin (2,4-D, naphthyl acetic acid, indole-3-acetic acid)
inducible enzyme. It is interesting that the structural requirements for induction are more specific than the substrate requirements. Since 2,4-D-Glu can be converted to 2,4-D-Asp in higher
concentrations than 2,4-D itself it suggests that the plant tissue
may possess an enzyme capable of catalyzing the direct conversion
of 2,4-D-Glu to 2,4-D-Asp. The hydroxylated metabolites are more
rapidly formed from 2,4-D-Glu than from 2,4-D thus raising the
question, can the amino acid conjugates be directly hydroxylated
and if so are they required for hydroxylation? To our knowledge
the in vitro hydroxylation of 2,4-D has not been demonstrated in
a cell free system and warrants further investigations employing
amino acid conjugates as substrates.
     Usually the hydroxylated 2,4-D metabolites are present as the
glucosides, however, small amounts of the free aglycone were
reported in bean plants (8). Amino acid conjugates of hydroxylated
2,4-D metabolites have not been reported, however, we do have
preliminary evidence for their occurrence.
     A significant quantity of 2,4-D is evidently present as a
glucose ester (7). The glucose esters of the amino acid conjugates have not yet been reported, however, it is quite possible
that they may exist. Small amounts of the amino acid conjugates
are often found in the water soluble fraction after Emulsin
treatment.
     The amino acid conjugates would undoubtedly possess different
permeabilities than the parent pesticide to cytoplasmic and sub-
cellular membranes. Thus, the biological activity and rapid
metabolism of the amino acid conjugates might be owing to their
more rapid penetration of the cell than the parent pesticide,
which results in an accumulation at the target sites of biological
activity and metabolism.
     Although only the amino acid conjugates of 2,4-D and indole-
3-acetic acid have so far been studied in depth, probably other
amino acid conjugates are equally important. Possibly all the
acidic auxin-like herbicides form amino acid conjugates and need
to be reexamined in light of current thinking.
     Additional studies are needed to determine if the different
herbicidal derivatives of 2,4-D, such as the amine salts, the
butyl ester and the butoxyethanol ester, also give rise to the

amino acid conjugates when applied to plants. The metabolism of
2,4-D has been examined in only a few plants and additional
plants should be investigated.

These data collectively demonstrate the wide distribution of
amino acid conjugates, their biological significance and plant
species specificity. Hopefully these results will stimulate
other investigators to be more aware of amino acid conjugates.
Analytical methods should be modified so that amino acid and
ester conjugates can be determined in residue work. Finally the
use of sterile plant tissue cultures offers many advantages for
metabolism studies.

Literature Cited

1. Knaak, J. B., Sullivan, L. J., J. Agr. Food Chem. (1968)
    16, 454.
2. Wit, J. G., Van Genderen, H., Biochem. J. (1966) 101, 698.
3. Lethco, E. J., Brouwer, E. A., J. Agr. Food Chem. (1966)
    14, 532.
4. Fieser, L. F., Fieser, M., "Organic Chemistry", p. 2,
    Reinhold Publishing Corp., New York, 1956.
5. Fruton, J. S., Simmonds, S., "General Biochemistry", John
    Wiley & Sons, New York, 1953.
6. Andrea, W. A., Good, N. E., Plant Physiol. (1961) 32, 566.
7. Klämbt, H. D., Planta (1961) 57, 339.
8. Feung, C. S., Hamilton, R. H., Witham, F. H., Mumma, R. O.,
    Plant Physiol. (1972) 50, 80.
9. Andrea, W. A., Good, N. E., Plant Physiol. (1955) 30, 380.
10. Andrea, W. A., van Ysselstein, M. W. H., Plant Physiol.
    (1956) 31, 235.
11. Andrea, W. A., van Ysselstein, N. W. H., Plant Physiol.
    (1960) 35, 225.
12. Good, N. E., Andrea, W. A., van Ysselstein, N. W. H., Plant
    Physiol. (1956) 31, 321.
13. Fang, S. C., Theisen, P., Butts, J. S., Plant Physiol. (1959)
    34, 26.
14. Bennet-Clark, T. A., Wheeler, A. W., J. Exp. Botany (1959)
    10, 468.
15. Thurman, D. A., Street, H. E., J. Exp. Botany (1962) 13, 369.
16. Wightman, F., Can. J. Botany (1962) 40, 689.
17. Zenk, M. H., Collow. Intern. Centre, Nat. Recherche Sci.,
    Paris (1963) 123, 241.
18. Sudi, J., Nature (1964) 201, 1009.
19. Sudi, J., N. Phytotologist (1966) 65, 9.
20. Winter, A., Thimann, K. V., Plant Physiol. (1966) 41, 335.
21. Olney, H. O., Plant Physiol. (1968) 43, 293.
22. Robinson, B. J., Forman, M., Addicott, F. T., Plant Physiol.
    (1968) 43, 1321.
23. Morris, D. A., Briant, R. E., Thomson, P. G., Planta (1969)
    89, 178.

24. Beyer, E. M., Morgan, P. W., Plant Physiol. (1970) 46, 157.
25. Lau, O. L., Murr, D. P., Yang, S. F., Plant Physiol. (1974) 54, 182.
26. Feung, C. S., Hamilton, R. H., Witham, F. H., J. Agr. Food Chem. (1971) 19, 475.
27. Feung, C. S., Hamilton, R. H., Mumma, R. O., J. Agr. Food Chem. (1973) 21, 637.
28. Frear, D. S., Swanson, H. R., Pesticide Biochem. and Physiol. (1973) 3, 473.
29. Shimabukuro, R. H., Lamoureux, G. L., Swanson, H. R., Walsh, W. C., Stafford, L. E., Frear, D. S., Pesticide Biochem. and Physiol. (1973) 3, 483.
30. Shimabukuro, R. H., Swanson, H. R., Walsh, W. C., Plant Physiol. (1970) 46, 103.
31. Shimabukuro, R. H., Frear, D. S., Swanson, H. R., Walsh, W. C., Plant Physiol. (1971) 47, 10.
32. Lamoureux, G. L., Stafford, L. E., Shimbukuro, R. H., Zaylskie, R. G., J. Agr. Food Chem. (1973) 21, 1020.
33. Shimabukuro, R. H., Walsh, W. C., Lamoureux, G. L., Stafford, L. E., J. Agr. Food Chem. (1973) 21, 1031.
34. Carter, M. C., Physiol. Plant (1965) 18, 1054.
35. Kaslander, J., Sijpesteijn, A. K., Van Der Kerk, G. J. M., Biochim. Biophys. Acta (1962) 60, 417.
36. Feung, C. S., Hamilton, R. H., Mumma, R. O., J. Agr. Food Chem. (1975) 23, 373.
37. Venis, M. A., Plant Physiol. (1972) 49, 24.
38. Feung, C. S., Mumma, R. O., Hamilton, R. H., J. Agr. Food Chem. (1974) 22, 307.
39. Hamilton, R. H., Feung, C. S., Myer, H. E., Mumma, R. O., Fifth Annual Meeting American Society Plant Physiologists, June 20, Cornell University, Ithaca, New York, 1974.
40. Wood, J. W., Fontaine, T. D., J. Org. Chem. (1952) 17, 891.
41. Wood, J. W., Fontaine, T. D., Mitchell, J. W., U. S. Patent No. 2,734,816 (1956).
42. Feung, C. S., Hamilton, R. H., Mumma, R. O., J. Agr. Food Chem. (1973) 21, 632.
43. Feung, C. S., Hamilton, R. H., Mumma, R. O., J. Agr. Food Chem. (1975) in press
44. Biemann, K., Cone, C., Webster, B. R., J. Amer. Chem. Soc. (1966) 88, 2597.
45. Arjmand, M., "Quantitative Gas-Liquid Chromatographic Analysis of 2,4-D Metabolites", Masters Thesis, 1975, The Pennsylvania State University, University Park, Pa.

6

Sulfate Ester Conjugates—Their Synthesis, Purification, Hydrolysis, and Chemical Spectral Properties

G. D. PAULSON

Metabolism and Radiation Research Laboratory, Agricultural Research Service, U.S. Department of Agriculture, Fargo, N. Dak. 58102

Since the early report that the dog readily metabolizes phenol to phenyl sulfate ($\underline{1}$), a large body of information has been developed about sulfate ester conjugation. Sulfoconjugates are a very diverse and widespread group of compounds that are found in microorganisms, plants ($\underline{2}$), insect ($\underline{3-5}$), mammals, birds, reptiles, amphibia, arthropods, and mollusks ($\underline{6}$). Dodgson and Rose ($\underline{7}$) have classified these compounds in the following way: 1) compounds with $P-O-SO_3^-$ linkages; 2) compounds with $C-O-SO_3^-$ linkages; 3) compounds with $N-SO_3^-$ linkages; 4) compounds with $N-O-SO_3^-$ linkages; and 5) compounds with $S-SO_3^-$ linkages. This review is restricted primarily to compounds with the $C-O-SO_3^-$ linkage, and particularly to aryl sulfates because these are most commonly encountered by the pesticide chemist and have been most extensively investigated. Some aspects of compounds with the $P-O-SO_3^-$ and $N-O-SO_3^-$ linkages will be discussed as they apply to xenobiotic metabolism.

Biosynthesis and Metabolic Fate

The early studies by DeMeio ($\underline{8}$), DeMeio and Tkacz ($\underline{9}$), DeMeio et al. ($\underline{10}$), DeMeio et al. ($\underline{11}$), Bernstein and McGilvery ($\underline{12}$, $\underline{13}$), Segal ($\underline{14}$), and others demonstrated that phenols were converted to aryl sulfates by the soluble fraction of rat liver homogenates when incubated with sulfate ions, ATP, and Mg^{2+} ions. Bernstein and McGilvery ($\underline{12}$, $\underline{13}$) and Segal ($\underline{14}$) discovered that an active sulfate was formed from ATP and sulfate ion and that the active sulfate reacted with a phenol to give phenyl sulfate. Robbins and Lipmann ($\underline{15-17}$) showed that the active sulfate was adenosine-3'-phosphate-5'-phosphosulfate (PAPS), and that two enzymes were involved in the formation of PAPS ($\underline{15}$, $\underline{16}$). The first enzyme, ATP-sulfurylase (ATP:sulfate adenyltransferase, 2.7.7.4) catalyzes the reaction of ATP and SO_4^{2-} ion to give adenosine-5'-phosphosulfate (APS) and pyrophosphate, and the second enzyme APS-kinase (ATP:adenyl sulfate 3'-phosphotransferase, 2.7.1.25) catalyzes the phosphorylation of APS to give PAPS. Baddiley et al. ($\underline{18}$, $\underline{19}$)

86

confirmed the structure of both APS and PAPS by synthesis. More detailed discussions of these and related studies are available (7, 20-24).

The thermodynamically unfavorable formation of APS ($\Delta F' = +11$ kCal) from ATP and sulfate is driven by the hydrolysis of pyrophosphate and by the rapid utilization of APS by APS-kinase (7, 20, 22). ATP-sulfurylase purified from yeast requires Mg^{2+} ion and its pH optimum is 7.5-9.0. However, Dodgson and Rose (7) discussed species differences and stressed the importance of several variables when assaying for the sulfate activating systems in other preparations. Roy (20, 21) reviewed the methods available for assay of ATP-sulfurylase activity, inhibitors of this enzyme, and studies on purification of this enzyme.

The conversion of APS to PAPS by the APS kinase isolated from yeast is essentially irreversible ($\Delta F' = -5$ Kcal) and requires Mg^{2+} ion; the optimum pH for this enzyme is 8.5-9.0 (17). There is apparently no detailed information on this enzyme in animal systems although it is presumed to be present in all tissues that form PAPS (21). Assay techniques and other studies on this enzyme, as well as the properties of APS and PAPS, have been reviewed by Roy (20, 21) and Dodgson and Rose (7). Recently, Wong (25) reported a new method for measuring the activity of the enzymes that generate PAPS and of the transferase enzymes; it was postulated that ATP acts as an allosteric modifier of one of the enzymes responsible for the synthesis of PAPS.

The ability to synthesize PAPS (an energetically expensive process for the organism) is common to a wide variety of plants, animals, and microorganisms. The involvement of PAPS in sulfate reduction, sulfate transport, sulfoconjugation of carbohydrates, and glycolipids and in many other diverse metabolic reactions (7, 20, 21) is beyond the scope of this review. Rather, this discussion will be restricted primarily to the involvement of PAPS and sulfotransferases in the biosynthesis of aryl sulfate esters and related compounds formed in the metabolism of xenobiotics. The biosynthesis of aryl sulfate esters is accomplished by transferring the sulfate group in PAPS to a receptor (ROH) to form $ROSO_3^-$. Evidence for the formation of disulfate conjugates of di- and tri-hydric phenols has been reported (26). In some cases, amines but not thiols can substitute for ROH as acceptors (21). Apparently there is no conclusive evidence for the biosynthesis of sulfate esters of hydroxylamines in vivo (which may be due to the inherent instability of these compounds); however, the evidence for the formation and transient existence of such compounds in vitro is convincing (27-31).

Apparently, no one has isolated a sulfotransferase in pure form (21), but it is well established that there are many different sulfotransferase enzymes (7, 20, 21, 32-39). Some of the sulfotransferases apparently have a high degree of substrate specificity but this conclusion must be verified with purified enzymes.

The intracellular location of the sulfate activating and sulfotransferase enzymes has not been studied for many tissues, but
in the liver they are present in the soluble fraction of the cell.
Although sulfate ester formation has been most extensively studied
in mammalian liver, many tissues, including the kidney, intestine,
brain, adrenal, mast cells, ovary, and testis, also have the ability to synthesize PAPS (21). Powell et al. (40) demonstrated that
the gut of the rat rapidly converted phenol to phenyl sulfate and
have questioned the belief of others that the liver is the major
organ involved in the metabolism of compounds of this nature.
Their results support the conclusion that phenols per se are not
transported from the gut but are conjugated before entering the
circulatory system.

The quantitative importance of sulfate ester conjugation varies with many factors which include the following: age of the
animal (35, 41); species of animal (42); tissue (42); sex of animal (43); size of dose (44, 45); sulfur nutritional state of animal (46, 47); time after dosing (47); disease state (48); substituent effects (49); and inhibitors (43, 50).

Roy (20, 21), Dodgson (51), Gregory and Robbins (22), Dodgson
and Rose (7), and Young and Maw (52) have reviewed the evidence
for the synthesis of sulfate esters by metabolic routes other than
those utilizing PAPS. Perhaps the best evidence for alternate
pathways has been obtained with lower animals such as mollusks,
but there are suggestions that other routes may occur in higher
animals as well. Ascorbic acid 3-sulfate and unknown sulfate
donors have been implicated. Some proposed mechanisms have been
discounted (21); but the possibility of alternate routes of sulfate ester biosynthesis has not been completely investigated.

The metabolic fate of some sulfate esters in animals has
been investigated. Studies, such as those reported by Flynn et al.
(53) and Hawkins and Young (54) demonstrated that many sulfate
esters are quickly eliminated in the urine with little or no metabolism. Park (6) reported that aryl sulfates were eliminated in
the urine by active transport. Curtis et al. (55) compared the
renal clearance of inulin and a series of aryl sulfates at different plasma concentrations and found that sulfate esters were se-
creted by the renal cells. The contribution of the renal secre-
tory process to overall urinary excretion ranged from 22 to 87%.
They concluded that the rapid elimination of the aryl esters studied was due to the rapid secretion of these compounds rather than
prevention of tubular resorption.

However, some sulfate esters are extensively metabolized to a
variety of products which include mercapturic acid derivatives
(56), doubly conjugated derivatives (53, 57), other compounds formed without removal of the sulfate group (58), and other unidentified metabolites (59-61). The degree of metabolism of sulfate
esters may vary with the sex of the animal (62). Studies have
shown that peritoneal barriers were permeable to some aryl sulfate
esters but other barriers were not; for instance, radioactivity

did not pass into the central nervous system when [^{35}S]aryl sulfate esters were given to rats (63).

Curtis et al. (55) reported that some aryl sulfate esters were bound to plasma proteins in vivo.

Biliary secretion, which is of quantitative importance with some sulfate esters, can be influenced by external factors. For instance, Powell et al. (64) reported that the biliary excretion of phenolphthalein disulfate by the rat was decreased by administration of diethylstilbestrol sulfate and diethylstilbestrol-monoglucuronide.

Sulfatase Enzymes

Because the pesticide chemist frequently uses sulfatase enzymes to cleave sulfate ester conjugates, a basic understanding of the kinetic characteristics and properties of the various members of this diverse group of enzymes is essential. At least six general groups of enzymes are responsible for the hydrolysis of sulfate esters which can be classified on the basis of their substrates as: aryl sulfatases, steroid sulfatases, mucopolysaccharide sulfatases (chondrosulfatase and heparin sulfatases), glycosulfatases, myrosulfatases, and alkyl sulfatases (22). Roy (65) and Dodgson and Rose (7) have used similar classification schemes. This discussion is restricted primarily to the aryl sulfatases because they have been studied in greatest detail and are of most interest to the pesticide chemist.

The studies that led to the discovery of aryl sulfatase enzymes have been reviewed (7). The general properties and classification of these enzymes, as well as kinetic and inhibitor studies, have been summarized (7, 20, 22, 65, 66). Enzymes for the hydrolysis of aryl sulfate esters are widespread in nature (22, 65), but the most detailed studies have been conducted with enzymes from mammals, mollusks, and microorganisms (22). The substrate specificity and properties of aryl sulfatases from different sources vary and failure to recognize this fact led to confusing and apparently contradictory results in the early studies (7). The more detailed studies (67-72) which clarified these points have been summarized (7).

It is now known that the mammalian liver contains two aryl sulfatases (designated A and B) in the lysosomes and a third form (designated C) in the microsome fraction. Aryl sulfatase A and B from mammalian lysosomes are inhibited by SO_4^{2-}, HPO_4^{2-}, and F^-, are not inhibited by CN^-, have a low pH optimum, and are most active in the hydrolysis of substrates such as nitrocatechol sulfate. These mammalian enzymes and aryl sulfatase enzymes from other sources that have similar substrate specificities and behavior toward inhibitors have been classified as "Type II enzymes." In contrast, aryl sulfatase C in the microsome fraction of mammalian liver has a pH optimum of 8, is most active on simple substrates such as p-nitrophenyl sulfate, and is inhibited by CN^-

but not by HPO_4^{2-}. This mammalian enzyme and aryl sulfatases from other sources with similar properties have been classified as "Type I Enzymes." The validity of the subdivision of the aryl sulfatases into Type I and Type II enzymes and inconsistencies that sometimes arise when the multicriteria classification system is used have been discussed (66). This classification system has shortcomings, but it is functional and should remind the pesticide chemist that the properties of aryl sulfatases from different sources may be quite dissimilar.

The aryl sulfatases catalyze the hydrolysis of the O-S bond, and the only known sulfate acceptor is water. There is no evidence that a metal is involved, and the reaction is apparently irreversible. Nicholls and Roy (66) suggested that the activation energy is probably about 12-14 Kcal/mole; however, the thermodynamics of the aryl sulfatase reaction apparently have not been studied in detail. The studies that have been reported on substrates and inhibitors of the aryl sulfatases, the active sites on these enzymes, and the kinetic and physical properties of these enzymes have been summarized (7, 66).

The procedures that have been used to assay aryl sulfatase activity have usually involved measuring the liberated phenol colorimetrically or determining the anionic form of the phenol in the visible or ultraviolet regions of the spectrum. Dodgson and Spencer (73) reviewed the methods, limitations, and problems that have been encountered with these procedures. More recent studies have been reported on direct cytochemical assay of aryl sulfatases (74), assays for sulfatases A and B (75, 76), the influence of the state of molecular aggregation on the enzymic hydrolysis of aryl sulfates (77), the tissue distribution of aryl sulfatases A and B (76), kinetic characteristics and inhibitors of aryl sulfatase A (78), electrophoretic separation and characterization of aryl sulfatase A and B (79), and evidence that cerebroside sulfates and aryl sulfates are degraded by the same enzyme (80). The latter report is of interest since humans with metachromatic leukodystrophy, a human sphingo-lipid storage disease, are deficient in aryl sulfatase A (81).

Induction of alkyl sulfatases in microorganisms has been reported (82, 83). Whether induction also occurs in higher animals and with other classes of compounds, such as aryl sulfates, has not been reported but may be worthy of further study.

Laboratory Synthesis

The most widely used methods for the synthesis of aryl sulfates employ sulfur trioxide or SO_3-amine adducts; the chemistry of sulfur trioxide, and its derivatives has been reviewed in detail (84). Many reagents have been used in this type of sulfation reaction including chlorosulfonic acid (45, 77, 85-89), triethylamine sulfur trioxide (90, 91), and pyridine sulfur trioxide (92, 93). Other reagents that have been used in the synthesis of sulfate

esters include pyrosulfate (94), fuming sulfuric acid, and sulfamic
acid (84). The use of chlor-[^{35}S]sulfonic acid in the preparation
of [^{35}S]-labeled aryl sulfate esters has been described (53, 54,
95).

Although not widely used by pesticide chemists, the reaction
of H_2SO_4 with a variety of compounds in the presence of dicyclo-
hexylcarbodiimide and a polar solvent (96-98) warrants careful
consideration. This method gives sulfate esters in good yield and
is especially useful in the preparation of sulfate esters of compounds that are unstable to reagents such as chlorosulfonic acid
and pyridine sulfur trioxide. Moreover, if the conditions are
judiciously adjusted, this procedure can be used for the selective
sulfation of polyfunctional molecules (98). This method is
particularly good for the synthesis of [^{35}S]-sulfate esters because
[$^{35}SO_4$] is readily available, relatively inexpensive, and is used
directly without conversion to chlorosulfonic acid or pyridine
sulfur trioxide.

Mumma (99) reported that ascorbic acid 2-sulfate and isopropo-
pylidene ascorbic acid sulfate acted as an *in vitro* sulfating
agent at elevated temperatures and/or in the presence of oxidiz-
ing agents. For example, alcohols such as 1-octanol and 3β-
cholestanol were readily sulfated when incubated with isopropyli-
dene ascorbic acid sulfate in the presence of bromine or when
incubated at 100°C. Quadri et al. (100) and Mumma et al. (101)
reported on the synthesis and characterization of L-ascorbic acid-
2-sulfate. The possible use of ascorbic acid sulfate and/or its
derivatives as a preparative method for sulfating phenols apparently has not been reported but may be worthy of further evaluation.

Recently, Nagasawa and Yoshidome (89) reported on the Cu(II)-
catalyzed reaction of 8-quinolyl sulfate in the synthesis of
D-galactose 6-sulfate, adenosine 5'-sulfate and dextran sulfate.
Whether this procedure can be used to prepare sulfate esters of
phenols, alcohols, and steroids waits further investigation.

Boyland and Nery (102) reported on the sulfation of phenyl-
hydroxylamine and related compounds with pyridine sulfur trioxide
and other reagents to form N-sulfonic and O-sulfonic acid derivatives (the product formed depended on the reaction conditions and
blocking groups used). These compounds were isolated as their
ammonium and potassium salts. Boyland and Nery (102) also made
the important observation that phenyl hydroxylamine-O-sulfonic acid
rearranged to 2-amino-phenyl sulfate.

The biosynthesis of sulfate esters with *in vitro* tissue preparations, fortified with PAPS or PAPS generating systems, has
been used by many workers (9, 12, 13, 32, 36, 37, 103-109). This
technique lends itself well to the synthesis of [^{35}S]-labeled sulfate esters.

The phenolsulfotransferase reaction is readily reversible
when the aryl sulfate ester involved is reactive (34, 110, 111).
For instance, Brunngraber (110) demonstrated the transfer of

sulfate from p-nitrophenyl sulfate to m-aminophenol in the presence
of phenol sulfotransferase and PAP. This fact has been exploited
as a convenient assay procedure. However, the possible use of
this technique for the synthesis of sulfate esters should be considered. This approach could be especially useful when the accept-
or molecule has one or more labile linkages.

Properties

Aryl sulfate esters are usually stable as their alkali salts,
especially when stored in the dark at low temperatures. Havinga
et al. (112) described the photochemical accelerated hydrolysis
of nitrophenyl sulfates. Aryl sulfate esters are highly soluble
in water and appreciably soluble in alcohols. An especially useful solvent is N-butanol because it can often be used to extract
aryl sulfate esters from aqueous solution. The fact that sulfate
esters form salts with organic bases such as p-toluidene (113),
p-bromoaniline (114), methylene blue (115, 116), and the amino-
acridines (72) is useful for the isolation of aryl sulfates because
most of them can be extracted from aqueous solutions with organic
solvents. Dodgson et al. (72) used 5-aminoacridine to isolate the
aryl sulfates excreted in the urine of rabbits fed p-chlorophenol
and related compounds. Roy and Trudinger (117) and Young and Maw
(52) discussed the application of this principle to the isolation
and identification of aryl sulfates.

The early work of Burkhardt (118, 119) and others established
that aryl sulfate esters are readily hydrolyzed by acids; the rate
of acid hydrolysis is increased when the sulfate moiety is attached to a position of low electron availability. For example, p-
nitrophenyl sulfate is easily hydrolyzed with acid. The mechanism
of acid hydrolysis of aryl sulfates has been studied (120-122).
Roy and Trudinger (117) discussed the problems with artifact formation when some sulfate esters are acid hydrolyzed. Batts (123)
reported that the rate of hydrolysis of sulfate esters was increased by a factor of 10^7 when the solvent was changed from pure
water to moist dioxane. Later, Goren and Kochansky (124) extend-
ed these studies and found that the solvolysis required initiation
by traces of impurities, presumably acting as an electrophile.
For instance, 2-octanol sulfate in clean teflon vessels was stable
to hot, moist dioxane.

In contrast to their lability under acid conditions, most
aryl sulfates are quite stable under basic conditions (7, 87, 117-
119). For example, Burkhardt and Lapworth (87) heated aryl sulfates to 150°C for 4 hours in strong alkali or half-concentrated
ammonia to bring about hydrolysis.

Separation and Purification Techniques

Assandri and Perazzi (125) reported on the separation of
phenolic O-glucuronides and phenolic sulfate esters by multiple

liquid-liquid partition. Their methods involved a counter current
technique with continuous flow of the solvents. However, the
isolation of sulfate esters from biological fluids, such as urine,
by this technique required pre-purification of the crude material
before the counter current fractionation procedure; impurities,
such as salts, interfered with the partition systems.

   Because of the polar nature of aryl sulfate esters, it is not
surprising that ion exchange chromatography has been used extensively in the purification of these compounds (12, 13, 45, 88,
126-131. Sephadex G-10 columns eluted with water (45, 88, 103,
130, 131), Sephadex G-15 columns eluted with water (132), and
Sephadex LH-20 columns eluted with either CH_3OH or H_2O (45, 103,
129-131) have been used for the purification of aryl sulfates. It
should be noted that mixed salt forms of sulfate esters are excreted by animals and that these different salt forms may be separated
on Sephadex LH-20 columns eluted with CH_3OH (103). Other chromatographic procedures that have been used to separate and purify
sulfate ester conjugates of pesticidal compounds include: Biogel
P-2 columns (131); XAD-2 columns (94, 116); Porapak Q columns (88,
129); and paper chromatography (88, 133, 134).

   Faakonmäki (135) reported on a direct gas chromatographic
analysis of steroid sulfates and glucuronides. Mass spectroscopy
showed that the steroid sulfates lost H_2SO_4 and a double bond was
formed giving a molecular ion 18 mass units lower than that of the
free sterol. In contrast, the glucuronic acid conjugates gave the
parent sterol. The applicability of this procedure, if any, to
aryl sulfates and glucuronides apparently has not been reported.
Preliminary studies with electron impact mass spectrometry at this
laboratory indicated that aryl sulfate esters (K salts) thermally
degrade to give a fragment corresponding to the phenol (usually
base peak) and fragments at lower masses.

Derivatization Procedures

   There apparently is little or no information in the literature concerning attempts to derivatize the sulfate group in aryl
sulfates. McKenna and Norymberski (136), Pasqualini et al. (137),
and Emiliozzi (138) reported on the formation of methylated derivatives when steroid sulfates were treated with diazomethane.
Studies at this laboratory indicated that aryl sulfates were not
methylated by diazomethane or, if they were, the products were
not stable; the latter explanation seems most likely. However,
further study is needed to clarify this point.

   In some cases, it is possible to derivatize other functional
groups in a molecule without cleaving or derivatizing the sulfate
group. For instance, Dodgson et al. (70) methylated the free
hydroxyl of a monosulfate ester of 4-chlorocatechol isolated from
rabbit urine. The methylation of the monosulfate esters of isopropyl 3,4-dihydroxycarbanilate with diazomethane (leaving the sulfate ester group intact) followed by replacement of the sulfate

ester with an acetoxy group made possible the characterization and
subsequent synthesis and identification of propham metabolites
(140).

A one-step method for replacing the sulfate group in aryl and
steroid sulfate esters with an acetoxy group (139) is useful for
characterization of compounds that are unstable to conventional
hydrolysis conditions. For example, utilization of this technique
made it possible to identify carbaryl metabolites in chicken urine
as conjugated forms of 1,5-dehydroxynaphthalene and 1,5,6-tri-
hydroxynaphthalene, compounds that are unstable to normal hydrolysis conditions (103). This technique was also used in the characterization of sulfate ester-containing metabolites of propham
(129, 140) and p-chlorophenyl N-methylcarbamate (45, 131).

Spectral Analysis

Hearse et al. (95) reported that a series of aryl sulfate
esters exhibited strong absorption from 240-280 mμ (maxima 250-275
mμ); whereas the parent phenols absorbed strongly from 270-310 mμ
(maxima 280-295 mμ). Moreover, the extinction coefficient of the
phenols was much greater than that of the sulfate esters. The
marked shift in the λmax and the increase in the extinction coefficient' associated with the conversion of the phenolic hydroxyl
to its ionized form were not shown by the corresponding aryl sulfate ester. This behavior has been exploited to develop assays
for the hydrolysis of aryl sulfates (73, 141).

Nuclear magnetic resonance studies (NMR) at this laboratory
have demonstrated that aryl sulfate esters shift the absorption
of ring protons downfield relative to their position in the spectrum of the parent phenol. As expected, the shift was greatest
for the proton ortho to the sulfate group. For instance, absorptions of the protons ortho and meta to the hydroxy in p-nitrophenol were at 6.93 and 8.10 ppm, respectively, (solvent-d_6-DMSO)
and these absorptions were shifted to 7.4 and 8.18 ppm, respectively, in the spectrum of p-nitrophenyl sulfate. NMR spectroscopy has been used in assigning structures to the mono and disulfate esters of 4-chlorocatechol (131). Further studies are
underway to more thoroughly investigate the effect of sulfate
esters on NMR absorption of aromatic compounds.

Since alkali salts of sulfate esters are solids and are only
slightly soluble in most organic solvents, the infrared spectra
of these compounds are usually measured in a KBr pellet or in a
Nujol mull. Chihara (142) reported that the spectra of sulfate
esters obtained from KBr pellets and mulls were not appreciably
different. However, it should be noted that the infrared spectra
of different salt forms of aryl sulfate esters are distinctly
different (103). Chihara (142) systematically studied the infrared spectra of a series of alkyl and aryl sulfate esters. He
assigned the two bands at 1210-1220 and 1240-1260 cm^{-1} to the SO_3
asymmetric stretching vibration; these absorptions were very

strong and not greatly shifted by a variety of substituents. He
assigned the strong absorption at 1040-1081 cm^{-1} to the SO_3 sym-
metric vibration and the less intense bands at 550-590 cm^{-1} (sometimes split) and 617-650 cm^{-1} to SO_3 bending vibrations. The absorption from 757 to 838 cm^{-1} was assigned to the S-O-C stretch.
Lloyd et al. (143) studied the infrared spectra of the sulfate
esters of alcohols, amino alcohols, and hydroxylated amino acids
and reported similar conclusions (absorption bands at 1210-1260
cm^{-1}, 1030-1050 cm^{-1}, and 770-810 cm^{-1}, assigned to the sulfate
ester). Related studies on the infrared spectra of polysaccharide
sulfates (144) and monosaccharide sulfates (145) have been reported. Hummel (146) discussed the infrared spectra of primary (absorption bands at 1220-1267 cm^{-1}, 1075-1100 cm^{-1}, and 834-840
cm^{-1}) and secondary (absorption bands 1228-1250 cm^{-1}, 1063-1075
cm^{-1}, and 926-945 cm^{-1}) sulfate esters in surfactant compounds.
Colthup et al. (147) presented the spectrum of n-dodecyl sulfate
which showed absorption bands at approximately 820-840 cm^{-1}, 1200-
1280 cm^{-1}, and 1060-1080 cm^{-1} which were assigned to the sulfate
moiety. Infrared spectroscopy has been used to characterize sulfate esters of drugs (94, 148), steroids (96, 138), and sulfate-
ester containing metabolites of pesticides which include mobam (88),
p-chlorophenylmethylcarbamate (45, 131), propham (129, 140),
carbaryl (103), chlorpropham (134), and barban (134).

The technique of laser ionization mass spectrometry (149) has
been used by Mumma and Vastola (150) to obtain the mass spectra
of the sodium and potassium salts of 1-hexyl, 1-decyl, and
1-octadecyl sulfate. The molecular species plus a cation ([M +
Na]$^+$ or [M + K]$^+$) was one of the more intense peaks in the spectra
but no other "organic ions" were observed. "Inorganic ions"
that were abundant in the spectra included [NaSO$_4$]$^+$, [Na$_3$SO$_4$]$^+$,
and the corresponding potassium-containing fragments. Approximately 1 mg of sample was used for these assays; but the authors reported "good spectra can be obtained on less sample." This procedure has been used to characterize a number of steroid, alkyl,
and aryl sulfate esters (97, 98).

Recently, Games et al. (151) published a brief report dealing
with the utility of field desorption mass spectrometry in the analysis of sulfate esters and related compounds. They found that
n-hexyl, n-decyl, and n-undecyl sulfates gave quasi molecular ions
at m/e 259, 315, and 329, respectively, but no fragment ions were
observed. Cyclohexylphenyl-4-sulfate gave a quasi-molecular ion
([M + K]$^+$) at m/e 333 but also gave an ion at m/e 176, presumably
resulting from cleavage and hydrogen transfer to form the parent
phenol. Thus, there is reason for optimism that field-desorption
mass spectrometry may be a useful tool. However, much additional
work needs to be done to determine if this approach will be of
practical importance in the identification of sulfate esters --
particularly sulfate esters from biological preparations where
complicating factors, such as mixed salt forms, may be a problem.
Hopefully, information to answer this and other questions about

field-desorption mass spectrometry of sulfate esters will soon be
available.

General Discussion

There is a large and growing body of information concerning
the biosynthesis, chemical synthesis, isolation, characterization,
and enzymology of the sulfate esters. However, a review of the
published literature dealing with pesticide metabolism reveals that
many pesticide chemists are not fully using the information and
technology that is available. For instance, in most of the reported studies, the sulfate esters were hydrolyzed either chemically or enzymatically and then only the "nonpolar" hydrolysis product was identified. Because some of these compounds, such as the
N-hydroxy sulfates, are potent biological agents (27-31, 152-154),
the isolation and identification of the intact molecule is important. The observation by Boyland and Nery (102) that the sulfate
ester of phenyl hydroxylamine rearranged to 2-amino phenyl sulfate
suggests that artifacts may be produced by conditions such as those
used to hydrolyze conjugated metabolites.

In many instances, structures of sulfate esters have been
assigned on the basis of enzyme hydrolysis studies and characterization of only the hydrolysis product. Incorrect assignment of
structure because of enzyme preparations that are contaminated with
other hydrolytic enzymes, as well as contaminates from nonezymatic
hydrolysis, are always possibilities that must be considered.
Often the markedly different properties of Type I and Type II sulfatases (see previous discussion) have been ignored in selecting
assay conditions and/or in selecting the type of enzyme to be used
for the hydrolysis of different classes of compounds.

Some workers have synthesized the suspected sulfate ester and
then characterized their unknown metabolites by co-chromatography
studies only. This is unfortunate because additional and usually
definitive data can be obtained by comparative UV, NMR, and IR
spectroscopy. These instrumental procedures are not destructive
and require only small samples (for UV and IR, a sample of 10 µg
or less is usually sufficient). We have found IR spectroscopy to
be especially useful in characterizing sulfate ester conjugated
pesticide metabolites; all of the sulfate esters that we have examined have shown the characteristic absorptions previously discussed
and the fingerprint region almost invariably showed sharp, intense
bands that are ideal for comparative IR spectroscopy studies.

Most pesticide chemists have not confirmed the structure of
suspected sulfate ester conjugated metabolites by synthesis. This
is surprising in light of the fact that there are several methods
in the literature for the synthesis of sulfate esters including
methods that are suitable for the sulfation of compounds with relatively labile linkages.

Information concerning the biological activity of sulfate
ester conjugates of pesticides and/or their primary metabolites is

very limited. Studied to determine the effect of dietary factors, drugs, hormones, disease states, and related factors on the sulfate ester conjugation of pesticides and their primary metabolites by animals would be of value. The fate of sulfate esters in soil and plant systems would also be of interest.

Finally, there is a need for better and faster methods of isolating sulfate esters from biological preparations. The existing techniques for characterization of sulfate esters need to be improved. Additional studies on NMR and field desorption mass spectrometry of this class of compounds may be especially fruitful.

Literature Cited

1. Bauman, E., Ber. Deut. Chem. Ges., 9: 54 (1876).
2. Salleh, N.A.M., El-Sissi, H. I., Nawwar, M.A.M., Phytochemistry, 14: 312 (1975).
3. Smith, J. N., Biol. Rev. Cambridge Phil. Soc., 30: 455 (1955).
4. Koolman, J., Hoffmann, J. A., Karlson, P., Hoppe-Seylers, Z. Physiol. Chem., 354: 1043 (1973).
5. Yang, R.S.H., Wilkinson, C. F., Biochem. J., 130: 487 (1972).
6. Parke, D. V., The Biochemistry of Foreign Compounds, Pergamon Press, New York (1968).
7. Dodgson, K. S., Rose, F. A., "Metabolic Conjugation and Metabolic Hydrolysis," (W. H. Fishman, ed.), Vol. 1, Academic Press, New York and London, 1970.
8. DeMeio, R. H., Acta Physiol. Latinoamer, 2: 195 (1952).
9. DeMeio, R. H., Tkacz, L., J. Biol. Chem., 195: 175 (1952).
10. DeMeio, R. H., Wizerkaniuk, M., Fabiani, E., J. Biol. Chem., 203: 257 (1953).
11. DeMeio, R. H., Wizerkaniuk, M., Schreibman, I., J. Biol. Chem., 213: 439 (1955).
12. Bernstein, S., McGilvery, R. W., J. Biol. Chem., 198: 195 (1952).
13. Bernstein, S., McGilvery, R. W., J. Biol. Chem., 199: 745 (1952).
14. Segal, H. L., J. Biol. Chem., 213: 161 (1955).
15. Robbins, P. W., Lipmann, F., J. Amer. Chem. Soc., 78: 2652 (1956).
16. Robbins, P. W., Lipmann, F., J. Amer. Chem. Soc., 78: 6409 (1956).
17. Robbins, P. W., Lipmann, F., J. Biol. Chem., 223: 681 (1958).
18. Baddiley, J., Buchanan, J. G., Letters, R., Proc. Chem. Soc. (London), p. 147 (1957).
19. Baddiley, J., Buchanan, J. G., Letters, R., J. Chem. Soc. (London), p. 1067 (1957).
20. Roy, A. B., Advances in Enzymology, 22: 205 (1960).
21. Roy, A. B., Handbuch der Experimentellen Pharmakologie, 28, p. 536-563 (1971).

22. Gregory, J. D., Robbins, P. W., Ann. Rev. Biochem., 29: 347 (1960).
23. Robbins, P. W. In "The Enzymes," P. D. Boyer, H. Lardy, and K. M. Myrback, eds., 2nd Ed., Vol. 6, p. 363, Academic Press, New York (1962).
24. Robbins, P. W., In "Methods in Enzymology," S. P. Colowick and N. O. Kaplan, eds., Vol. 6, p. 766, Academic Press, New York (1963).
25. Wong, K. P., Anal. Biochem., 62: 149 (1974).
26. Vestermark, A., Boström, H., Experientia, 16: 408 (1960).
27. Irving, C. C., "Metabolic Conjugation and Metabolic Hydrolysis," (W. H. Fishman, ed.), Vol. 1, Academic Press, New York and London, 1970.
28. Irving, C. C., Xenobiotica, 1: 387 (1971).
29. Miller, J. A., Miller, E. C., In "Physico-Chemical Mechanisms of Carcinogenesis," (E. D. Bergmann and B. Pullman, eds.), Vol. 1, p. 237, Israel Academy of Sciences and Humanities, Jerusalem, 1969.
30. DeBaun, J. R., Miller, E. C., Miller, J. A., Cancer Research, 30: 577 (1970).
31. Weisburger, J. H., Yamamoto, R. S., Williams, G. M., Grantham, P. H., Matsushima, T., Weisburger, E. K., Cancer Research, 32: 491 (1972).
32. Spencer, B., Biochem. J., 77: 294 (1960).
33. Nose, Y., Lipmann, F., J. Biol. Chem., 233: 1348 (1958).
34. Gregory, J. D., Lipmann, F., J. Biol. Chem., 229: 1081 (1957).
35. Carroll, J., Spencer, B., Biochem. J., 94: 20p (1965).
36. Holcenberg, J. S., Rosen, S. W., Arch. Biochem. Biophys., 110: 551 (1965).
37. Adams, J. B., Edwards, A. M., Biochem. Biophys. Acta, 167: 122 (1968).
38. Banerjee, R. K., Roy, A. B., Mol. Pharmacol., 2: 56 (1966).
39. Meek, J. L., Foldes, A., Frontiers Catecholamine Res., p. 167 (1973).
40. Powell, G. M., Miller, J. J., Olavesen, A. H., Curtis, C. G., Nature, 252: 234 (1974).
41. Wengle, B., Acta Soc. Med. Upsalien, 69: 105 (1963).
42. DeMeio, R. H., Arch. Biochem., 7: 323 (1945).
43. Miller, J. J., Powell, G. M., Olavesen, A. H., Curtis, C. G., Xenobiotica, 4: 285 (1974).
44. Minck, K., Schupp, R. R., Illing, H.P.A., Kahl, G. F., Netter, K. J., Naumym-Schmiedeberg Arch. Pharmacol., 279 347 (1973).
45. Paulson, G. D., Zehr, M. V., J. Agr. Food Chem., 19: 471 (1971).
46. Slotkin, T., Distefano, V., Biochem. Pharmacol., 19: 125 (1970).
47. Kurzynske, J. S., Smith, J. T., Fed. Proc., 34, 882 (1975).
48. Gessner, T., Biochem. Pharmacol., 23: 1809 (1974).

49. Sato, T., Suzuki, T., Fukuyama, T., J. Biochemistry, 43: 421 (1956).
50. Mulder, G. J., Pilon, A.H.E., Biochem. Pharmacol., 24: 517 (1975).
51. Dodgson, K. S., Proceedings of the Fourth International Congress of Biochemistry, Vienna, 13: 23 (1958).
52. Young, L., Maw, G. A., "The Metabolism of Sulphur Compounds," Methuen, London (1958).
53. Flynn, T. G., Dodgson, K. S., Powell, G. M., Rose, F. A., Biochem. J., 105: 1003 (1967).
54. Hawkins, J. B., Young, L., Biochem. J., 56: 166 (1954).
55. Curtis, C. G., Hearse, D. J., Powell, G. M., Xenobiotica, 4: 595 (1974).
56. Clapp, J. J., Young, L., Biochem. J., 118: 765 (1970).
57. Gatehouse, P. W., Roy, A. B., Dodgson, K. S., Powell, G. M., Lloyd, A. G., Olavesen, A. H., Biochem. J., 127: 661 (1972).
58. Denner, W.H.B., Olavesen, A. H., Powell, G. M., Dodgson, K. S., Biochem. J., 111: 43 (1969).
59. Dodgson, K. S., Tudball, N., Biochem. J., 74: 154 (1960).
60. Powell, G. M., Curtis, C. G., Biochem. J., 99: 34p (1966).
61. Hearse, D. J., Powell, G. M., Olavesen, A. H., Biochem. Pharmacol., 18: 197 (1969).
62. Hearse, D. J., Powell, G. M., Olavesen, A. H., Dodgson, K. S., Biochem. Pharmacol., 18: 181 (1969).
63. Hearse, D. J., Powell, G. M., Olavesen, A. H., Dodgson, K. S., Biochem. Pharmacol., 18: 205 (1969).
64. Powell, G. M., Gregory, P. A., Olavesen, A. H., Jones, A. G., Biochem. Soc. Trans., 1: 1165 (1973).
65. Roy, A. B., "The Enzymes," (P. E. Boyer, ed.), 3rd Ed., Vol. 5, p. 1-19, Academic Press, Inc., New York (1971).
66. Nicholls, R. G., Roy, A. B., "The Enzymes," (P. D. Boyer, ed.), Vol. 5, p. 21, Academic Press, Inc., New York (1971).
67. Roy, A. B., Biochem. J., 53: 12 (1953).
68. Roy, A. B., Biochem. Biophys. Acta, 14: 149 (1954).
69. Dodgson, K. S., Spencer, B., Thomas, J., Biochem. J., 57: 21 (1954).
70. Dodgson, K. S., Spencer, B., Thomas, J., Biochem. J., 59: 29 (1955).
71. Dodgson, K. S., Spencer, B., Wynn, C. H., Biochem. J., 62: 500 (1956).
72. Dodgson, K. S., Rose, F. A., Spencer, B., Biochem. J., 60: 346 (1955).
73. Dodgson, K. S., Spencer, B., Methods of Biochemical Analysis, Vol. 4, p. 211 (1957).
74. Hanker, J. S., Thornburg, L. P., Yates, P. E., Romanovicz, D. K., Histochemistry, 41: 207 (1975).
75. Worwood, M., Dodgson, K. S., Hook, G.E.R., Rose, F. A., Biochem. J., 134: 183 (1973).
76. Hook, G.E.R., Dodgson, K. S., Rose, F. A., Worwood, M., Biochem. J., 134: 191 (1973).

77. Baxter, T. H., Kostenbauder, H. B., J. Pharm. Sci., $\underline{58}$: 33 (1969).
78. Stinshoff, K., Biochim. Biophys. Acta, $\underline{276}$: 475 (1972).
79. Dubois, G., Baumann, N., Biochem. Biophys. Res. Comm., $\underline{50}$: 1129 (1973).
80. Stinshoff, K., Jatzkewitz, H., Biochim. Biophys. Acta, $\underline{377}$: 126 (1975).
81. Austin, J., Bolasubramanian, A., Pattabiraman, T., Saraswathi, S., Basu, D., Bachhawat, B., J. Neurochem., $\underline{10}$: 805 (1963).
82. Fitzgerald, J. W., Dodgson, K. S., Payne, W. J., Biochem. J., $\underline{138}$: 63 (1974).
83. Dodgson, K. S., Fitzgerald, J. W., Payne, W. J., Biochem. J., $\underline{138}$: 53 (1974).
84. Gilbert, E. E., "Sulfonation and Related Reactions," Interscience Publishers, New York, New York (1965).
85. Feigenbaum, J., Neuberg, C. A., J. Amer. Chem. Soc., $\underline{63}$: 3529 (1941).
86. Burkhardt, G. N., Wood, H., J. Chem. Soc., 141 (1929).
87. Burkhardt, G. N., Lapworth, A., J. Chem. Soc., p. 684 (1926).
88. Robbins, J. D., Bakke, J. E., Feil, V. J., J. Agr. Food Chem., $\underline{18}$: 130 (1970).
89. Nagasawa, K., Yoshidome, H., J. Org. Chem., $\underline{39}$: 1681 (1974).
90. Dusza, J. P., Joseph, J. P., Bernstein, S., Steroids, $\underline{12}$: 49 (1968).
91. Hardy, W. B., Scalera, M., J. Amer. Chem. Soc., $\underline{74}$: 5212 (1952).
92. Paterson, J.Y.F., Klyne, W., Biochem. J., $\underline{43}$: 614 (1948).
93. Sobel, A. E., Spoerri, P. E., J. Amer. Chem. Soc., $\underline{63}$: 1259 (1941).
94. Fujimoto, J. M., Haarstad, V. B., J. Pharmacol. Exp. Therap., $\underline{165}$: 45 (1969).
95. Hearse, D. J., Olavesen, A. H., Powell, G. M., Biochem. Pharmacol., $\underline{18}$: 173 (1969).
96. Mumma, R. O., Lipids, $\underline{1}$: 221 (1966).
97. Hoiberg, C. P., Mumma, R. O., Biochim. Biophys. Acta, $\underline{177}$: 149 (1969).
98. Hoiberg, C. P., Mumma, R. O., J. Amer. Chem. Soc., $\underline{91}$: 4273 (1969).
99. Mumma, R. O., Biochim. Biophys. Acta, $\underline{165}$: 571 (1968).
100. Quadri, S. F., Seib, P. W., Deyoe, C. W., Carbohydrate Res., $\underline{29}$: 259 (1973).
101. Mumma, R. O., Verlangieri, A. J. Weber, W. W., Carbohydrate Res., $\underline{19}$: 127 (1971).
102. Boyland, E., Nery, R., J. Chem. Soc., $\underline{5217}$ (1962).
103. Paulson, G. D., Zaylskie, R. G., Zehr, M. V., Portnoy, C. E., Feil, V. J., J. Agr. Food Chem., $\underline{18}$: 110 (1970).
104. Goldberg, I. H., Delbrück, A., Fed. Proc., $\underline{18}$: 235 (1959).
105. Layton, L. L., Frankel, D. R., Arch. Biochem., Biophys., $\underline{31}$: 161 (1951).

106. Raub, H. R., Hobkirk, R., Can. J. Biochem., 44: 657
     (1966).
107. Raub, H. R., Hobkirk, R., Can. J. Biochem., 46: 749
     (1968).
108. Raub, H. R., Hobkirk, R., Can. J. Biochem., 46: 759
     (1968).
109. Torday, J., Hall, G., Schweitzer, M., Giroud, C.J.P.,
     Can. J. Biochem., 48: 148 (1970).
110. Brunngraber, E. G., J. Biol. Chem., 233: 472 (1958).
111. Fendler, E. J., Fendler, J. H., J. Org. Chem., 33: 3852
     (1968).
112. Havinga, E., DeJongh, R. O., Dorst, W., Recueil, 75: 378
     (1956).
113. Barton, A. D., Young, L., J. Amer. Chem. Soc., 65: 294
     (1943).
114. Laughland, D. H., Young, L., J. Amer. Chem. Soc., 65:
     657 (1944).
115. Goertz, G. R., Crepy, O. C., Judas, O. E., Longchampt,
     J. E., Jayle, M. F., Clin. Chim. Acta, 51: 277 (1974).
116. Miyabo, S., Kornel, L., J. Steroid Biochem., 5: 233 (1974).
117. Roy, A. B., Trudinger, P. A., "The Biochemistry of
     Inorganic Compounds of Sulphur," University Press,
     Cambridge (1970).
118. Burkhardt, G. N., Evans, A. G., Warhurst, E., J. Chem. Soc.,
     p. 25 (1936).
119. Burkhardt, G. N., Ford, W.G.K., Singleton, E., J. Chem.
     Soc., p. 17 (1936).
120. Kice, J. L, Anderson, J. M., J. Amer. Chem. Soc., 88: 5242
     (1966).
121. Batts, B. D., J. Chem. Soc. (B), p. 551 (1966).
122. Benkovic, S. J., Dunikoski, L. K., Biochemistry, 9: 1390
     (1970).
123. Batts, B. D., J. Chem. Soc. (B), p. 547 (1966).
124. Goren, M. B., Kochansky, M. E., J. Org. Chem., 38: 3510
     (1973).
125. Assandri, A., Perazzi, A., J. Chromatog., 95: 213 (1974).
126. Jenner, W. N., Rose, F. A., Nature, 252: 237 (1974).
127. Anders, M. W., Latoree, J. P., J. Chromatog., 55: 409 (1971).
128. Hobkirk, R., Davidson, S., J. Chromatog., 54: 431 (1971).
129. Paulson, G. D., Jacobsen, A. M., Zaylskie, R. G., Feil,
     V. J., J. Agr. Food Chem., 21: 804 (1973).
130. Paulson, G. D., Jacobsen, A. M., J. Agr. Food Chem., 22:
     629 (1974).
131. Paulson, G. D., Zehr, M. V., Docktor, M. M., Zaylskie,
     R. G., J. Agr. Food Chem., 20: 33 (1972).
132. Slotkin, T. A., Distefano, V., Au, W.Y.W., J. Pharmacol.
     Exp. Ther., 173: 26 (1970).
133. Tocco, D. J., Buhs, R. P., Brown, H. D., Matzuk, A. R.,
     Mertel, H. E., Harman, R. E., Trenner, N. R., J. Med. Chem.,
     7: 399 (1964).

134. Grunow, W., Böhme, C., Budczies, B., Fd. Cosmet. Toxicol.,
     8: 277 (1970).
135. Faakonmäki, P. I., Acta Endocr. (Kbh), Suppl. 119, p. 118
     (1967).
136. McKenna, J., Norymberski, J. K., J. Chem., Soc., p. 3893
     (1957).
137. Pasqualini, J. R., Zelnik, R., Jayle, M. F., Bull. Soc.
     Chim. Fr., p. 1171 (1962).
138. Emiliozzi, R., Societe Chimique De France, p. 911 (1960).
139. Paulson, G. D., Portnoy, C. E., J. Agr. Food Chem., 18: 180
     (1970).
140. Paulson, G. D., Docktor, M. M., Jacobsen, A. M., Zaylskie,
     R. G., J. Agr. Food Chem., 20: 867 (1972).
141. Dodgson, K. S., Spencer, B., Biochem. J., 53: 444 (1953).
142. Chihara, G., Chem. Pharm. Bull. (Tokyo), 8: 988 (1960).
143. Lloyd, A. G., Tudball, N., Dodgson, K. S., Biochim. Biophys.
     Acta, 52: 413 (1961).
144. Lloyd, A. G., Dodgson, K. S., Price, R. G., Rose, F. A.,
     Biochim. Biophys. Acta, 46: 108 (1961).
145. Lloyd, A. G., Dodgson, K. S., Biochim. Biophys. Acta, 46:
     116 (1961).
146. Hummel, D., "Identification and Analysis of Surface Active
     Agents," Interscience Publishers, New York, London, Sydney
     (1962).
147. Colthup, N. B., Daly, L. H., Wiberly, S. E., "Introduction
     to Infrared and Raman Spectroscopy," Academic Press, New
     York and London (1964).
148. Yeh, S. Y., Woods, L. A., J. Pharm. Sci., 60: 148 (1971).
149. Vastola, F. J., Mumma, R. O., Pirane, A. J., Organic Mass
     Spectrometry, 3: 101 (1970).
150. Mumma, R. O., Vastola, F. J., Organic Mass Spectrometry, 6:
     1373 (1972).
151. Games, O. E., Games, M. P., Jackson, A. H., Olavesen, A. H.,
     Rossieter, M., Tet. Letters, p. 2377 (1974).
152. Lotikar, P. D., Scribner, J. D., Miller, J. A., Miller,
     E. C., Life Sci. 5: 1263 (1966).
153. DeBaun, J. R., Rawley, J. Y., Miller, E. C., Miller, J. A.,
     Proc. Soc. Expt. Biol. Med., 129: 268 (1966).
154. Lin, M. S., Walden, D. B., Exp. Cell Res., 86: 47 (1974).

Glutathione Conjugates

D. H. HUTSON

Shell Research Ltd., Tunstall Laboratory, Sittingbourne Research Centre, Sittingbourne, Kent, England

Almost 100 years ago, two groups of German scientists, studying the fate of halobenzenes in mammals, unwittingly initiated the study of glutathione (GSH) conjugation. Baumann and Preusse (1) isolated a cysteine derivative from the acidifed urine of mammals treated with bromobenzene. Jaffé (2) isolated a similar metabolite of chlorobenzene. Their results were published in the same volume of Chemisches Berichte in 1879. The derivatives were called mercapturic acids and were shown later to be S-aryl-N-acetyl-L-cysteines (Figure 1). The mechanism of formation of this type of mercapturic acid is not immediately obvious from a consideration of the structures of precursor and product and almost 100 years elapsed before the reaction pathway was fully elucidated. Waelsch (3), and Brand and Harris (4) suggested in the 1930's that glutathione (GSH) was the source of the cysteine for the biosynthesis of mercapturic acids, but as late as the 1940's and 1950's, dietary protein and tissue cysteine (5)(6)(7) were still being considered as sources. Glutathione was shown to be the source in 1959 by James and her co-workers (8)(9) in England. As far as the mammal is concerned, mercapturic acid formation and GSH conjugation are inextricably linked and the former must be included in this discussion.

In most cases the initial reaction in mercapturic acid biosynthesis is the enzyme-catalysed reaction between GSH and either the foreign compound itself or a metabolite of that foreign compound. The metabolic activation of a compound to the ultimate reactant with GSH is an important facet of this type of conjugation and its significance will be discussed later. It is now clear that mercapturic acid biosynthesis from a precursor R-X (for example benzyl chloride) proceeds in four stages which are shown in Figure 2 . The final product, benzyl mercapturic acid, is excreted in the urine but it is the first step, conjugation with GSH, which is crucial in the destruction

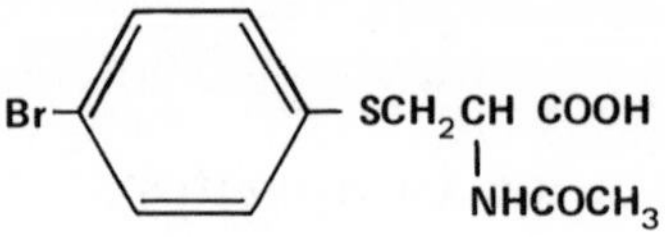

Figure 1. Mercapturic acid derived from bromobenzene

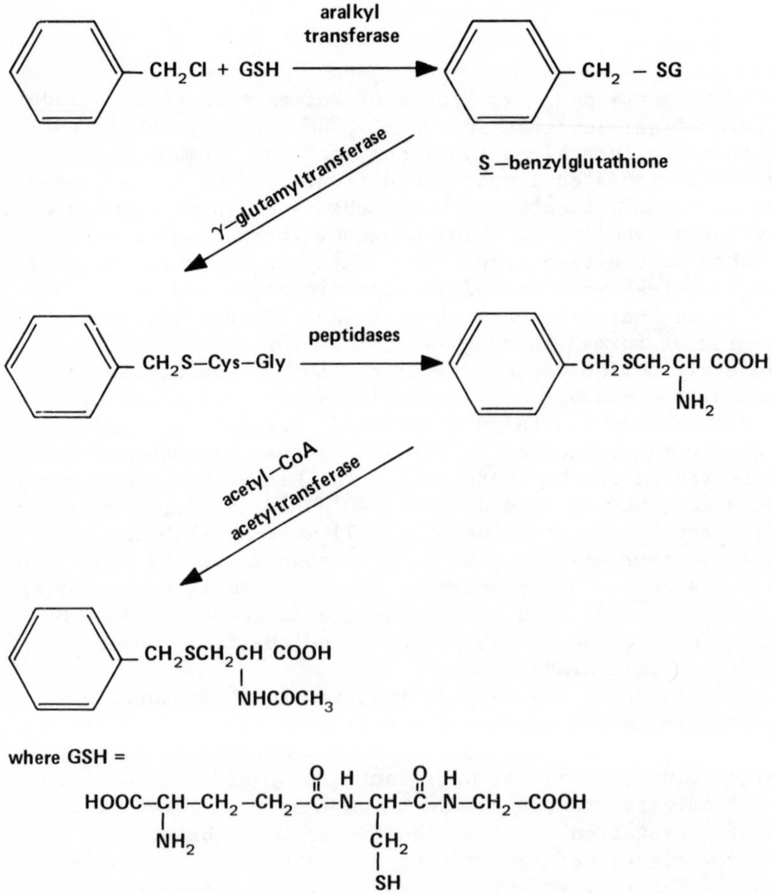

Figure 2. Biosynthesis of mercapturic acids

of the benzyl chloride.

Most of the background information available up to 1973
on GSH conjugation and mercapturic acid formation is obtainable
from reviews by Boyland and Chasseaud (10), Wood (11) and
Hutson (12). In addition Chasseaud has published two reviews
(13)(14) dealing specifically with the enzymes which catalyse
the initial transfer - the glutathione transferases.

1. The Chemical Requirements For Glutathione Conjugation

Much of the structure-activity data on this mode of conjugation has been acquired via observations of mercapturic acid
excretion in the urine of animals treated with the precursor.
A disadvantage of this approach has been that the yields of
mercapturic acids in the urine are often a very poor measure
of GSH conjugation. These conjugates are excreted in the bile
and they may also be extensively metabolised in the liver, and
intestinal tract, or during enterohepatic circulation. An advantage of the approach however, is that it reveals mercapturic
acid formation from substrates which do not themselves react
with GSH and transferase. Undoubtedly the best way to investigate GSH conjugation by the mammal in vivo is to analyse
metabolites excreted in the bile. This can be done relatively
simply by cannulating the bile duct under general anaesthesia
prior to dosing the animal. Such experiments may be short-term,
e.g. 3 hours, without recovery from the anaesthetic, or long-
term, e.g. several days, during which time the bile can be collected in a surgically implanted vessel and the animal can
be allowed to live normally in a metabolism cage fitted for
the collection of urine and faeces.

The use of these various techniques over the last 20 years
or so has demonstrated that several general types of compounds
are substrates for GSH conjugation. We face the usual xenobiochemical dilemma of how to make a useful chemical classification
of these substrates. Classification by function of compound
is not very enlightening. Classification on the basis of the
group transferred (e.g. alkyl, aryl, etc) fails to acknowledge
the importance of the activation effected by the leaving group.
Thus in the generalised reaction shown in Figure 3 the nature
of X is as important as that of R in determining the rate of
reaction. However, the notion of group transfer is well
entrenched and will be used below, but the reader is cautioned
(i) that the specificity does not lie solely with the group
which is transferred and (ii) certain classifications of substrates are only important special cases of group transfer
(e.g. aromatic 'epoxide transfer' is only a special case of
aryl transfer).

Alkyl Transfer

Alkyl Halides. The simple fumigants such as methyl bromide, ethylene dibromide and many other halogenated paraffins form GSH conjugates (15)(16). The sequence of reactivity lies in the predictable order I > Br > Cl > F. The efficiency of dechlorination depends on activation of the carbon-chlorine bond by electron withdrawing groups e.g. carbonyl and nitrile. The reaction rates of some halides are shown in Figure 4.

Alkyl Alkylsulphonates. The alkylating agents methyl methanesulphonate (17) and ethyl methanesulphonate (18) (Figure 5) are metabolised via GSH conjugation. Both the alkyl halides and the sulphonates react spontaneously with GSH in vitro but the reaction rate is considerably enhanced by the presence of hepatic GSH alkyl transferase.

Organophosphate Insecticides. Methyl parathion (19), methyl paraoxon (20), tetrachlorvinphos and several other dimethyl phosphoric acid triesters (21) (Figure 6) are demethylated by the action of GSH transferase. The reaction is very important in limiting the acute toxicity of these compounds in certain species. The pathway is usually only effective for methyl groups. In a recent study using a mixed methyl ethyl alkenylphosphate (temivinphos, Figure 6) we have found that demethylation was the only observable reaction in vitro. However, the O-desethylation of the nematacide Mocap (Figure 7) by rabbit liver cytosol has been detected (22), the ethyl group being transferred to GSH. Methyl glutathione and methyl mercapturic acid (in bile and urine, respectively) are barely detectable in the metabolism of the organophosphate insecticides because the former is rapidly further catabolised, the methyl group being eliminated as CO_2 (23). However we have detected small quantities of methyl mercapturic acid, S-methyl cysteine and S-methyl cysteine oxide in the urine of rats dosed with [^{14}C-methyl]dichlorvos (24). Dimethyl organophosphate triesters are weak methylating agents; they undoubtedly methylate GSH but our studies in vitro (21) revealed only very low rates of reaction with GSH in the absence of enzyme.

Aralkyl Transfer

Although aralkyl transfer may just be a special case of alkyl transfer, there is a tendency for stabilisation of the carbonium ion derived from these substrates, and there may therefore be an element of SN1 reaction possible with some substrates, for example benzyl chloride.

Aralkyl Halides. Benzyl chloride is a mercapturic acid precursor (Figure 2; Figure 8) and Boyland and Chasseaud (25)

In the reaction :

$$R-X + GSH \longrightarrow RSG + HX$$

the nature of X is as important
as that of R in determining the
possibility of reaction   *Figure 3.*

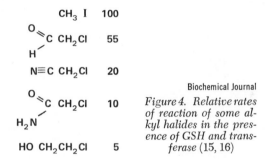

CH$_3$ I	100
	55
N≡C CH$_2$Cl	20
	10
HO CH$_2$CH$_2$Cl	5

Biochemical Journal

*Figure 4. Relative rates
of reaction of some al-
kyl halides in the pres-
ence of GSH and trans-
ferase* (15, 16)

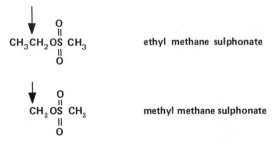

CH$_3$CH$_2$OS CH$_3$ ethyl methane sulphonate

CH$_3$ OS CH$_3$ methyl methane sulphonate

Figure 5. Substrates of GSH alkyl transferase

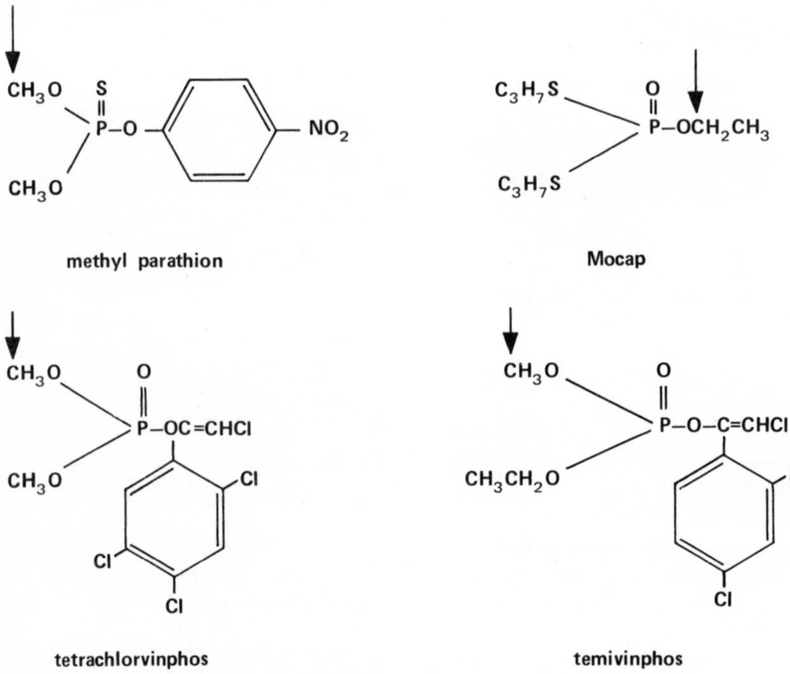

Figure 6. *Organophosphate substrates for GSH alkyl transferase*

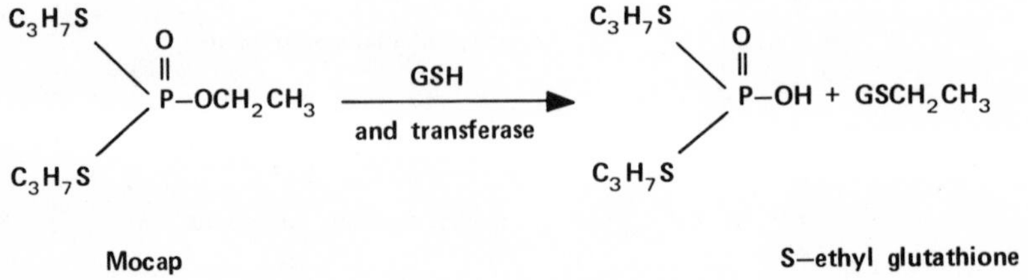

Biochemical Pharmacology

Figure 7. (22)

have carried out some studies on the enzyme which catalyses
the formation of S-benzyl glutathione. It has been distinguished
from the other GSH transferases by heat inactivation, precipitation, and species distribution.

    Aralkyl Sulphates. Benzyl, 1-menaphthyl (Figure 8) and
phenanthr-9-ylmethyl sulphates are substrates for the transferase. The corresponding aralkyl alcohols (which are not
substrates) are mercapturic acid precursors in vivo and it would
seem likely that O-sulphate conjugation is an obligatory intermediate step in this process (26). This is an unusual role for
O-sulphates; they are usually terminal metabolites.

Aryl Transfer

    Substrates for aryl transfer to GSH include 1,2-dichloro-4-
nitrobenzene, 1,2,4,5-tetrachloro-3-nitrobenzene, 4-nitro-
pyridine N-oxide and sulphobromophthaleine (a dye used to test
liver function). The structures, and points of attack of GSH,
are shown in Figure 9. Paraoxon and methyl paraoxon are de-
arylated by a GSH transferase (20). We see in this situation
(Figure 10), GSH acting at alternative points on the same
molecule. If R = methyl, demethylation predominates, if R =
ethyl, de-arylation predominates.
    The GSH-dependent cleavage of a diphenyl ether has recently
been reported. Fluorodifen (2,4'-dinitro-4-trifluoromethyl
diphenyl ether), one of a new class of herbicides used for the
control of broad-leaved weeds in soya beans, peanuts, cotton
and rice, is de-arylated by a GSH aryl transferase. The enzyme
has been isolated from the epicotyl tissue of pea seedlings
(27) and is one of the few examples of plant GSH transferases
reported to date. The products of reaction have been identified
as S(2-nitro-4-trifluoromethyl phenyl) glutathione and p-
nitrophenol (Figure 11). Limited substrate specificity studies
have shown that substitutents causing a large decrease in
electron density at C - 1 are necessary for enzyme activity.
This is commensurate with nucleophilic attack of glutathione
sulphur at this carbon atom.
    Bromobenzene, the precursor of the first known mercapturic
acid, is not a substrate for glutathione aryl transferase. This
indicates that the reaction in vivo is not a hydride ion displacement but that another reaction precedes GSH conjugation.

Epoxide Transferases

    When rats and rabbits are dosed with bromobenzene, careful
extraction of the urine affords N-acetyl-S-(4-bromo-1,2-dihydro-
2-hydroxyphenyl)-L-cysteine (Figure 12). This was isolated as
its cyclohexylamine salt (28) which could be decomposed in acid
solution to the mercapturic acid shown in Figure 1. This

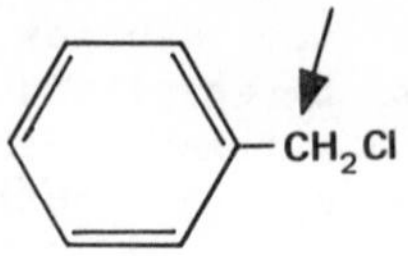

benzyl chloride

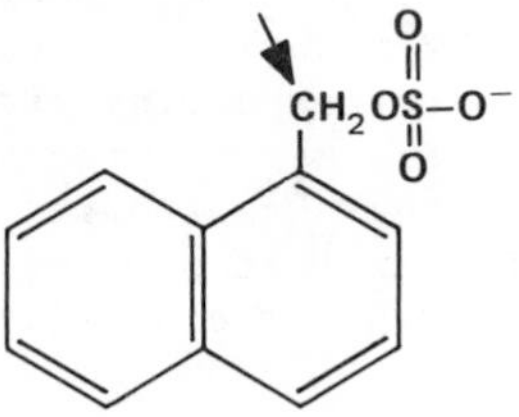

Figure 8. Substrates for GSH aralkyl transferase

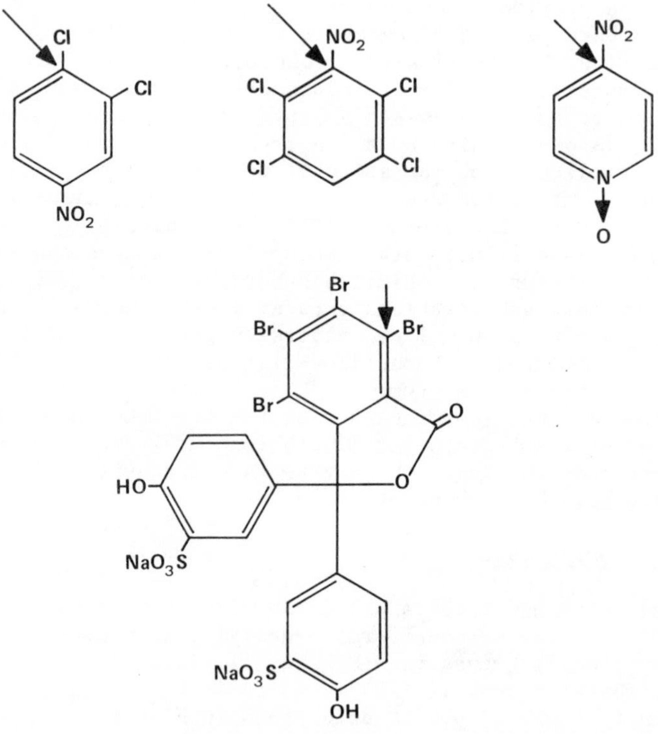

menaphthyl sulphate

Figure 9. Substrates for GSH aryl transferase

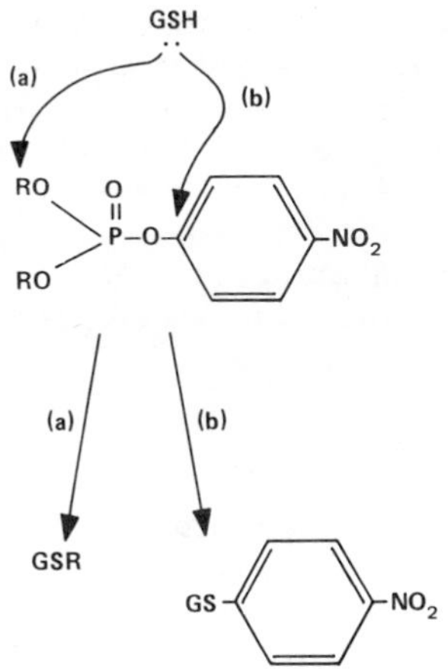

Life Sciences

Figure 10. Alternative GSH-dependent pathways for the degradation of a dialkyl phosphate (20)

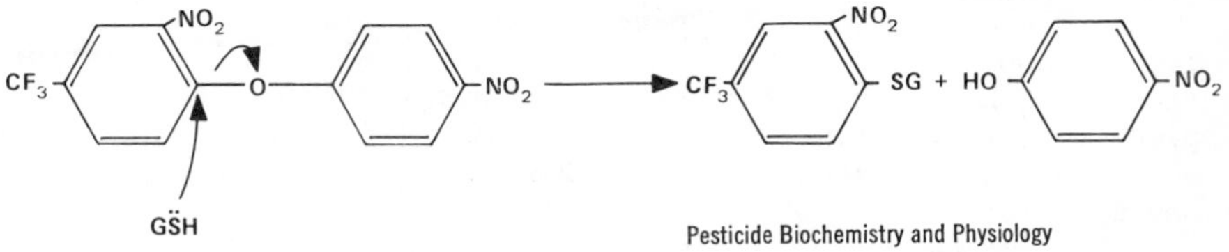

Pesticide Biochemistry and Physiology

Figure 11. GSH-dependent cleavage of a diphenyl ether (27)

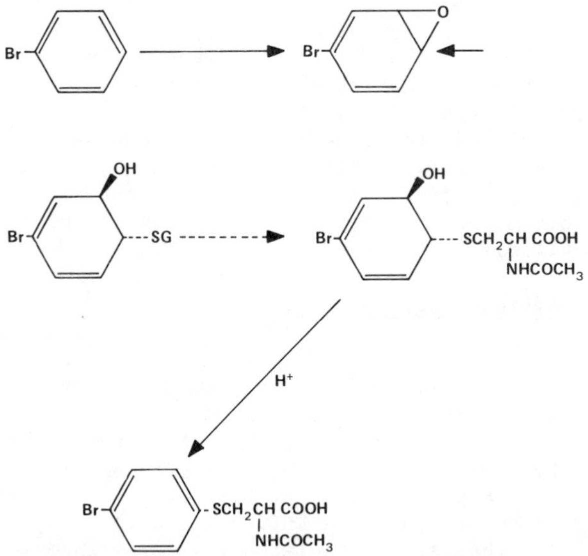

Biochemical Journal

Figure 12. Mercapturic acid formation from aromatic hydrocarbons (28)

observation, together with the now classic work on arene oxides
at N.I.H. (29), and the discovery of GSH epxoide transferase
by Boyland and coworkers (30), has led to the recognition that
aromatic hydrocarbons form GSH conjugates and mercapturic acids
only after oxygenation to epoxides. A typical reaction pathway
is shown in Figure 12.

Aliphatic epoxides, e.g. styrene oxide (29) and 1-phenoxy-
prop-2-enyl oxide (31) (Figure 13), are also substrates for
a GSH transferase and form mercapturic acids.

Alkene Transfer

Compounds containing activated carbon-carbon double bonds,
with the general formula shown in Figure 14 are conjugated with
GSH by the enzyme GSH alkene transferase which is present in
the livers of various mammalian species and various tissues
of the rat (32). Chasseaud (33) has studied the enzyme(s) in
some depth. Efficiency of the reaction of the substrate depends
on the electron-withdrawing power of the group X and on electron
repulsion or attraction exerted by groups R, R' and R''. Reaction
occurs at the β carbon atom of the substrate to yield the
product shown in Figure 14. Some effective substrates of the
enzyme are illustrated in Figure 15 which shows that group X
may be carbonyl, nitrile, sulphone or nitro.

Alkyl Mercapto Transfer

The first example of a toxicating (bioactivation) reaction
mediated by a GSH transferase was discovered by Casida and coworkers (34). It is catalysed by a soluble liver enzyme and
involves the attack of GSH at the thiocyanate sulphur of an
alkyl thiocyanate resulting in the liberation of HCN (Figure 16).
The existence of this reaction leads to an unusual protective
effect which we can now explain. The insecticidal dimethyl
phosphate triester, fenitrothion, reduces the toxicity of several
organic thiocyanates to mice, either by lowering hepatic GSH
or by competitively inhibiting the glutathione transferase.

Transfer of Nitrogen Heterocycles

A paper published in 1970 reported the conjugation of
atrazine (Figure 17) with GSH in sorghum leaf sections (35).
This paper was important because not only did it describe the
first example of a conjugate of a sym-triazine with GSH, but
it demonstrated for the first time that plants possessed the
capability for GSH conjugation. At this time, during a study
of the metabolism of another sym-triazine herbicide, cyanazine
(Figure 17), in the rat, we isolated the first example of a
sym-triazinyl mercapturic acid (36) (of N-desethyl cyanazine)
and subsequently identified the predictable GSH conjugates in

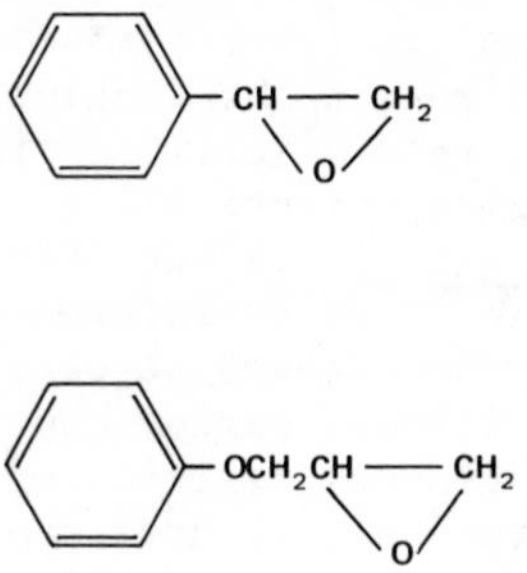

Figure 13. Substrates for GSH epoxide transferase

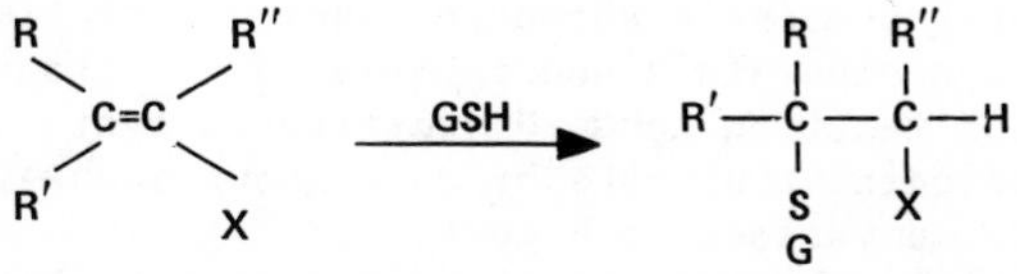

Figure 14. The action of GSH alkene transferase

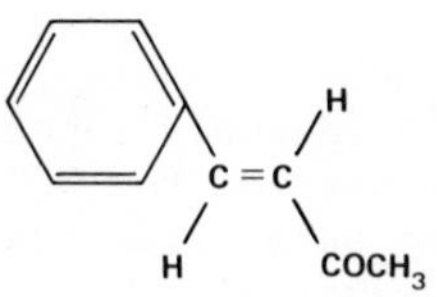

trans—benzylidene acetone

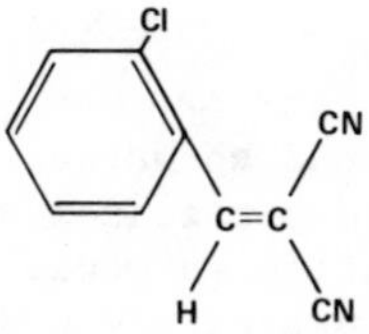

CS riot control agent

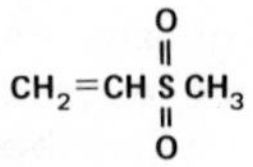

methyl vinyl sulphone

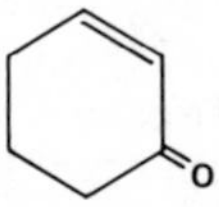

cyclohex—2—ene—l—one

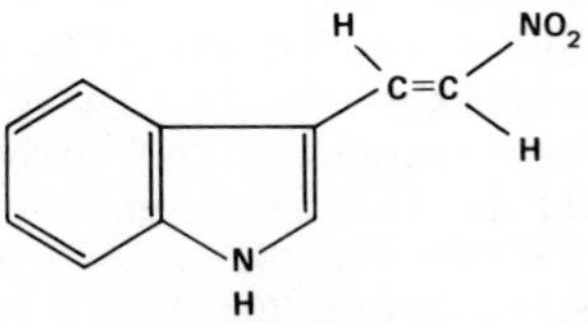

(3—β—nitrovinyl)indole

Biochemical Journal

Figure 15. Substrates for GSH alkene transferase

(33)

bile and the "triazinyl transferase" in liver cytosol (37).
Atrazine and its metabolites which retain the 2-chloro group
have been shown recently to conjugate similarly in vitro (38).
It was puzzling to find that Simazine (Figure 17) in our hands,
though yielding GSH conjugates in vitro, did not afford
mercapturic acids in the urine when dosed to rats.

The conjugation in plants has been found to occur in sugar
cane leaves, corn leaves, sorghum leaves, barley shoots, Johnson
grass and Sudan grass and to be a general reaction of 2-chloro-
sym-triazines (39)(40). Further studies (41) have shown that
the "atrazinyl glutathione" initially formed is catabolised
in plants first by loss of glycine, then glutamic acid (cf the
reverse order for mammals shown in Figure 2) to yield the S-
substituted cysteine. This undergoes an S → N trans-
triazinylation reaction to an N-substituted cysteine which then
forms a lanthionine conjugate by an unknown mechanism. These
reactions are summarised in Figure 18.

A possibly analogous reaction of diazinon (Figure 19)
has been reported (42), however, this may be an analogue of
aryl transfer.

Miscellaneous

Casida and co-workers (43) have recently reported that
thiocarbamate herbicides are also activated for GSH conjugation
by S-oxygenation in mouse liver. The sulphoxide can be isolated
but the conjugates are unstable (Figure 20).

General Comment

This brief review has demonstrated the wide variety of
structures which may be involved in GSH conjugation. There are
undoubtedly other reactions to be discovered. There is however,
an important unifying feature about all of the substrates of
the GSH transferases. They all possess, to a greater or lesser
extent, an electrophilic carbon atom which has been marked in
many of the figures so far. The reactions all possess a marked
similarity to an SN2 reaction of a sulphur nucleophile with
electrophilic carbon. This is clearly the main driving force
for the reaction.

In this property, GSH conjugation differs from the other
two general types of conjugation, in that the activation for
the reaction derives mainly from the intrinsic reactivity of
the xenobiotic. Thus we can recognise the three major types
of conjugation (Figure 21) as (a) in which the foreign acceptor
substrate reacts enzymatically with an endogenous reactive donor,
as with glucuronide formation from UDPGA (b) in which the
foreign substrate becomes activated to a donor via an endogenous
mechanism and then reacts with an acceptor, as with the formation of hippurates from benzoic acids and (c) in which the

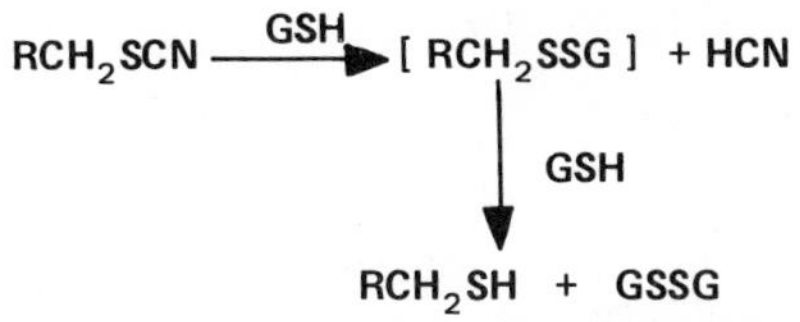

Pesticide Biochemistry and Physiology

Figure 16. Bioactivation of alkyl thio-cyanates by GSH alkyl transferase (34)

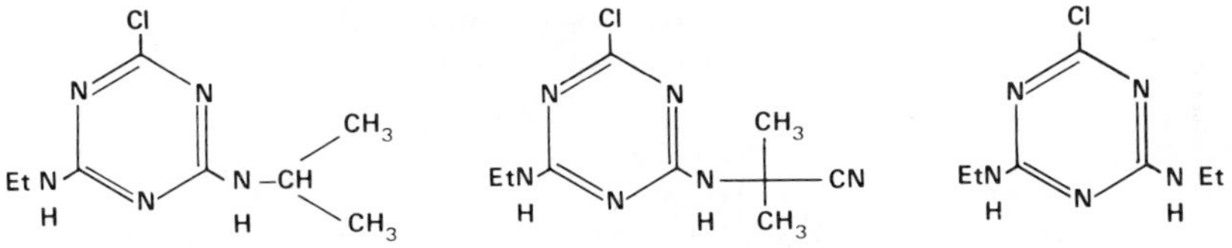

atrazine   cyanazine   simazine

Figure 17. Substrates of GSH triazinyl transferase

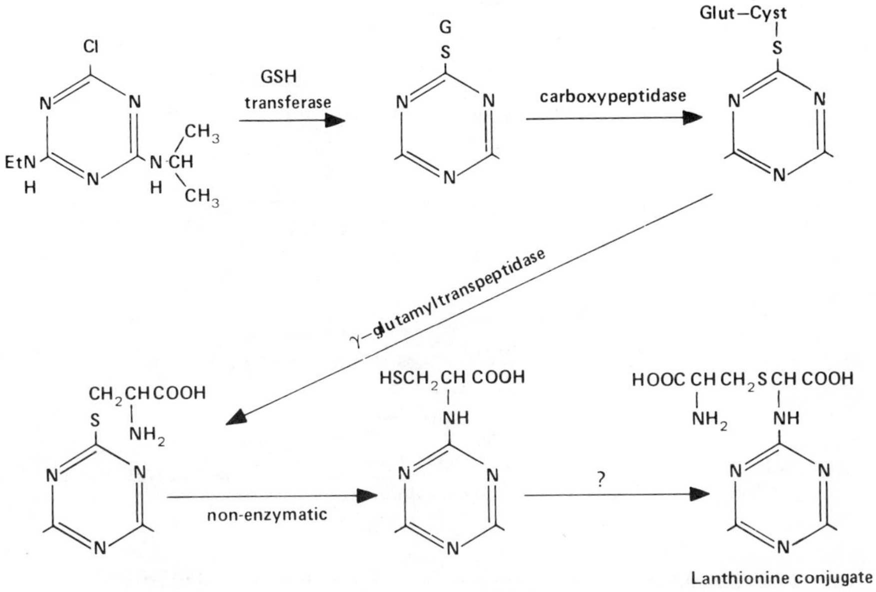

Lanthionine conjugate

Journal of Agricultural and Food Chemistry

Figure 18. Catabolism of a sym-triazine conjugate in plants (41)

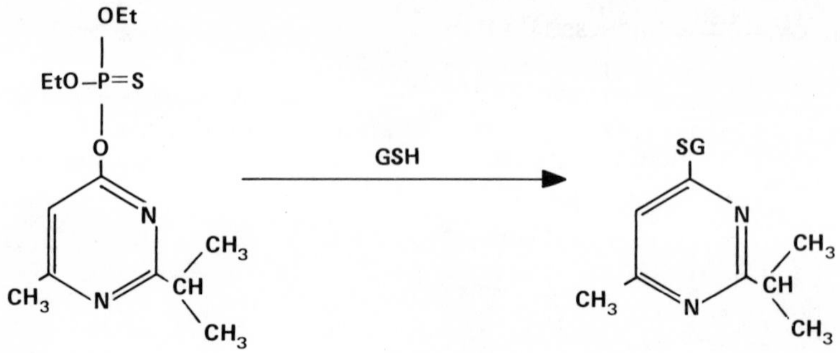

Pesticide Biochemistry and Physiology

Figure 19. Detoxication of diazinon (42)

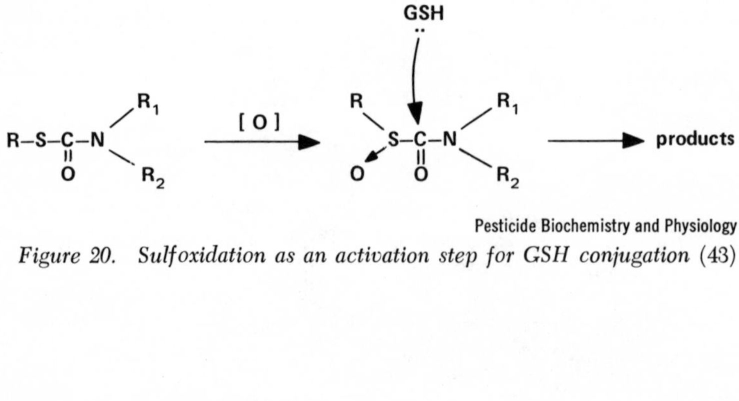

Pesticide Biochemistry and Physiology

Figure 20. Sulfoxidation as an activation step for GSH conjugation (43)

(a) PHENOL —— active donor UDPGA + enzyme ——▶ PHENYL GLUCURONIDE

(b) BENZOIC ACID —— activation via CoA ——▶ BENZOYL CoA (active donor) —— Glycine + enzyme ——▶ BENZOYL GLYCINE

(c) ALKYL HALIDE —— GSH + enzyme ——▶ ALKYL GLUTATHIONE

no further activation needed

Figure 21. Classes of conjugation reaction

foreign substrate itself possesses the required reactivity.

2. Some Techniques Used To Investigate Cysteine and GSH Conjugates

Mercapturic Acids

These present no special problems as they can usually be extracted from urine at pH 2-3 with ether, or from freeze-dried urine by extraction with methanol. They can then be methylated and subjected to the usual physical methods such as nmr spectrometry and mass spectrometry. If the carbon - sulphur bond is stable to acid, it is useful to remember that a mercapturic acid of a neutral compound is only anionic before hydrolysis, but amphoteric after hydrolytic removal of the N-acetyl group. These changes can be monitored by paper electrophoresis at pH 7 and pH 2. A sensitive microtest for a carbon - sulphur bond (sodium formate fusion test and detection of H_2S with lead acetate paper) is also an aid to identification. This test works at the 5 μg level. This combination of tests was used to identify the mercapturic acid derived from cyanazine (36), with the exception that acid hydrolysis split the C-S bond. A general test for mercapturic acid formation is to label the sulphur pool of rats by treatment with [^{35}S]cysteine or by feeding on a diet containing [^{35}S]yeast, and then challenging some of the animals with the test compound. Mercapturic acids in the urine can be distinguished readily from the background of sulphate etc.

Synthesis. These compounds may be readily synthesised by refluxing the sodio derivative of N-acetyl-L-cysteine with the precursor. If the true precursor is an active metabolite, this method will not be applicable if the metabolite is not to hand. However, in this situation it is often possible to devise other routes utilising chemically synthesised intermediates.

Biosynthesis. This is not usually practical in a controlled manner in vitro because there are too many stages in the overall reaction.
Certain of the techniques discussed below for GSH conjugates are, however, applicable to the mercapturic acids.

Glutathione Conjugates

These conjugates present more problems than do the mercapturic acids because of their high polarity and amphoteric nature.

Isolation. The conjugates will usually have to be isolated from aqueous media composed of preparations from insects, plants, mammalian liver or mammalian bile. They are best isolated from

freeze-dried preparations (from which lipophilic materials have
been extracted) by extraction with methanol containing 5 to
25% of water.

Purification. The concentrated methanol extract may be
purified by thin-layer chromatography, column chromatography
using silicic acid and very polar solvents (37)(41), e.g.
butanol:formic acid:water (70:7:7 v/v), or by paper chromatography using butanol:acetic acid:water mixtures. Methods of
detection will depend on the compound but radiochemical methods
and UV absorbance will be common. The GSH conjugates are
ninhydrin-positive and this provides a sensitive and very helpful indicator, particularly if another method (e.g. radio-
chemistry) is simultaneously available for the xenobiotic portion
of the conjugate. Ion-exchange radiochromatography as described
by Lamoureux et al (35) can be a very powerful tool but may
require much investment of time to optimize. As a glutathione
conjugate is progressively purified, preparative paper electrophoresis utilising the amphoteric nature of the molecules may
be useful as a penultimate purification step. The buffer ions
will then have to be removed by a further chromatographic step.

Structural Identification. One of the most complete
identifications of a glutathione conjugate published in recent
years is that of the plant atrazine conjugate by Lamoureux
et al (35). This may be a reflection of the surprise at dis-
covering the mechanism in plants. Figure 22, adapted from the
publication of these workers, illustrates the steps taken to
effect the identification.

(a) Chemical Hydrolysis. Acid hydrolysis (6N HCl, 100°, 20 h)
followed by routine application of an amino acid analyzer serves
to identify glycine and glutamic acid. Cysteine will be identified (as cystine) if the C-S bond is acid-labile, but normally
it will appear as an S-substituted cysteine. This can be
derivatised as the N-2,4-dinitrophenyl derivative and methylated
to afford a derivative which may well be suitable for mass
spectrometry. S-Aryl glutathiones are labile to base (e.g.
N NaOH, 40°, 2 h) yielding the thiophenol (as dimer if oxidised).
During this process the cysteinyl moiety undergoes β-elimination
to yield the tripeptide, γ-glutamyldehydroalanylglycine, which
can be hydrolysed in acid to glutamic acid, glycine and pyruvic
acid (44).

(b) Hydrogenolysis. Refluxing with Raney nickel in aqueous
ethanol can be an excellent and simple way of demonstrating
the site of sulphur substitution on the xenobiotic portion of
the molecule. The latter should then be amenable to mass
spectrometry.

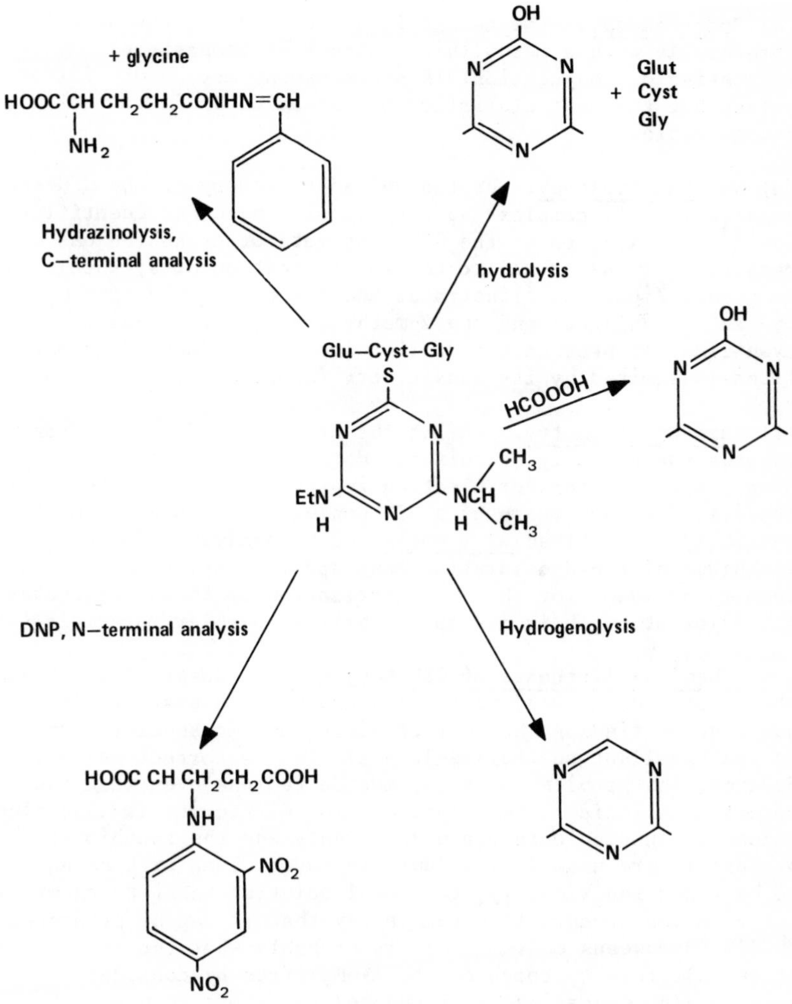

Journal of Agricultural and Food Chemistry

Figure 22. Identification of a glutathione conjugate (35)

(c) Terminal Analyses. Classic C and N-terminal amino acid
analysis are clearly necessary for unequivocal identification
when this is required.

(d) Spectroscopy. The change in UV spectrum of the xenobiotic
(preferably with a methylthio analogue in comparison) may be
suggestive of conjugation. IR spectroscopy may be of limited
value, but the characteristics of the amide bonds may be
recognisable.

(e) NMR Spectroscopy. Proton NMR spectroscopy of the GSH conjugates is very complex and not normally used for identification.
The ^{13}C NMR spectra of the GSH conjugate of cyanazine has been
measured (45) as an aid to the identification of synthetic
compound. Figure 23 illustrates the spectra of GSH, the
cyanazine conjugate and the 2-methylmercapto analogue of
cyanazine. At present the application of this method to metabolites is limited by its sensitivity (about 1-5 mg).

(f) Mass Spectrometry. Whilst the fragments of the GSH conjugates can normally be suitably derivatised and presented to
a mass spectrometer for electron impact and chemical ionization
studies, I am not aware of a GSH conjugate which has been successfully derivatised as a whole and so analysed. The new
technique of field-desorption mass spectrometry (FDMS) would
seem to be ideal for the characterisation of these conjugates.
Its major strength is its applicability to polar materials (46).

Chemical Synthesis of GSH Conjugates. Chemical synthesis
should provide no real problems because it is essentially a
question of finding the correct electrophilic species to present
to the excellent sulphur mucleophile in the correct medium.
However, the problem can sometimes be tedious even when the
parent xenobiotic is reactive. A major difficulty is that glutathione is insoluble in organic solvents and the xenobiotic
precursors are usually insoluble in water. Each problem has
to be faced individually, the ideal solution being to chemically
activate the precursor in such a way that it can be presented
to GSH in aqueous medium (this is probably what the transferase
achieves). This approach can be exemplified by considering cyanazine again. Cyanazine was incubated for about 24 h at 40°
with a saturated solution of GSH in DMSO in the presence of
solid sodium bicarbonate. A poor yield of the conjugate was
laboriously purified from the reaction mixture. However, when
cyanazine is reacted with trimethylamine in acetone, a 2-
trimethylammonium chloride analogue is formed in excellent yield
(37). The product is stable, freely soluble in water, and reacts
very rapidly with GSH in the presence of a mole of sodium
bicarbonate. The reaction is equally suitable for the synthesis
of gram quantities and sub-milligram quantities (e.g.

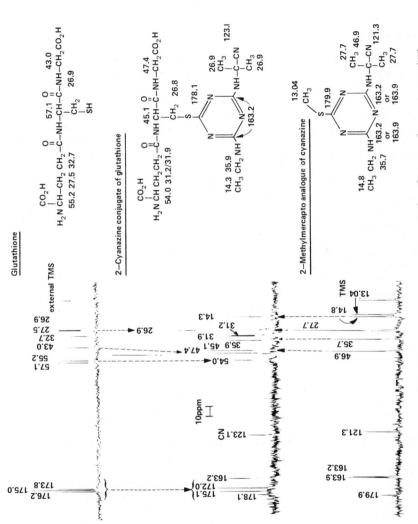

Figure 23. ^{13}C Nuclear magnetic resonance spectra of glutathione and a conjugate (45)

radiosynthesis). The reactions are shown in Figure 24.

3. Enzymic Synthesis and Degradation of Glutathione Conjugates

The biosynthesis and biodegradation of these conjugates
are helpful in their characterisation and to an understanding
of the mechanisms of their formation.

Biosynthesis

Crude transferase may be prepared from rat or rabbit liver
cytosol (100,000 g supernatant) and stored frozen for long
periods. If it is dialysed before storage (to remove GSH) and
partially purified by ammonium sulphate precipitation (21)(37),
the requirement for GSH can be demonstrated by including a
control reaction containing no GSH during the biosynthesis.
GSH should be added at a concentration of about $10^{-3}\underline{M}$ and at
pH 7.5. The GSH conjugate of cyanazine can be synthesised
directly by this method (37). Some conjugates cannot be so prepared however because bioactivation steps (e.g. epoxidation)
occur in vivo. These reactions must be initiated in vitro by
incubating with the transferase in the presence of liver microsomes and cofactors. The post-mitochondrial supernatant will
often substitute for the latter mixture, but it is preferable
to carry out the various stages separately to understand what
events are occurring. The first studies of GSH conjugation in
plants were effectively in vitro studies, utilising the action
of excised sorghum leaf sections on atrazine (35). The enzyme
has been purified 8-fold from corn leaves by extraction, centrifugation, ammonium sulphate fractionation and gel filtration.
(40).

Biodegradation

γ-Glutamyltranspeptidase should remove the glutamic acid
from a GSH conjugate and carboxypeptidase should remove glycine.
Few controlled studies have been carried out however. An
exception is the hydrolysis of the S-atrazine derivative of
glutamyl-cysteine isolated from sorghum (41). The metabolite
was incubated with hog kidney γ-glutamyltranspeptidase and the
product was identified as the S-atrazine conjugate of cysteine.
We have also observed the hydrolysis of the GSH conjugate of
cyanazine under similar conditions, but using alanine as the
glutamate acceptor.
The use of C - S lyase in the characterisation of the
lanthionine conjugate of atrazine (41) is also of great value
in the specialised area of sym-triazine biochemistry.

Some Properties of the Glutathione S-transferases

Virtually all of the mammalian transferases reported are located in the cytosol of the liver cell. The main impression gained from working in this area is that there is an enzyme for every substrate. The careful work of Chasseaud in differentiating between several alkene transferases (33) has been followed by the purification of three other transferases from rat liver cytosol to homogeneity. Six reactions were used to monitor the purifications (31). The results were not encouraging in terms of substrate specificity. The three proteins possessed a considerable overlap in specificity and they were specific neither for the leaving group nor for the carbon skeleton of the transferred group.

However, the proteins are true enzymes. They obey Michaelis-Menton kinetics, they are very specific in their requirement for GSH. Glutathione methyl transferase (46-fold purified) is stereospecific in its demethylation of dimethyl 1-naphthyl phosphorothionate (47). A very thorough kinetic study has been carried out on 76-fold pure GSH aralkyltransferase (menaphthyl sulphate) (48). The reaction product activates the enzyme when GSH is saturating and the concentration of substrate is low. The enzyme may exist in two sub-units separable by isoelectric focussing. The thiol group of GSH does not possess any special nucleophilic properties (towards benzene oxide) which would not be predicted from its pKa (49), therefore the specificity of GSH relative to other endogenous thiols must be conferred by the enzyme(s). It has been suggested (50) that ligandin, a soluble protein in liver cells which binds several organic ions, may be identical with GSH aryltransferase. This protein comprises about 4% of that in the liver cell. The multiplicity and overlapping substrate specificity of these transferases is difficult to acknowledge, and it may yet prove possible to rationalise the activities as being due to isoenzymes. The situation is somewhat analogous to that of cytochrome P450, the terminal mono-oxygenase of liver microsomes. This material is present in relatively massive amounts and its identity as one enzyme, or as a plethora of related enzymes, is still the subject of much discussion.

The mechanism of action of the enzyme(s) is not yet known. It is thought that both substrates are bound to protein. It is likely that this interaction allows the highly polar, solvated GSH molecule into a reasonably dipolar aprotic environment in which an efficient SN2 reaction with the lipophilic foreign compound can take place. Spectacular increases in the rates of SN2 reactions have been observed when reaction conditions are changed from protic solvents to dipolar aprotic solvents (e.g. acetone, DMSO) (51).

5. The Occurrence of GSH Transferases

The transferases appear to be wide spread in nature.

Mammals

Investigations of inter-species variations are patchy, and have not routinely been carried out, but what information exists suggests that the cytosol of all mammalian liver contain the enzyme(s). Chasseaud (33) made a six-substrate comparison of GSH alkene transferases in livers of rat, mouse, ferret, rat, dog, rabbit, guinea-pig, hamster, human adult and human foetus. Levels were generally of the same order, but those for humans were low on average. The methyl transferase (organophosphate insecticides) is present in the livers of rat, mouse, rabbit, dog and pig (21). The methyl transferase (methyl iodide) is present in the livers and kidneys of rat, mouse, guinea-pig, rabbit, lamb (very high), ox, monkey, pig and cat(16). The aryl transferase is widely distributed in mammals, including sheep (52). Organ variation of the enzyme is now being studied increasingly. A ten-substrate study of alkene transferase in rat liver and kidney showed that the activity in liver exceeded that in kidney by factors of about one and a half to seven fold depending on the substrate used for assay. Methyl transferase (methyl iodide) has been found in the rat liver, kidney and adrenal but not in the heart, lung, spleen, blood or brain (16). However, brain, spleen, lung, heart, kidney and muscle are reported to be active in the GSH-dependent demethylation of methyl parathion (19).

Epoxide transferase is also present in extra-hepatic tissues including those of the foetus (53). A recent study of GSH-epoxide transferase using naphthalene-1,2-oxide and styrene oxide as substrates and [^{35}S]GSH for quantitation has revealed that this enzyme is widely distributed in rat tissues, though liver and kidney contain about 10 and 7 times the mean value found in the other tissues (54). The same technique was used to show that the transferase to naphthalene-1,2-oxide was present in livers from a variety of species as follows: (activity as nmoles conjugate formed/g wet wt/min. in parentheses) sheep (463); horse (300); cattle (330); pig (86); monkey (96); rabbit (152); guinea-pig (448); male rat (228); female rat (187); male mouse (665) and female mouse (327) (54). Aryl transferase has been found in the lung of rabbit at about one-fifth of the specific activity found in liver.

Birds

Alkene transferase is present in pigeon liver (33). Aryl transferase has been found in the livers of several wild birds in the order pheasant ⩾ gull ⩾ coot ⩾ duck ⩾ eider ⩾ grebe ⩾

goosander (55). Chicken liver and kidney contain low alkyltransferase activity (16).

Insects

Alkyltransferase occurs in the midgut region of the horn
beetle, in silk worm larvae (19), and in houseflies (56). Aryl
transferase occurs in grass grubs (52). Aralkyl transferase
occurs in the locust (57), housefly, flourbeetle, cockroaches,
cattle tick, cotton stainer and turnip beetle (58). The insect
enzyme can often be distinguished from the mammalian enzyme
by differential inhibition (52)(58).

Plants

We do not yet know the generality of GSH conjugation in
plants but aryl transfer (27)(44) and several triazinyltransfers
(39) have now been characterised at the USDA Laboratory at Fargo.
Some organophosphate insecticides are O-dealkylated in plants
(59) and this reaction may well be mediated by such a transferase.

The Normal Role of the Enzymes

In mammals GSH transferases play a role in the metabolism
of steroidal estrogens in vivo (13). Two conjugates of 17-β-
estradiol have been biosynthesised (60). The 2,3-unsaturated
acylcoenzyme A thiol esters are also substrates for a transferase (61). Their reaction with GSH could be the initiating
mechanism for the excretion of the normal S-(carboxyalkyl)
cysteines found in animal and human urine.

6. The Significance of GSH Conjugation

General

The formation of a GSH conjugate effects a dramatic change
in the physical properties of a molecule. Thus a small, lipophilic, neutral molecule may be altered in one or two bio-
synthetic steps into a molecule which is about twice as large,
very lipophobic and possessing both anionic and cationic properties. It is therefore highly likely that, regardless of the
chemical changes that have taken place, the physical changes
will destroy the biocidal properties of the parent molecule.
The formation of the conjugates however, may have different
consequences in plants, insects and animals.

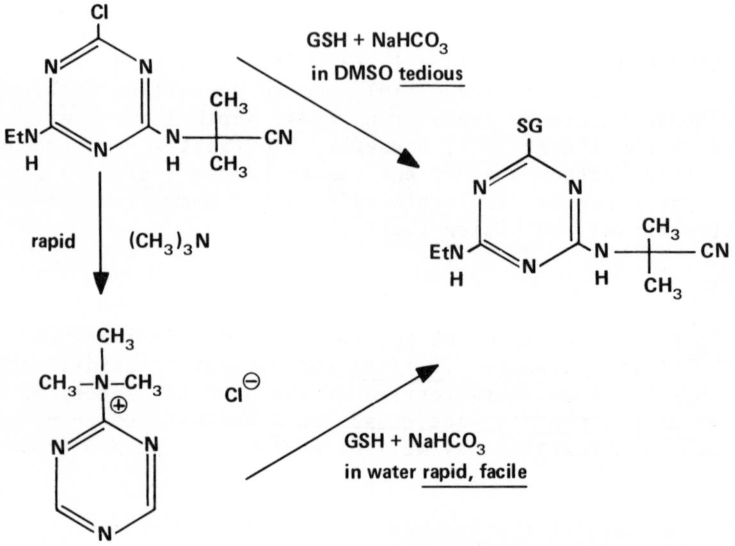

Pesticide Biochemistry and Physiology
Figure 24. Alternative syntheses of GSH conjugate (37)

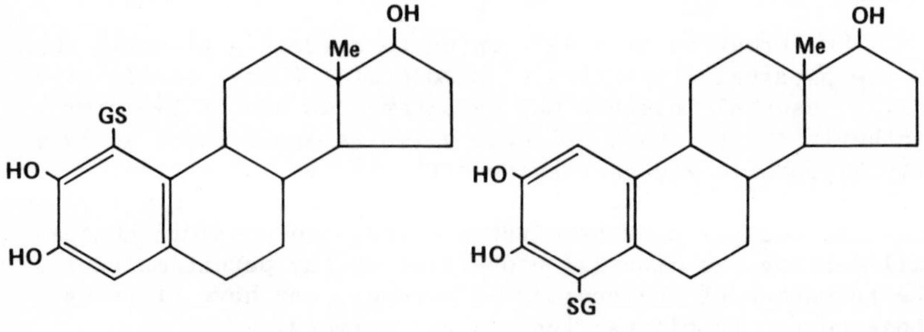

Figure 25. Glutathione conjugates of 17 β-estradiol

Plants

Plants do not have efficient excretory mechanisms for GSH
conjugates, however, the reaction normally leads to destruction
of biological activity of a compound. Therefore the possession
of an appropriate transferase by a particular species may be
expected to confer a degree of resistance to a foreign compound.
This has been shown to be the case with atrazine in susceptible
and resistant plant species (62)(63). However, it seems that
plants, like animals, possess a range of enzymes with various
specificities. The transferase that catalyses the cleavage of
fluorodifen is different from the enzyme that catalyses the
metabolism of atrazine. Therefore, the finding that fluoro-
difen- resistant plants like cotton, soyabean and peanut are
susceptible to atrazine was not surprising (27). The binding
of certain herbicides to plant components has been reported.
The catabolism of GSH conjugates and subsequent biochemical
incorporation of a substituted cysteine may be a mechanism of
such binding. Alternatively the incorporation may be due to
the direct interaction of the precursor with thiol groups of
proteins. In view of the specificity of the protein-
synthesising systems, the former mechanism would seem unlikely.

Insects

Certain of the transferases effect the detoxication of
organophosphate insecticides in insects and there is some
evidence that resistance to these insecticides in the tobacco
budworm is associated with higher levels of alkyl transferase
than are present in susceptible strains(64). The excretion
processes for GSH conjugates in insects have not been studied
in detail.

Mammals

There are three major consequences of GSH conjugation in
the mammalian liver:
(i)   specific bioactive properties are lost;
(ii) the conjugate is ideally structured for biliary secretion
and the compound is therefore efficiently removed from the
organ;
(iii) electrophilic compounds/metabolites are scavenged.

The last of these consequences is probably the most
important one. It is now established that many foreign compounds
exert their long-term toxic effects by reaction of electrophilic
centres in the parent molecules (or of centres generated by
metabolism) with DNA, RNA, proteins and other critical sites
in the cell. Glutathione, being present in all cells in
relatively high concentrations and containing the nucleophilic

thiol group, is a protective agent par excellence. Gillette
and coworkers have clearly demonstrated this protective effect
against the hepatic necrosis induced by massive doses of the
drug, acetaminophen. The degree of necrosis correlates with
the extent of oxidative metabolism of acetaminophen (65). The
severity of the necrosis is proportional to the amount of
covalent binding of a metabolite to liver protein (66). The
binding is mediated by cytochrome P450 (67). Acetaminophen
causes a dose-related decrease in hepatic GSH. Experimental
depletion of GSH potentiates the necrosis and increases the
covalent binding. Administration of cysteine reverses these
effects (68). A similar situation exists with bromobenzene,
which is oxidatively metabolised to its 3,4-oxide (Figure 12)
which is then detoxified by the action of GSH and GSH epoxide
transferase. However, at high doses, or under conditions of
GSH depletion, the centrilobular regions of the liver are
subject to necrosis (69). These regions are close to the portal
blood supply and therefore subjected to the highest concentration of compound. Glutathione can be progressively depleted
from these cells as it is utilised in reaction with the bio-
actived bromobenzene (the epoxide). When GSH levels are
severely depleted (>90%), the epoxide reacts with other components of these cells.

The apparently wide occurrence of the GSH transferases
in other mammalian organs is also an important aspect of the
protection afforded by GSH. Generally speaking, the enzyme which
plays the largest part in the metabolic generation of
electrophiles (the microsomal mono-oxygenase), is less active
in extra-hepatic tissues, in comparison with the liver. Nevertheless, GSH and the transferase(s) are present in these tissues
which may, therefore, be better protected from covalent
interactions than the liver. It would be surprising, however,
if there were not many exceptions to this generalisation
because the balance between oxygenase and the transferase will
vary with substrate and tissue.

Literature Cited

1. Baumann, E. and Preusse, C. Chem. Ber. (1879), 12, 806.
2. Jaffé, M. Chem. Ber. (1879), 12, 1092.
3. Waelsch, H. Arch. exp. Pathol. Pharmakol. (1930), 156, 356.
4. Brand, E. and Harris, M. M. Science, (1933), 77, 589.
5. Stekol, J. A. J. biol. Chem. (1939), 128, 199.
6. Smith, J. N., Spencer, B. and Williams, R. T. Biochem. J.
   (1950), 47, 284.
7. Mills, G. C. and Wood, J. L. J. biol. Chem. (1956), 219, 1.
8. Barnes, M. M., James, S. P. and Wood, P. B. Biochem. J.
   (1959), 71, 680.

9. Bray, H. G., Franklin, T. J. and James, S. P. Biochem. J.
    (1959), 71, 690.
10. Boyland, E. and Chasseaud, L. F. Adv. Enzymol. (1969), 32,
    173.
11. Wood, J. L., in "Metabolic Conjugation and Metabolic
    Hydrolysis", p. 261, Ed. W. H. Fishman, Academic Press,
    New York, Vol. 2, 1970.
12. Hutson, D. H., in "Foreign Compound Metabolism in Mammals",
    The Chemical Society, London; Vol. 3, 1975, p.537; Vol. 2,
    1972, p.385; Vol. 1, 1970, p.364.
13. Chasseaud, L. F., in "Glutathione; Proceedings of the 16th
    Conference of the German Society of Biological Chemistry,
    Tübingen, 1973", Eds. L. Flohé, H. Ch. Benöhr, H. Sies,
    H. D. Waller and A. Wendel. G. Thieme, Stuttgart, 1974,
    p.90.
14. Chasseaud, L. F. Drug Metabolism Reviews, (1974), 2, 185.
15. Johnson, M. K. Biochem. J. (1966), 98, 38.
16. Johnson, M. K. Biochem. J. (1966), 98, 44.
17. Pillinger, D. J., Fox, B. W. and Jackson, H., in "Isotopes
    in Experimental Pharmacology", p.415, Ed. L. J. Roth,
    University of Chicago Press, 1965.
18. Booth, J., Boyland, E. and Sims, P. Biochem. J. (1961),
    79, 516.
19. Fukami, J. and Shishido, J. J. econ. Entomol. (1966), 59,
    1338.
20. Hollingworth, R. M., Alstott, R. L. and Litzenberg, R. D.
    Life Sci. (1973), 13, 191.
21. Hutson, D. H., Pickering, B. A. and Donninger, C.
    Biochem. J. (1972), 127, 285.
22. Iqbal, Z. M. and Menzer, R. E. Biochem. Pharmacol. (1972),
    21, 1569.
23. Hollingworth, R. M., in "Biochemical Toxicology of
    Insecticides", p.75, Ed. R. D. O'Brien and I. Yamamoto,
    Academic Press, New York, 1970.
24. Hutson, D. H. and Hoadley, E. C. Xenobiotica, (1972), 2,
    107.
25. Boyland, E. and Chasseaud, L. F. Biochem. J. (1969), 115,
    985.
26. Gillham, B. Biochem. J. (1971), 121, 667.
27. Frear, D. S. and Swanson, H. R. Pest. Biochem. Physiol.
    (1973), 3, 473.
28. Gillham, B. and Young, L. Biochem. J. (1968), 109, 143.
29. Daly, J. W., Jerina, D. M. and Witkop, B. Experientia,
    (1972), 28, 1129.
30. Boyland, E. and Williams, K. Biochem. J. (1965), 94, 190.
31. Pabst, M. J., Habig, W. H. and Jakoby, W. B. Biochem.
    Biophys. Res. Commun. (1973), 52, 1123.
32. Boyland, E. and Chasseaud, L. F. Biochem. J. (1967), 104,
    95.

33. Chasseaud, L. F. Biochem. J. (1973), 131, 765.
34. Ohkawa, H., Ohkawa, R., Yamamoto, I. and Casida, J. E. Pest. Biochem. Physiol. (1972), 2, 95.
35. Lamoureux, G. L., Shimabukuro, R. H., Swanson, H. R. and Frear, D. S. J. Agric. Fd. Chem. (1970), 18, 81.
36. Hutson, D. H., Hoadley, E. C., Griffiths, M. H. and Donninger, C. J. Agric. Fd. Chem. (1970), 18, 507.
37. Crayford, J. V. and Hutson, D. H. Pest. Biochem. Physiol. (1972), 2, 295.
38. Dauterman, W. C. and Muecke, W. Pest. Biochem. Physiol. (1974), 4, 212.
39. Lamoureux, G. L., Stafford, L. E. and Shimabukuro, R. H. J. Agric. Fd. Chem.. (1972), 20, 1004.
40. Frear, D. S. and Swanson, H. R. Phytochemistry, (1970), 9, 2123.
41. Lamoureux, G. L., Stafford, L. E., Shimabukuro, R. H. and Zaylskie, R. G. J. Agric. Fd. Chem. (1973), 21, 1020.
42. Shishido, T., Usui, K., Sato, M. and Fukami, J. Pest. Biochem. Physiol. (1972), 2, 51.
43. Casida, J. E., Kimmel, E. C. Ohkawa, H. and Ohkawa, R. Pest. Biochem. Physiol. (1975), 5, 1.
44. Shimabukuro, R. H., Lamoureux, G. L., Swanson, H. R., Walsh, W. C., Stafford, L. E. and Frear, D. S. Pest. Biochem. Physiol. (1973), 3, 483.
45. Leworthy, D. P. (1975). Sittingbourne Research Centre, unpublished work.
46. Schulten, H. R., Prinz, H., Beckey, H. D., Tomberg, T. W., Klein, W. and Korte, F. Chemosphere, (1973), 2, 23.
47. Hutson, D. H. Med. Fac. Landbouwwet. Gent, (1973), 38, 741.
48. Gillham, B. Biochem. J. (1973), 135, 797.
49. Reubens, D. M. R. and Bruice, T. C. J.C.S. Chem. Comm. 1974, 113.
50. Kaplowitz, N., Percy-Robb, I. W. and Javitt, N. B. J. Exp. Med. (1973), 138, 483.
51. Miller, J. and Parker, A. J. J. Am. Chem. Soc. (1961), 83, 117.
52. Clark, A. G., Darby, F. J. and Smith, J. N. Biochem. J. (1967), 103, 49.
53. Juchau, M. R. and Namkung, M. J. Drug Metab. Disposit. (1974), 2, 380.
54. Hayakawa, T., Lemahieu, R. A. and Udenfriend, S. Arch. Biochem. Biophys. (1974), 162, 223.
55. Wit, J. G. Europ. J. Pharmacol. (1968), 5, 100.
56. Ishida, M. and Dahm, P. A. J. econ. Entomol. (1965), 58, 602.
57. Cohen, A. J. and Smith, J. N. Biochem. J. (1964), 90, 449.
58. Cohen, A. J., Smith, J. N. and Turbert, H. Biochem. J. (1964), 90, 457.
59. Beynon, K. I., Hutson, D. H. and Wright, A. N. Residue Rev. (1973), 47, 55.

60. Kuss, E. Z. physiol. Chem. (1969), 350, 95.
61. Speir, T. W. and Barnsley, E. A. Biochem. J. (1971), 125, 267.
62. Shimabukuro, R. H., Swanson, H. R. and Walsh, W. C. Plant Physiol. (1970), 46, 103.
63. Shimabukuro, R. H., Lamoureux, G. L., Frear, D. S. and Bakke, J. E. IUPAC Internat. Symp. on Terminal Residues, Tel Aviv, Ed. A. S. Tahori, Butterworths, London, 1971, p.323.
64. Bull, D. L. and Whitten, C. J. J. Agric. Fd. Chem. (1972), 20, 561.
65. Mitchell, J. R., Jollow, D. J., Potter, W. Z., Davis, D. C., Gillette, J. R. and Brodie, B. B. J. Pharmacol. Exptl. Therap. (1973), 187, 185.
66. Jollow, D. J., Mitchell, J. R., Potter, W. Z., Davis, D. C., Gillette, J. R. and Brodie, B. B. J. Pharmacol. Exptl. Therap. (1973), 187, 195.
67. Potter, W. Z., Davies, D. C., Mitchell, J. R., Jollow, D. J., Gillette, J. R. and Brodie, B. B. J. Pharmacol. Exptl. Therap. (1973), 187, 203.
68. Mitchell, J. R., Jollow, D. J., Potter, W. Z., Gillette, J. R. and Brodie, B. B. J. Pharmacol. Exptl. Therap. (1973), 187, 211.
69. Reid, W. D. and Krishna, G. Exp. Mol. Pathol. (1973), 18, 80.

8

Miscellaneous Conjugates—Acylation and Alkylation of Xenobiotics in Physiologically Active Systems

JORG IWAN

Schering AG, Berlin/Bergkamen, Plant Protection Division,
D-1 Berlin 65, Postfach 650311, West Germany

"Conjugation", as used by the pesticide chemist, represents a collective term for reactions usually catalyzed by enzymes and in which endogenous substrates
are linked to xenobiotics. Binding to the foreign
compounds takes place at functional groups that may
already be present in the parent molecules or added
in the course of the metabolic process. Formation of
conjugates can affect the biological activity of a
pesticide or drug in two ways: Through decrease of
lipid solubility and through the alteration of molecular structures essential for the exertion of physiological effects. Therefore, conjugation normally results in detoxification but, as far as acylations and
alkylations are concerned, this is not always the
case.

1. ACYLATION

1.1. Acetylation. Transfer of acetate from
acetyl-coenzyme A (whose biosynthesis, as a reminder,
is shown in fig. 1) to an amino group of a xenobiotic,
is certainly the best understood acylation reaction
observed in biotransformations of foreign compounds.
This reaction, which is catalyzed by arylamine
N—acetyltransferases, represents a general metabolic
pathway of aromatic amines, sulfonamides and
hydrazino compounds as well as nonaromatic amines in
mammals (1 - 4). There are strong indications that
generation of the N-acetyl derivatives proceeds via a
simple "ping-pong" mechanism involving two consecutive
steps: Formation of acetylacetyltransferase through a
reaction between acetyl-CoA and acetyltransferase and,
secondly, reaction of the enzyme complex with a suitable substrate to produce the N-acetate and acetyltransferase (5 - 7).

Studies on detoxification of 4,6-dinitro-o-cresol in the rabbit, conducted by Smith, Smithies and Williams in 1952 (8), belong to the early investigation revealing N-acetylation of a pesticide or of one of its degradation products (fig. 2).

Among the DNOC metabolites extracted from urine these same workers found 6-amino-4-nitro-o-cresol and 6-acetamido-4-nitro-o-cresol, the N-acetyl compound (as an O-glucuronide) being the most abundant conversion product. Detection of these substances, whose identification was based on paper chromatography as well as on UV spectrophotometry, and comparison with authentic standards demonstrated that in the rabbit, reduction and subsequent acetylation of dinitro-o-cresol constituted the predominant mechanism of inactivation.

Intensive research on microbial decomposition of aniline derivatives occurring as degradation products from a variety of pesticides in the soil environment started approximately in the middle sixties. In the course of these studies which received a special momentum by the discovery of microbiological azobenzene formation, N-acetylation of anilines was recognized as a metabolic conversion common in a variety of fungi, bacteria and algae (9 - 15). Some of the compounds studied are shown in fig. 3.

During their investigations on the metabolism of metobromuron by selected soil microorganisms Tweedy and coworkers detected the rapid and quantitative acetylation of p-bromoaniline (9) (fig. 4).

These authors suggested that conversion of anilines to their respective N-acetyl derivatives may be competitive with the concentration dependent oxidative coupling to azobenzenes in soil.

Identification of the acetanilides was accomplished through cochromatography of soil and culture extracts with known standard compounds or thinlayer and gas chromatography combined with mass spectrometry. A characteristic of the mass spectra obtained using electron impact ionization was the loss of ketene from the molecular ions thus giving rise to the appearance of the corresponding aniline radical ions in high relative abundance.

Needless to say these modern studies were carried out employing carbon 14 labeled parent compounds.

Highly interesting as to the way of its formation, its mass spectrometric fragmentation pattern and its ultimate metabolic fate was the detection of 4-chloro-2-hydroxy-acetanilide in cultures of Fusarium oxysporum fed with p-chloroaniline (13) (fig. 5).

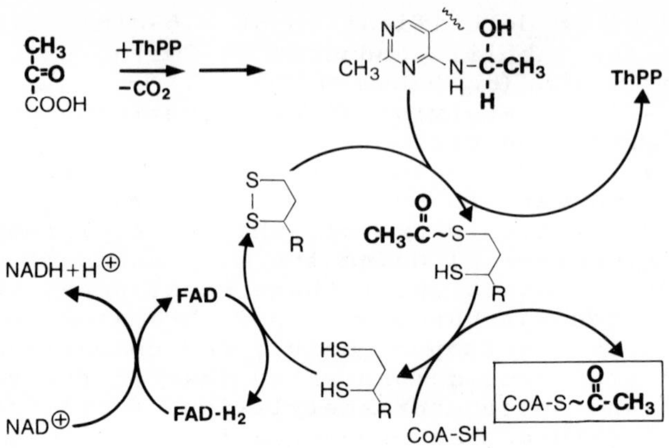

Figure 1.

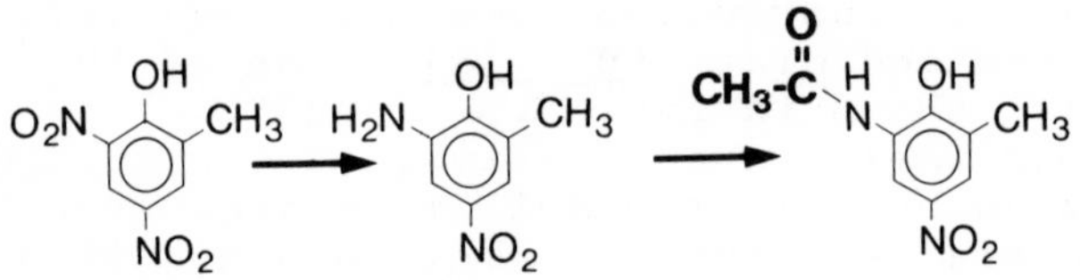

Figure 2.

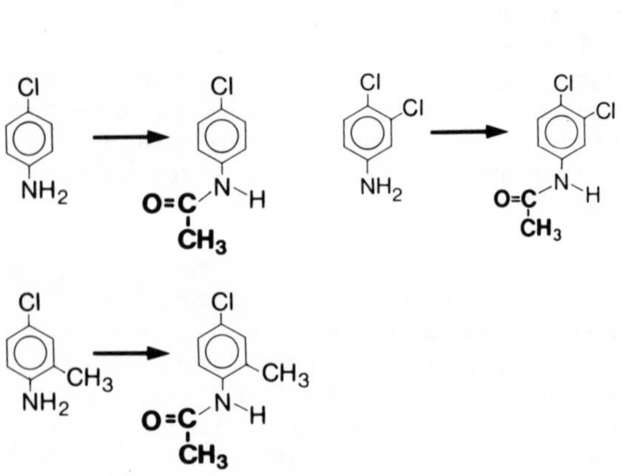

Figure 3.

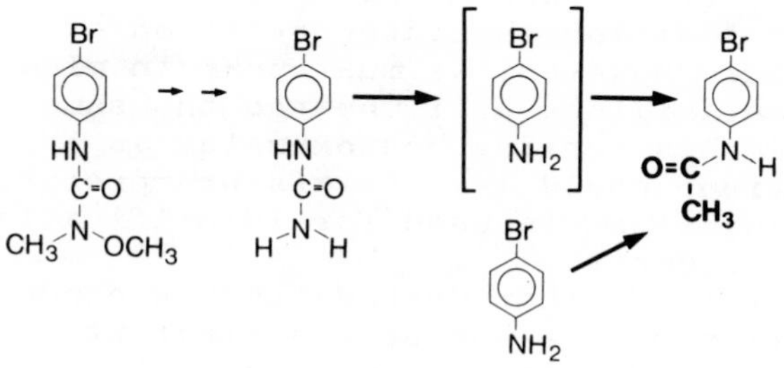

Figure 4.

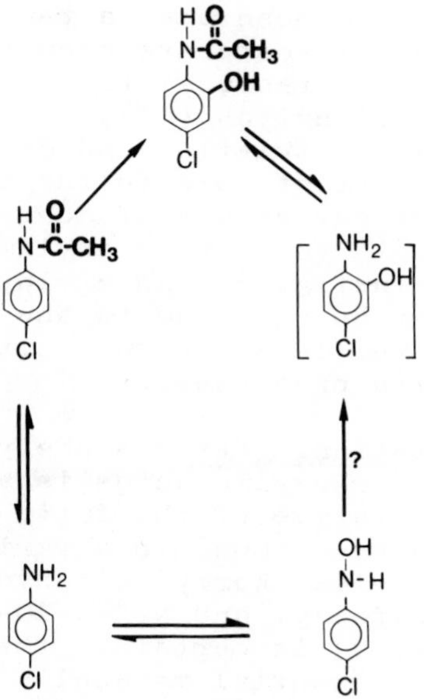

Figure 5.

Upon electron impact, generation of a benzoxazole
radical ion by ortho-elimination of water was observed (fig. 6).

Discovery of such a fragment may be a useful
diagnostic tool for the interpretation of respective
mass spectra; however, we must bear in mind that o-
hydroxy-acetanilides will undergo the same cyclization
on heating thus forming benzoxazoles as artefacts.
This behaviour could lead to misinterpretations if a
glc-ms combination is used for identification of
metabolic products.

As far as further degradation is concerned, o-
hydroxylation may be one of the first steps toward
microbial ring fission of anilines.

Conversion of m-aminophenol to m-hydroxy-acet-
anilide on sugar beet leaves after spraying with
desmedipham provides an example of the participation
of higher plants in the acetylation of a foreign
compound (16) (fig. 7).

However, the N-acetyl derivative was detected
among the chloroform soluble materials recovered from
the leaf rinse of treated beets. Microbial interaction
at the plant surface thus effecting N-acetylation,
therefore, cannot be excluded.

Acetyl transfer from 4-(N-hydroxyacetamido)-
biphenyl to 4-aminoazobenzene, a reaction catalyzed
by a special acetyltransferase occurring in rat liver,
may illustrate that transacylations do not necessarily
bring about detoxification (17).

In this case acetylation and deacetylation apparently play an important role in the carcinogenic activity of arylamines. By way of acetyl transfer from
the hydroxamic acid to another arylamine - the metabolic step shown in fig. 8 - an arylhydroxylamine is
released that can be oxidized to the corresponding
nitroso derivative. This, in turn, is more carcinogenic than any of its precursors.

1.2. Formylation. Besides acetylation, conjugation with other carboxylic acids is also possible. The
simplest representative of the fatty acid homologues,
formic acid, has been found in a variety of activated
forms in living cells. Formyltetrahydrofolic acid
occuring as N-10-formyl and N-5-N-10-methenyl-tetra-
hydrofolate (fig. 9) is certainly the most important
formate carrier. Essential metabolic steps in the
biosynthesis of purines (fig. 10) depend on the availability of this formyl source.

Other biologically active compounds may also be
considered as formate donors. Formyl-CoA, for in-

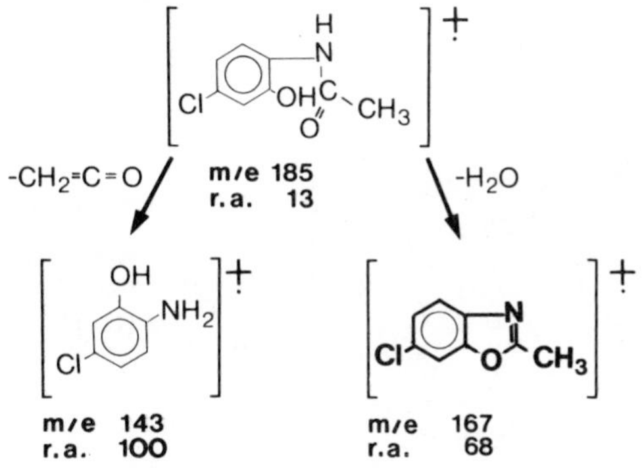

Figure 6.

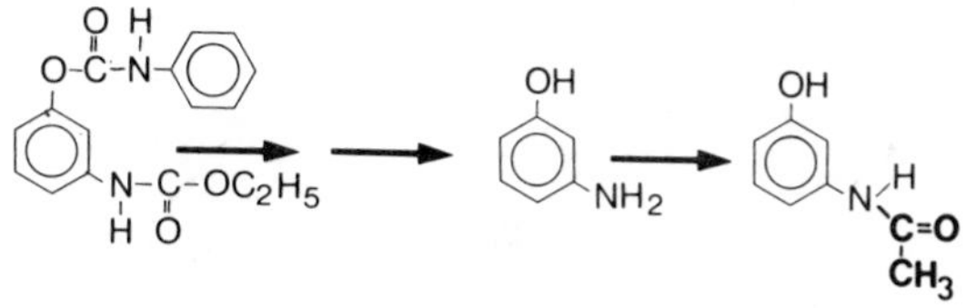

Figure 7.

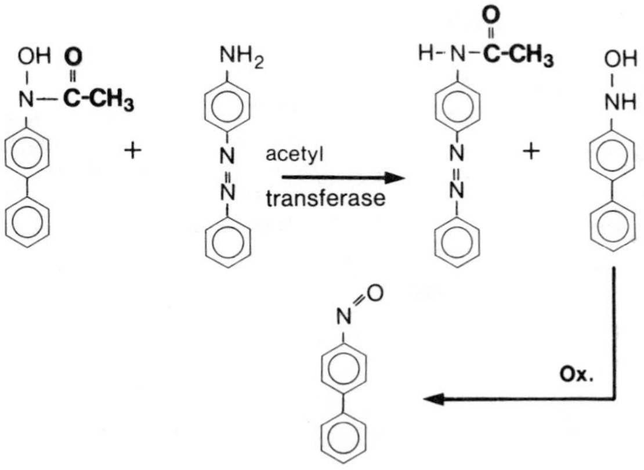

Figure 8.

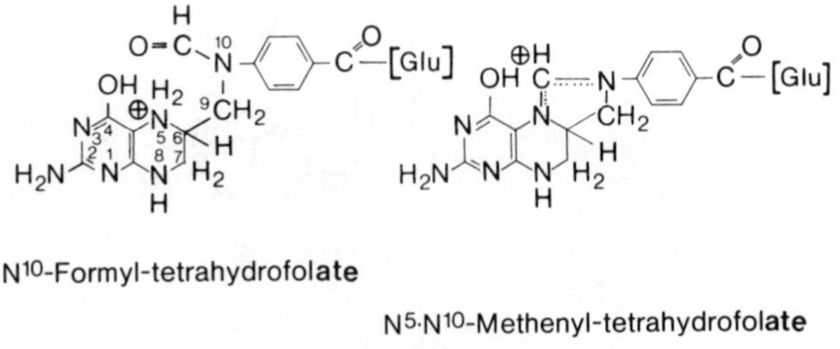
N10-Formyl-tetrahydrofolate
N5·N10-Methenyl-tetrahydrofolate

Figure 9.

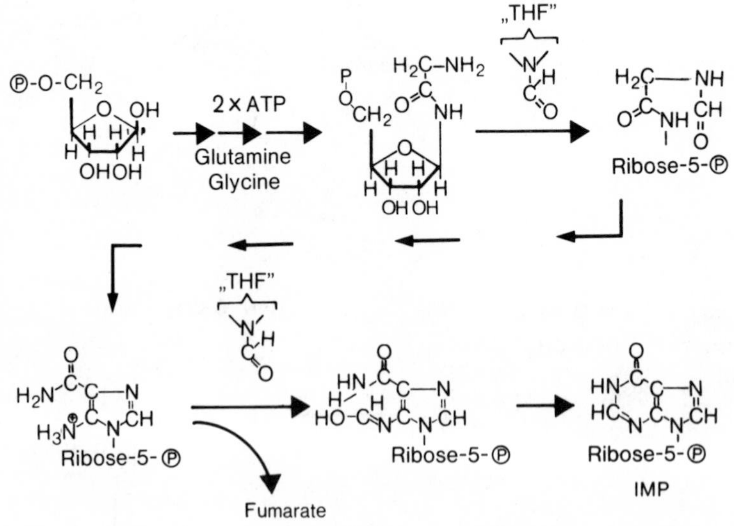
„THF"
2 × ATP
Glutamine
Glycine
Ribose-5-Ⓟ
Fumarate
IMP

Figure 10.

stance, has been demonstrated in microorganisms and higher plants as shown in fig. 11 which gives a partial view of the glyoxylate cycle.

N-Formyl-L-kynurenine, an intermediate in tryptophane metabolism, was found to be capable of transferring formate to aniline, naphthylamine and anthranilic acid in vitro (18) (fig. 12).

The transformylation was catalyzed by kynurenine formamidase from guinea-pig liver. 2-Formamido-1-naphtyl hydrogen sulphate, detected as a metabolite of 2-naphthylamine in the urine of dogs and rats, may have received its formyl group by reaction with this enzyme system (19).

Microbial transformation of anilines to formanilides in soil and pure fungal cultures was observed by Kaufman, Kearney and Plimmer (12, 13, 14, 20) (fig. 13).

3,4-Dichloroformanilide, as a metabolite of 3,4-dichloroaniline in soil, could be identified after purification by column and preparative thinlayer chromatography using infrared and mass spectrometry. Loss of 28 mass units from the parent ion was indicative of the elimination of a formyl group as carbon monoxide, a fragmentation analogous to the loss of ketene from acetanilides. Interpretation of the spectra was confirmed by synthesis of authentic 3,4-dichloroformanilide obtained from 3,4-dichloroaniline through reaction with acetic formic anhydride or refluxing in ethyl formate under atmospheric pressure.

1.3. Malonic acid conjugation. Conjugation with malonic acid has been observed as a mechanism for detoxification of D-amino acids in higher plants (21, 22). Malonyl-CoA which plays a central role in the biosynthesis of fatty acids (fig. 14) presumably mediates this reaction.

Malonyl transfer to a foreign compound by microorganisms was reported by Ross and Tweedy (23). Studies on the fate of chlordimeform in mixed microbial cultures revealed that 4-chloro-o-toluidine, occurring through stepwise hydrolytical cleavage of the parent compound, was transformed to the respective malonanilic acid derivative (fig. 15).

In some experiments this metabolite amounted to as much as 46 % of the total administered radioactivity thus suggesting that malonic acid conjugation may be an important mechanism for detoxification of aromatic amines.

Identification of the conjugate was accomplished using thinlayer chromatographic purification and mass

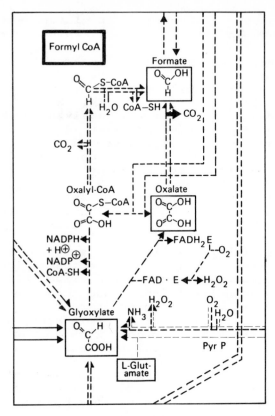

Figure 11.

Figure 12.

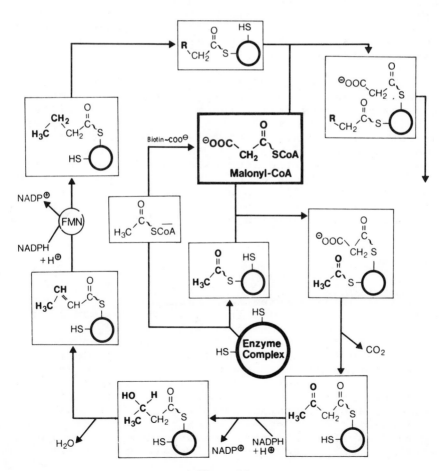

Figure 13.

Figure 14.

spectrometry. Characteristic of the mass spectrum was
a loss of 44 mass units from the parent ion to $\frac{m}{e}$ 183/
185 which indicated the elimination of carbon dioxide.
From this point downward in the direction of lower
mass numbers the spectrum became identical to that of
4-chloro-o-acetotoluidide the base peak representing
the toluidine radical ion at $\frac{m}{e}$ 141/143. Synthesis of
authentic 4'-chloro-2'-methylmalonanilic acid confirmed these results.

Evidence for the capacity of higher plants to
conjugate aniline derivates with malonic acid was recently obtained in studies on the metabolism of 2,6-
dichloro-4-nitroaniline in soybeans (24) (fig. 16).

The metabolic pathway involved reduction of DCNA
to the corresponding p-phenylenediamine and subsequent
malonyl transfer to yield N-(4-amino-3,5-dichloro-
phenyl)-malonamic acid as the major conversion product. Spectrometric identification could be confirmed
by synthesis from 2,6-dichloro-p-phenylenediamine
through reaction with ethyl chloroformylacetate followed by mild hydrolysis.

  1.4. Miscellaneous acylations. Various other
carboxylic acids are known to exist as activated substrates in living cells. Assuming the availability of
a suitable transferase, all of them might be regarded
as potential acylation reagents for xenobiotics if
these foreign compounds can penetrate to the site of
enzymatic action. The studies of Smith showing that
Streptomyces venezuelae can produce acetyl, propionyl
and butyryl as well as probably pentanoyl and hexanoyl
analogues of chloramphenicol, if the chloride ion concentration in the growth medium is limited, may illustrate this point (25).

2. ALKYLATION

  2.1. Methylation. As early as in 1894 it was
suggested by Hofmeister that methylation of organic
and inorganic compounds in animal tissues may occur
via transfer of an intact methyl group (26). A particular compound acting as a possible methyl source, however, was not mentioned. In 1933, 39 years later,
Challenger, Higginbottom and Ellis succeeded in eluci-
dating the nature of the so-called Gosio-gas, a volatile arsenic compound produced by several fungi growing on arsenic containing media (27). These authors
demonstrated that Gosio-gas entirely consisted of
trimethylarsine, and they proposed a mechanism involving stepwise addition of formaldehyde to arsenious

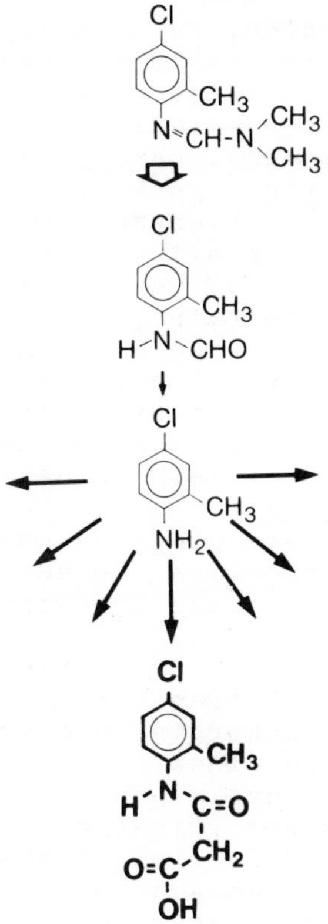

Figure 15.

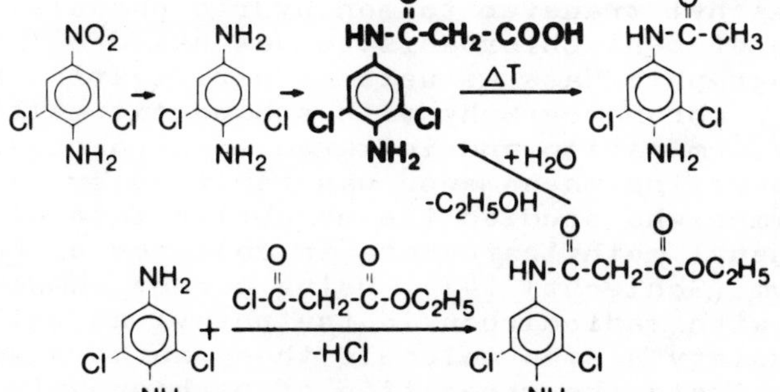

Figure 16.

acid, followed by reduction, for the formation of this
substance. 15 Years later, in 1954, Challenger and coworkers obtained experimental evidence that methylation of arsenic and other metalloids was mediated by
an active form of methionine, namely S-adenosyl-
methionine (28). These findings had been made possible
through the research of du Vigneaud and his school,
who between 1939 and 1942 provided the first scientific proof that methionine was the methyl source for
transmethylations in white rats (29). Other workers
had demonstrated the importance of methionine for the
dimethylamino group of hordenine in barley seedlings
(30, 31) and the methoxylation of barley lignin (32,
33). Cantoni and colleagues, finally, have been able
to explain the mechanism of methionine activation, as
shown in fig. 17 (34 - 37).
    Adenosine triphosphate plays an essential role
in this reaction as it provides the adenosine moiety
as well as the energy necessary for stimulation of
the process.
    A large number of papers describing enzyme sys-
tems which catalyze methyl transfer from S-adenosyl-
methionine to naturally occurring and foreign compounds has been published since then, and in a review
article by Greenberg de novo synthesis of the
methionine methyl group was illustrated in detail
(38). The scheme outlined in fig. 18 displays most of
the steps involved in the one-carbon transfer which
ultimately results in methylation of homocysteine.
    Biological methylation of a pesticide or rather
one of its metabolic products was reported by Loos
and coworkers who detected 2,4-dichloroanisole formation in the 2,4-D containing growth medium of an
Athrobacter sp. (39) Cserjesi and Johnson observed
pentachloroanisole production from pentachlorophenol
in cultures of Trichoderma virgatum (40) (fig. 19).
    Both reactions are remarkable as they represent
two examples of the, in other respects rarely encountered, methyl transfer to monohydric phenols. Identification of pentachloroanisole was based on gas-
chromatographic data as well as purification by
thinlayer chromatography with subsequent melting
point determination and infrared spectroscopy.
    A puzzling phenomenon was reported by Kaufman
and Plimmer who studied the metabolic fate of p-
chlorophenyl methylcarbamate in cultures of Fusarium
oxysporum (Schlecht) (41). Using parent compounds
labeled with radiocarbon in the phenyl as well as the
methyl moiety of the molecule these authors were able
to demonstrate the generation of p-chloroanisole which

CH₃
|
S
|
CH₂
| +ARP*∼P*∼P $\xrightarrow{\text{GSH}}$
CH₂   Mg⁺⁺
|
CH-NH₂
|
COOH

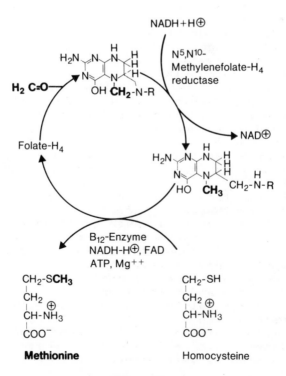

+P*∼P*+iP

Figure 17.

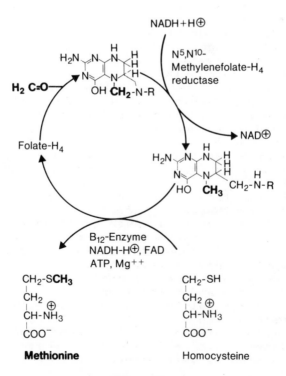

Methionine   Homocysteine

Figure 18.

contained both the methyl and the ring label. One may
speculate that the methyl carbon entered into one-
carbon metabolism and was reattached to the phenolic
oxygen in a transmethylation process (fig. 20).
    However, no p-chloroanisole could be detected
when p-chlorophenol was fed to the fungal cultures.
    Contrary to the usual finding that methylation
of phenolic compounds decreases their toxicity, microbial methyl transfer to 2,5-dichloro-4-methoxy-
phenol, a conversion product of chloroneb, results in
the opposite effect. Wiese and Vargas recently showed
that a variety of fungi could decompose chloroneb to
the corresponding dichloromethoxyphenol and independ-
ently resynthesize the fungicide in considerable
quantities (42) (fig. 21).
    The latter reaction is very important as it sug-
gests that the persistence of chloroneb in soil may
be explained better by simultaneous decomposition and
resynthesis rather than by an inherent inertness of
the compound.
    Gas chromatography was employed to detect
chloroneb and its degradation products in the microbial cultures. Authenticity of these substances was
ascertained by mass spectrometry.
    Metabolism of carbaryl in lactating cows in-
volved hydroxylation as well as methylation and sul-
phuric acid conjugation thus yielding 1-methoxy-5-
(methylcarbamoyloxy)-2-naphthyl sulphate among other
metabolic compounds (43) (fig. 22).
    Upon treatment with sulphatase this conjugate
released 5-methoxy-6-hydroxy-1-naphthol methylcarbamate whose identity was confirmed by comparison of
infrared and mass spectral data with those of authentic material.
    Similar alterations of a carbamate were described
in connexion with isopropyl carbanilate metabolism in
the chicken (44).
    After purification by ion exchange and gel
chromatography the 4-sulphate ester of isopropyl
3-methoxy-4-hydroxycarbanilate could be isolated from
the urine of the propham treated birds (fig. 24). The
structure of this compound was elucidated through infrared spectroscopy as well as derivatizations followed by mass spectrometric analysis. Synthesis of an
authentic sample and comparison of its IR-data with
those of the metabolite afforded final confirmation.
    Transformation of pentachloronitrobenzene to
pentachlorothioanisole was first reported in 1969 (45,
46) and later again in 1971 (47). The mechanism of
this conversion is very interesting, as it occurs in

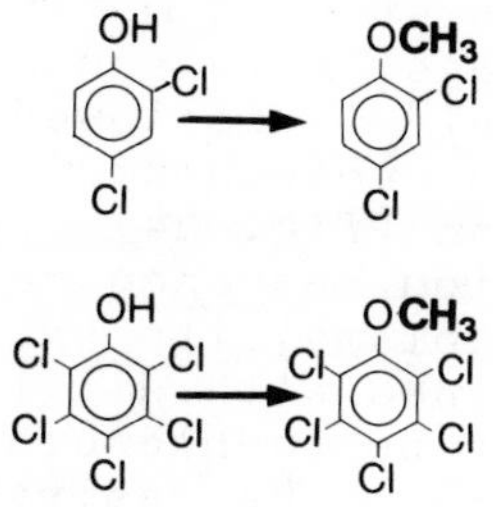

Figure 19.

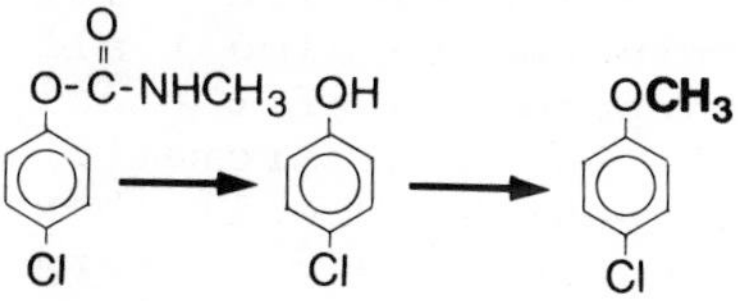

Figure 20.

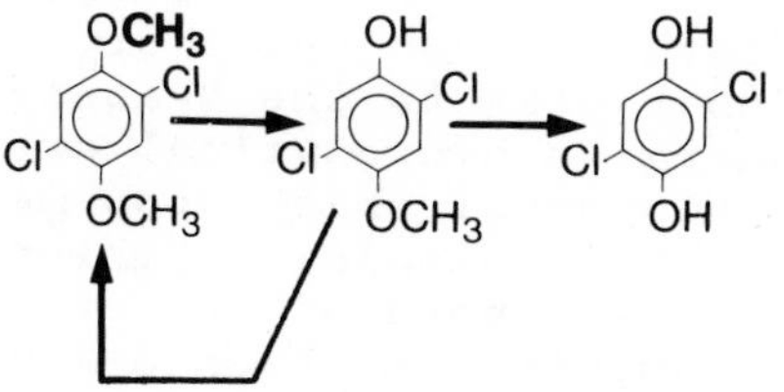

Figure 21.

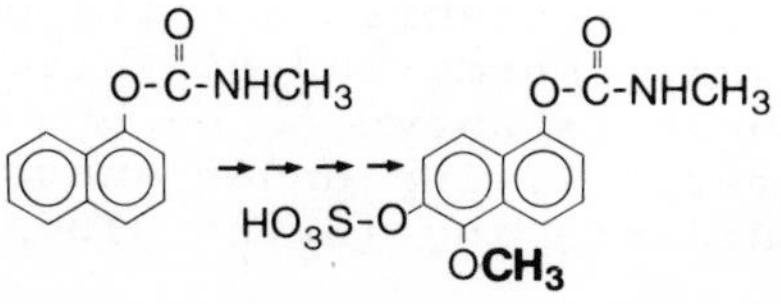

Figure 22.

mammals, higher plants and microorganisms as well.
Kuchar and coworkers who studied metabolism of pentachloronitrobenzene in dogs, rats and cotton plants
suggested alkaline hydrolysis followed by the formation of a thiophenol and subsequent methylation as a
possible way of formation (fig. 24).

While such a reaction sequence may be conceivable
in mammals and microorganisms, its existence in higher
plants can only be imagined with difficulty. Search
for a general pathway of pentachlorothioanisole generation, therefore, appears to be desirable. A hint at
the nature of this pathway might be obtained from the
works of Betts, Bray and coworkers, who were able to
isolate respective mercapturic acids from the urine of
tetra and pentachloronitrobenzene treated rabbits (48,
49). Mercapturic acid formation, however, proceeds via
reaction with glutathione, stepwise hydrolysis to a
cysteine derivative and acetylation. Glutathione, in
turn, is widely distributed in almost all living organisms, so that incorporation of the sulphur atom of
pentachlorothioanisole into the aromatic moiety, by
enzymatic reaction of pentachloronitrobenzene with
GSH, seems highly probable. The discovery of enzyme
catalyzed glutathione conjugation of triazines in
higher plants supports this view (50, 51).

Based on these observations the mechanism of
pentachlorothioanisole formation may be explained as
shown in fig. 25.

The first step in this process would involve
displacement of the labile nitro group by glutathione
followed by generation of a mercapturic acid at least
in mammals. The glutathione conjugate or, any one of
the intermediates "en route" to the N-acetylcysteine
derivative, then undergoes hydrolytical or reductive
cleavage to form a thiophenol which upon methylation
yields the respective thioanisole. A "thionase" capable of splitting the C - S bond of the cysteine moiety in glutathione conjugates has been demonstrated in
rats, rabbits and dogs (52).

Pentachlorothioanisole was identified by mass
spectrometry after initial purification using steam
distillation, thinlayer, column and gas chromatography
as clean-up procedures (45 - 47).

Biological methylation of inorganic, divalent
mercury to methyl and dimethylmercury is a noteworthy
reaction because of its mechanism and its possible
ecological significance. Formation of these volatile
alkyl mercury compounds is known to be brought about
by methanogenic bacteria living under anaerobic conditions in river sediments and in the sludge of

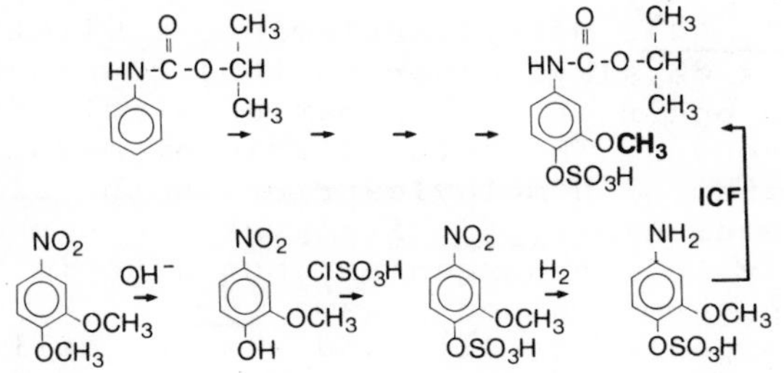

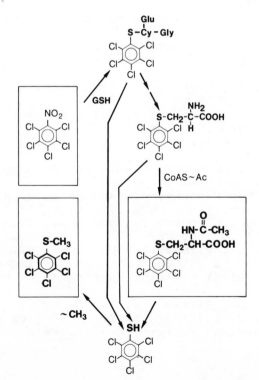

Figure 23.

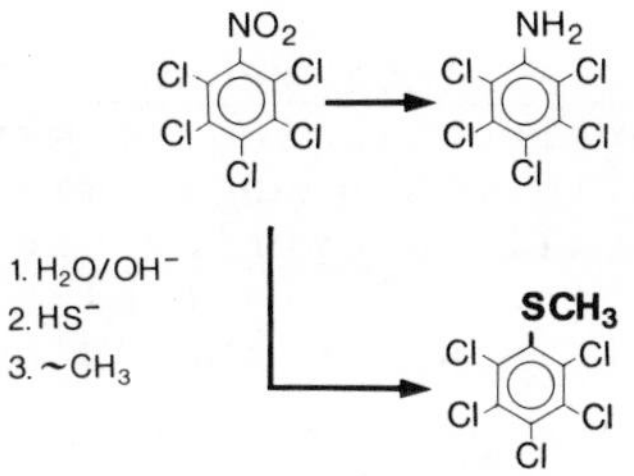

Figure 24.

Figure 25.

sewage beds (53, 54). Methylcobalamine, a vitamin B_{12} analogue, serves as alkyl donor in this transmethylation which can occur as an enzymatic as well as a nonenzymatic process. The natural role of methylcobalamine is that of a methylcarrier in the biosynthesis of methionine (55); its presence was demonstrated not only in microorganisms but also in mammalian tissues and human blood plasma (56).

In model experiments with and without cell free extracts of methanogenic bacteria the alkyl mercury compounds derived from the reaction of divalent mercury ions with methylcobalamin were characterized by thinlayer and gas chromatography.

2.2. Transfer of larger alkyl groups. Transfer of propyl groups from propylcobalamine to mercury ions was shown to proceed in vitro under mild reducing conditions (57). Naturally occurring biological processes involving migration of intact alkyl groups larger than methyl are not known to the author.

Literature Cited

1. Chon, T. C., and Lipmann, F., (1952); J. Biol. Chem. 196, 89
2. Tabor, H., Mehler, A. H., and Stadtman, E. R., (1953); J. Biol. Chem. 204, 127
3. Lipmann, F., (1953); Bacteriol. Rev. 17, 1
4. Colvin, L. B., (1969); J. Pharm. Sci. 58, (12) 1433
5. Cleland, W. W., (1963); Biochim. Biophys. Acta 67, 104
6. Steinberg, M. S., Cohen, S. N., and Weber, W. W., (1971); Biochem. Biophys. Acta 235, 89
7. Riddle, B., and Jencks, W. P., (1971); J. Biol. Chem. 246, 3250
8. Smith, J. N., Smithies, R. H., and Williams, R. T., (1953); Biochem. J. 54, 225
9. Tweedy, B. G., Loeppky, C., and Ross, J. A., (1970); J. Agr. Food. Chem. 18, (5), 851
10. Tweedy, B. G., Loeppky, C., and Ross, J. A., (1970); Science 168, 482
11. Kaufman, D. D., Plimmer, J. R., Iwan, J., Klingebiel, U. I.; 162nd National Meeting of the American Chemical Society, Washington D. D., Sept. 1971
12. Kaufman, D. D., Plimmer, J. R., Iwan, J., Klingebiel, U. I.; 163rd National Meeting of the American Chemical Society, Boston, Mass., April 1972

13. Kaufman, D. D., Plimmer, J. R., and Klingebiel, U. I., (1973); J. Agr. Food Chem. 21, (1), 127
14. Kaufman, D. D.; International Symposium, The Interaction of Herbicides, Microorganisms and Plants; Wroclaw, Poland, Aug./Sept. 1973
15. Iwan, J., and Goller, D.; The Third International Congress of Pesticide Chemistry (IUPAC), Helsinki, July 1974
16. Knowles, C. O., and Sonawane, B. R., (1972); Bull. Env. Contam. Toxicol. 8, (2), 73
17. Booth, J., (1966); Biochem. J. 100, 745
18. Santti, R. S. S., and Hospsu-Havu, V. K., (1968); Biochem. Pharmacol. 17, 1110
19. Boyland, E., and Manson, D., (1966); Biochem. J. 99, 189
20. Kearney, P. C., and Plimmer, J. R., (1972); J. Agr. Food Chem. 20, (3), 584
21. Zenk, M. H., and Scherf, H., (1964); Planta 62, 350
22. Rosa, N., and Neish, A. C., (1968); Can. J. Biochem. 46, 797
23. Ross, J. A., and Tweedy, B. G., (1973); Bull. Env. Contam. Toxicol. 10, (4), 234
24. Kadunce, R. E., Stolzenberg, G. E., and Davis, D. G., (1974); Abstr. Papers, Am. Chem. Soc. 168 Meet. Pest. 5
25. Smith, Ch. G., (1958); J. Bacteriol. 75, 577
26. Hofmeister, (1894); Arch. exp. Path. Pharm., 33, 209, 213
27. Challenger, F.. Higginbottom, C., and Ellis, L., (1933); J. Chem. Soc. 1933, 95
28. Challenger, F., Lisle, D. B., and Dransfield, P. B., (1954); J. Chem. Soc. 1954, 1760
29. du Vigneaud; "A Trial of Research", Cornell Univ. Press, Ithaca, N. Y. 1952
30. Kirkwood and Marion, (1951); Can. J. Chem. 29, 30
31. Dubeck and Kirkwood, (1952); J. Biol. Chem., 199, 307
32. Byerrum and Flokstra, (1952); Fed. Proc. 11, 193
33. Byerrum, Dewey, and Ball, (1953); Fed. Proc. 12, 186
34. Cantoni, G. L., (1952); J. Amer. Chem. Soc., 74, 2942
35. Baddiley, Cantoni, and Jamieson, (1953); J. Chem. Soc. 1953, 2662
36. Cantoni, G. L., (1953); J. Biol. Chem., 204, 403
37. Cantoni, G. L., and Durrel, J., (1957); J. Biol. Chem. 225, 1033
38. Greenberg, D. M., (1963); Adv. Enzymol. 25, 395

39. Loos, M. A., Roberts, R. N., and Alexander, M.,
    (1967); Can. J. Microbiol. 13, 691
40. Cserjesi, A. J., and Johnson, E. L., (1972); Can.
    J. Microbiol. 18, 45
41. Kaufman, D. D., and Plimmer, J. R.; Biodegrada-
    tion of p-Chlorophenyl Methylcarbamate
42. Wiese, M. V., and Vargas, J. M., (1973); Pestic.
    Biochem. Phys. 3, 214
43. Dorough, H. W.; Pesticide Terminal Residues,
    Invited Papers from the International Symposium
    on Pesticide Terminal Residues, Tel-Aviv, Israel,
    Febr. 1971
44. Paulson, G. D., Docktor, M. M., Jacobsen, A. M.
    and Zaylskie, R. G., (1972); J. Agr. Food Chem.
    20, (4), 867
45. Kuchar, E. J., Geenty, F. O., Griffith, W. P.,
    and Thomas, R. J., (1969); J. Agr. Food Chem. 17,
    (6), 1237
46. Nkanishi, T., and Oku, H., (1969); Phytopathology
    59, 1761
47. Borzelleca, J. F., Larson, P. S., Crawford, E.
    M., Hennigar, G. R., Kuchar, E. J., and Klein, H.
    H., (1971); Toxicol. Appl. Pharmacol. 18, 522
48. Bray, H. G., Hybs, Z., James, S. P., and Thorpe,
    W. V., (1953); Biochem. J. 53, 266
49. Betts, J. J., James, S. P., and Thorpe, W. V.,
    (1955); Biochem. J. 61, 611
50. Frear, D. S., and Swanson, H. R., (1970);
    Phytochemistry 9, 2123
51. Lamoureux, G. L., Stafford, L. E., Shimabukuro,
    R. H., (1972); J. Agr. Food Chem. 20, (5), 1004
52. Colucci, D. F., and Buyske, D. A., (1965);
    Biochem. Pharmacol. 14, 457
53. Wood, J. M., Kennedy, F. S., and Rosen, C. G.,
    (1968); Nature 220, 173
54. Imura, N., Sukegawa, E., Pan. S.-K., Nagao, K.,
    Kim, J.-Y., Kwan, T., and Ukita, T., (1971);
    Science 172, 1248
55. Guest, J. R., Friedman, S., Woods, F. R. S.,
    (1962); Nature (London) 195, 340
56. Lindstrand, K., (1964); Nature 204, 188
57. Westöö, G., (1967); Acta Chem. Scand. 21, 1790
    (1967)

9

Nature of Propanil Bound Residues in Rice Plants as Measured by Plant Fractionation and Animal Bioavailability Experiments

M. L. SUTHERLAND

Monsanto Agricultural Co., 800 N. Lindbergh Blvd., Saint Louis, Mo. 63166

In 1968 Still applied ^{14}C carbonyl labeled propanil (3,4 dichloropropionanilide) to the roots of rice and 6 days later could account for only 25% of applied ^{14}C. Yih applied ring ^{14}C labeled propanil to leaves of rice plants and 21 days later could account for only 46% of applied ^{14}C.

MONSANTO employed ring ^{14}C propanil in metabolism chambers that allowed $^{14}CO_2$ and other volatile radioactive products to be trapped. Treatments consisted of 5 lb/A applications of propanil phenyl-^{14}C to: 1) rice foliage, 2) soil containing rice plants, and 3) soil only. The percent ^{14}C distribution and accountability at the end of 27 days is shown below:

FRACTION		CHAMBER NUMBER 1 (%)		2 (%)	3 (%)
COLD TRAP		12.5		0.2	0.3
$^{14}CO_2$		0.2		0.9	0.5
SOIL		9.3		88.5	85.8
PLANT					
CHCl$_3$ SOLUBLE	10.2		10.5		
AQUEOUS SOLUBLE	18.1	70.4	17.0	1.9	---
INSOLUBLE	71.7		72.5		
^{14}C ACCOUNTABILITY		92.4		91.5	86.6

The above study showed that: 1) acceptable ^{14}C accountability could be attained when ring labeled propanil was utilized and all possible fractions analyzed, 2) the major portion of the volatile ^{14}C trapped from chamber 1 occurred in days 3 through 5, 3) most of the ^{14}C in rice plants harvested from propanil foliar or soil treatments was not extractable, 4) the foliar applied propanil-phenyl-^{14}C must have been absorbed by rice leaves since it appeared to have undergone a similar degree of plant insolubilization as that absorbed by rice plants harvested from propanil treated soil, 5) the detection of $^{14}CO_2$ suggested the possibility that the insoluble plant ^{14}C could be due to incorporation of $^{14}CO_2$ by rice plants into polymeric natural products.

Rice plants harvested at 3 and 12 weeks after soil applications of 5 lb/A propanil-phenyl-^{14}C were exposed to a plant
separation scheme. Radioactivity was spread throughout all of
the fractions for both 3 rd and 12 th week harvests. TLC examination and enzymatic hydrolysis of 3 rd week rice plant fractions failed to reveal the presence of any natural products.
Only trace quantities of low specific activity ^{14}C natural products (starch, hemicellulose, ligin, and cellulose) were found
in the 12 th week rice fractions. Of greater significance was
the fact that alkaline hydrolysis of each plant fraction revealed that the major portion of ^{14}C was associated with the
3,4-dichloroaniline-phenyl-^{14}C moiety. Shown below is the
distribution of ^{14}C in the various 12 th week plant fractions
and the percent 3,4-dichloroaniline-phenyl-^{14}C in each fraction:

FRACTION	^{14}C (%)	3,4-DICHLOROANILINE (% of ^{14}C)
HEXANE	4.8	78.2
BLIGH/DYER		
CHCl$_3$	11.0	48.4
AQUEOUS	20.6	38.8
STARCH REMOVAL		
ACIDIC ETHANOL	22.7	63.8
STARCH	1.6	----
PROTEIN	1.8	----
HEMICELLULOSE	8.7	48.4
LIGNIN	24.4	63.2
CELLULOSE	2.0	47.3
^{14}C ACCOUNTABILITY	97.6	----

The insoluble character of the major portion of the propanil
rice residue suggests that the animal digestibility of these non-
extractable residues would be low. If so, such insoluble residues would have neglible toxicological effect. To this end
bioavailability studies were undertaken to determine if monogastric animals could release the non-extractable ^{14}C in treated
rice plants. White albino rats were dosed with propanil-phenyl-
^{14}C treated plant samples via gavage and maintained in Aerospace
metabolism cages. The highly restrictive Aerospace metabolism
cage provides good feces/urine separation, eliminates coprophagy,
and allows monitoring of ^{14}CO$_2$ and other volatile products.
When 4 albino rats were dosed with pre-extracted 12 th week rice
plant filter cake, no ^{14}CO$_2$ and other ^{14}C volatile products
could be detected, while 76.1% of the initial ^{14}C appeared in
the feces with only 2.4% in the urine. When the study was repeated in the Aerospace cages utilizing non-extracted 12 th week
rice, the excretion of ^{14}C was 6.5% in the urine and 78.2% in
the feces. Since the 12 th week rice contained 36% extractable
^{14}C it appears that the rat was less efficient in removing the
^{14}C propanil residues from the rice plant than solvent extraction. To overcome the low ^{14}C accountability the rat bioavail-

ability studies were repeated in less restrictive Econo metabolism cages with 12 th week non-extracted rice. The percent ^{14}C
recovery rose to 88.8% in the feces, 11.3% in urine, with an
accountability of 99.9%. Combustion of tissue in all studies
conducted, revealed less than 0.05 ppm in fat, kidney, liver
and muscle.

The highly restrictive nature of Aerospace metabolism
cages causes constipation and/or low food intake resulting in
low ^{14}C recoveries due to incomplete passage of the dose. Normal rat metabolism such as excretion of products via urine is
also slowed down.

It should be pointed out that not all absorbed components
are excreted exclusively via the urine. Some absorbed components can be eliminated by the feces due to bile excretion into
the digestive tract. At the request of FDA, biliary fistula
recycling experiments were carried out by Bio-Test Laboratories
on dog and mice. Rice plants harvested 3 weeks after 10 lb/A
propanil-phenyl-^{14}C soil treatment were extracted via Bligh/
Dyer and the resulting filter cake was used in the recycling
experiments. Bio-Test found less than 0.05% of the ^{14}C dose
appearing in the bile of dogs or mice. The distribution of ^{14}C
in the urine and feces was very similar to that found in the rat
bioavailability studies. There was no significant tissue retention of ^{14}C by either dogs or mice.

In summary, the rapid insolubilization of ^{14}C in propanil-
phenyl-^{14}C treated rice plants can be accounted for by the
association of the 3,4-dichloroaniline moiety with various plant
fractions. The degree of absorption of bound propanil residues
in rice by monogastric animals can be monitored directly by
urinary excretion activity. Due to the low absorption by monogastric animals, the bound propanil residues should be considered
as low toxicological concern.

Literature Cited:

1. Still, G. G., Science (1968) 159, 992-993.

2. Yih, R. Y., McRae, D. H., Wilson, H. F., Science (1968)
    161, 376.

10

Solubilization of Bound Residues from
3, 4-Dichloroaniline–^{14}C and Propanil–Phenyl–^{14}C
Treated Rice Root Tissues

GERALD G. STILL—Agricultural Research Service, U.S. Department of Agriculture, Metabolism and Radiation Research Laboratory, Fargo, N. Dak. 58102

FRANK A. NORRIS—BASF-Wyandotte Corp., 100 Cherry Hill Rd., Parsippany, N.J. 07054

JORG IWAN—Schering AG, Berlin/Bergkamen, West Germany

Rice plants, with their aryl acylamidase, rapidly cleave the herbicide, propanil (3,4-dichloropropionanilide), to 3,4-dichloroaniline and propionic acid (Fig. 1). Rice plants treated with either propanil or 3,4-dichloroaniline incorporate the aromatic nucleus in a form that is not easily extracted with aqueous or organic solvents. We have reported that greater than 80% of the aromatic moiety of propanil is incorporated into root treated rice plants as a Bligh-Dyer bound residue (1).

In our laboratory, the primary extraction is based on the procedure described by Bligh-Dyer (2, 3) which uses a ternary mixture of chloroform:methanol:water (Fig. 2) yielding Bligh-Dyer insoluble residues and soluble polar and nonpolar phases. As an example, the extraction of rice plant roots treated with 3,4-dichloroaniline-^{14}C yielded only 8.3% of the administered radiolabel in the soluble fractions, with 91.7% retained in the insoluble or bound residue fraction. Radiocarbon distribution studies showed that 12.5% of the radiolabel was in shoot tissue and 79.2% was in the roots.

Catalytic reduction has been used successfully to remove plant cuticle components. When bound residues from propanil-^{14}C treated rice roots were subjected to low pressure catalytic reduction (platinum on charcoal as catalyst and acetic acid as solvent), 15% to 25% of the bound radiolabel was solubilized. Two major components were isolated and identified as 3,4-dichloroacetanilide and N-3,4-dichlorophenylcyclohexylamine (Fig. 3). Subsequent investigation showed that N-3,4-dichlorophenylcyclohexylamine was a product of the reduction and was also formed by the reduction of 3,4-dichloroaniline. The importance of this product is that all of the solubilized radiolabel contained the intact 3,4-dichloroaniline moiety (4).

In order to better characterize the insoluble 3,4-dichloroaniline-^{14}C residues in rice roots, the residues were extracted as described in Fig. 4. Exhaustive soxhlet extraction with a ternary mixture of acetone:benzene:ethanol (1:1:1) followed by boiling the residues in water both yielded 3,4-dichloroaniline.

Cl
NH
CH2
CH3
C
O
Cl
Propanil
Propionic Acid
CO2
Cl
NX2
Cl
N-(3,4-Dichlorophenyl)-Glucosamine
Bound Residues
Cl
NH2
Cl
3,4-Dichloroaniline

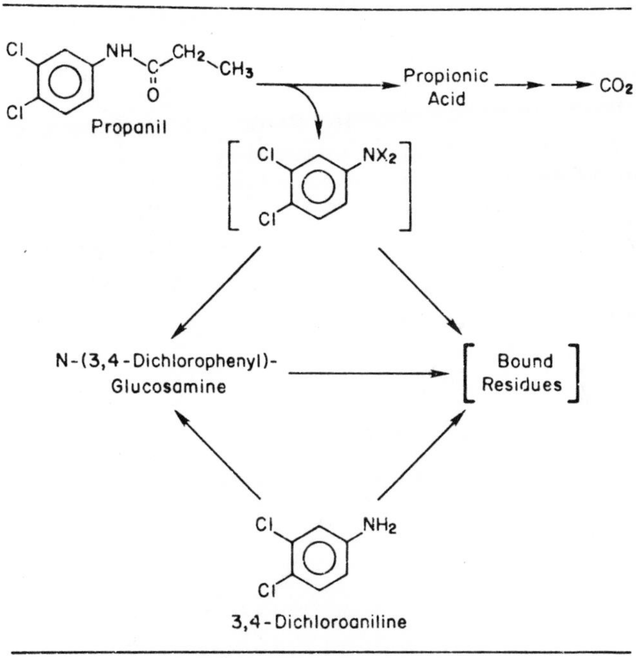

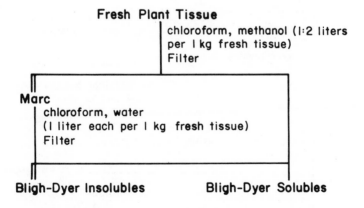

G. G. Still, Plant Physiol., 43, 543 (1968).
G. G. Still, Science, 161, 992 (1968).

Figure 1. Propanil metabolism in rice

Fresh Plant Tissue
chloroform, methanol (1:2 liters per 1 kg fresh tissue)
Filter
Marc
chloroform, water
(1 liter each per 1 kg fresh tissue)
Filter
Bligh-Dyer Insolubles
Bligh-Dyer Solubles

Bligh and Dyer, Can. J. Biochem. Physiol., 37, 911 (1959).
Still and Mansager, Pest. Biochem. Physiol., 3, 87 (1973).

Figure 2.

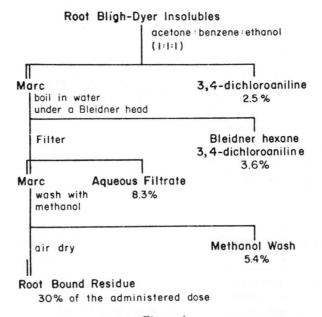

Figure 3.

Figure 4.

The marc (solids) from the boiling water treatment were filtered, extracted with methanol and the resulting "root bound residues" were used in the studies described in the remainder of this communication.

The Bleidner distillation/extraction head (Fig. 5) was used to isolate 3,4-dichloroaniline from the boiling water extraction and was also used in other studies described in this report (5).

The liberation of 3,4-dichloroaniline by boiling water was further studied. It was found that N-(3,4-dichlorophenyl)-glucosamine was easily hydrolyzed in neutral boiling water (Fig. 6). In rice N-(3,4-dichlorophenyl)-glucosamine is a polar metabolite of propanil (1). These data suggest that some of the isolated 3,4-dichloroaniline was released by hydrolysis of a bound N-aryl glucosylamine.

Table 1 summarizes the results from a number of exploratory experiments designed to find a method to free the 3,4-dichloro-aniline bound residues in rice root tissues. Each digestion was carried out at reflux for 16 hours using the Bleidner ex-traction head. The percentages correspond to the total bound radiocarbon in the "root bound residues". The fraction reported as volatile was defined as the radiocarbon extracted into hexane using the Bleidner extraction head. The percentage of organic soluble residue was the radiocarbon extracted into ethyl ether by partition of the aqueous phase after removal of the marc. The marc was the material found to be insoluble after extrac-tion.

Strong oxidative digestion methods proved fruitless. Once the digestion proceeded, the products were oxidized to carbon dioxide. Treatment with aqueous alkali did not release apprec-iable amounts of radiocarbon from the bound residues. The vol-atile radiolabeled products from alkali hydrolysis were iden-tified as 3,4-dichloroaniline. Aqueous alkali should have dissolved the polyuronide hemicelluloses.

Hydrochloric, hydrobromic, and perchloric acids each solu-bilized about 40% of the bound radiocarbon from the insoluble residue. This may imply that 40% of the component(s) of the insoluble residue were acid labile. Hot dilute acids should dissolve noncellulosic polysaccharides, polyuronide, hemicellu-loses, and pectic substances.

A volatile product was isolated from the Bleidner hexane fraction after hydrobromic acid digestion. This material was tentatively identified as N-(3,4-dichlorophenyl)-furfurylidimine (Fig. 7). These data again suggest the presence of a bound N-(3,4-dichlorophenyl) substituent.

Other solubilization methods were tested (Table 2). Ex-traction with dioxane:HCl (9 parts dioxane:1 part 1 N HCl) removed most of the radiolabeled material from the "root bound residue" but the products from this digestion were not stable. Although many different attempts were made to stabilize these products, all were found to be unsatisfactory. Solubilization

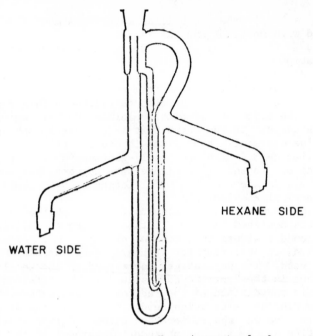

Figure 5. Bleidner distillation/extraction head

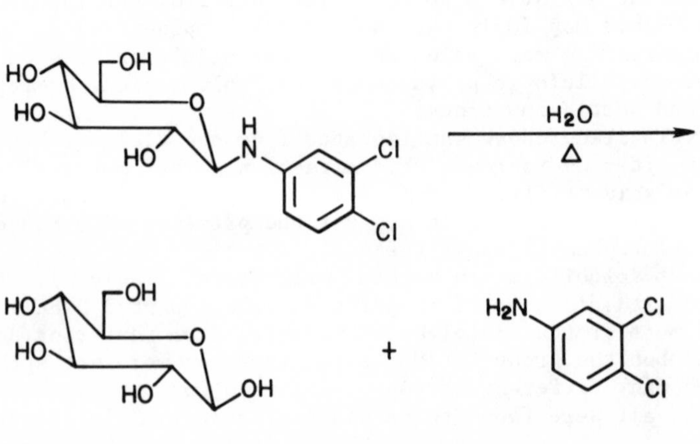

Figure 6.

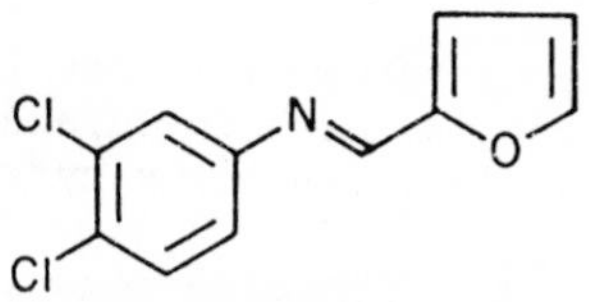

II

Figure 7.

Table 1

PERCENT OF ^{14}C RECOVERED IN FRACTIONS

Digestion	Marc	Water	Organic	Volatile
KM_nO_4 (basic)	0	–	–	> 90*
CrO_3	0	–	–	> 90*
H_6IO_6	0	<10	–	> 90*
H_2O_2 (neutral)	0	<10	–	> 90*
NaOH (0.1N, 1.0N, 5.0N)	82-90	–	–	–
HCl (pH 1)	62	4	15	1
HBr (pH 1)	62	22	5	6
$HClO_4$ (pH 1)	60	9	4	1

*This volatile fraction was not soluble in hexane.

Table 2

PERCENT OF ^{14}C RECOVERED IN FRACTIONS

Digestion	Marc	Water	Organic	Volatile
Dioxane : HCl	2	30	60	–
Methanol : HCl	30	–	70	–
HNO_3 (5%–10%)	1-3	3-11	22-25	21-33
HNO_3 (pH 1-3)	24-95	2-20	32	1-24
$H^{15}NO_3$ (5%)	–	–	–	23
HNO_3 (pH 1) + Urea	51	19	28	1
HCl + $NaNO_2$ (1N)	66	32	9	3
$NaNO_3$ (0.1 M)	89	7	3	1
NO_2/Water	15	12	50	25
NO_2/Ether	97	–	1	2
NO_2/Ethanol	60	–	40	–

with methanol:HCl (9 parts methanol:1 part 1 N HCl) was also
used to digest lignin but, as with the dioxane:HCl system, the
results were not acceptable because the hydrolysis products were
unstable.

Ten percent nitric acid has been used in lignin analysis
and was tested in our studies. It proved to be a very efficient
reagent for digesting the radiolabeled components from "root
bound residue." Nitric acid plus urea was a poor reagent. Urea
is known to destroy the oxides of nitrogen. Nitrous acid was
generated by the addition of hydrochloric acid to sodium nitrite,
plus a trace of nitric acid. This was a poor solubilizing reagent. However, nitrogen dioxide in water or ethanol was an
efficient solubilizing agent. These studies suggest that the
reactive species in the nitric acid system was not the hydrogen
ion, the nitrate ion, the nitrous ion, nor nitric acid itself.
The low reactivity of the nitrogen dioxide in ether and the
increased activity of the more polar solvents suggested a reactive polar species. In concentrated nitric acid, nitrous acid
is almost completely ionized to the nitrosonium ion (NO^+) (Fig.
8).

These nitric acid digestions produced a significant amount
of volatile hexane-soluble, radiolabeled products (20-33% of
the bound radiolabel). Figure 9 reports the structures of four
of the radiolabeled components isolated in the volatile fraction
from nitric acid solubilization of rice "root bound residues."
Isolated compounds included; 1,2-Dichlorobenzene (12%), 1,2-
dichloro-4-nitrobenzene (17%), 1,2-dichloro-3-nitrobenzene (3%),
and a compound tentatively identified as 5,6-dichlorobenzofurazan
(5,6-dichlorobenzo-2,1,3-benzoxodiazole) (19%). Again, these
data indicate that the dichlorophenyl moiety of the bound residues was intact in at least one-third of the bound residues.

The assumption that the aniline nitrogen of 3,4-dichloro-
aniline-^{14}C was still present in these volatile products must be
challenged because of the 1,2-dichloro-3-nitrobenzene component.
The isolation of this compound led us to believe that one of the
nitrogens of the 5,6-dichlorobenzofurazan resulted from nitra-
tion. To test this hypothesis, solubilization experiments were
conducted using nitric acid-^{15}N and bound root residues. These
experiments yielded 1,2-dichloro-3-nitrobenzene-^{15}N and 1,2-
dichloro-4-nitrobenzene-[15]N. Therefore, the nitrogen atom of
the nitro moiety must come, in part, from the acid (nitrosonium
ion). There was insufficient material to determine an isotopic
ratio in the dichlorobenzofurazan from the root bound residue.

The digestion of model compounds with nitric acid is summarized in Fig. 10. When 3,4-dichloroaniline, 3,4-dichloro-
acetanilide, 3,4-dichlorophenol, and 4,5-dichloro-2-nitroaniline
were refluxed with 1-10% nitric acid, none of the volatile
hexane soluble products observed from the digestion of the bound
residue were found. However, when N-ethyl-3,4-dichloroaniline
was treated with nitric acid, the volatile products were the

$$N_2O_4 \rightleftharpoons 2NO_2 \rightleftharpoons NO^+ + NO_3^-$$ *Figure 8.*

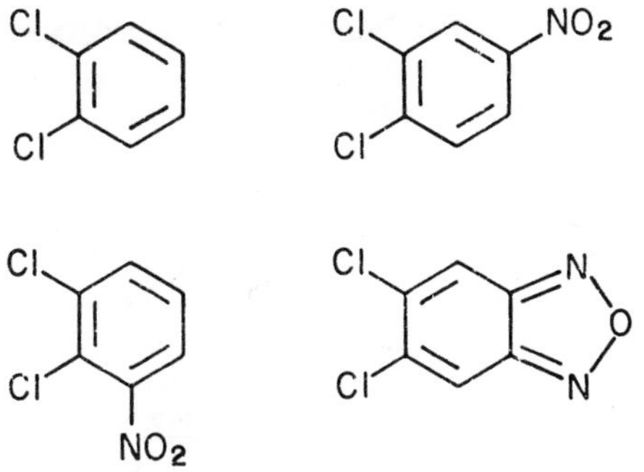

Figure 9.

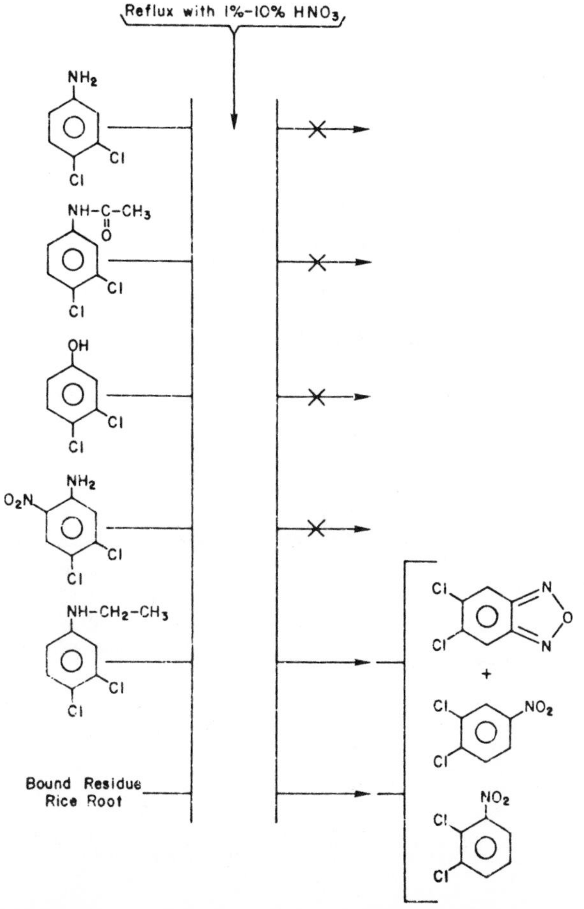

Figure 10.

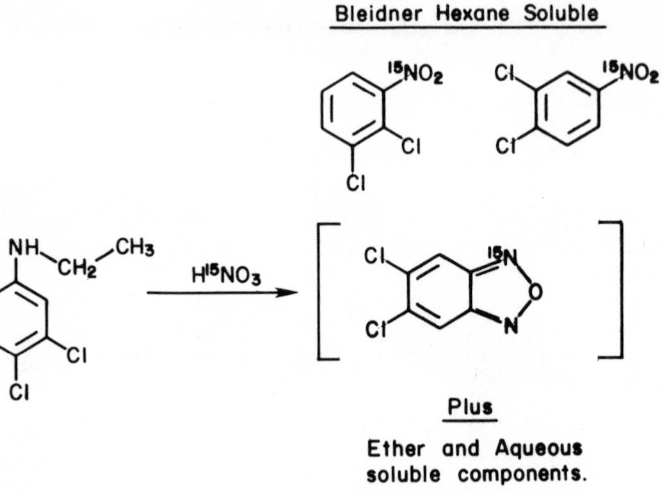

Bleidner Hexane Soluble

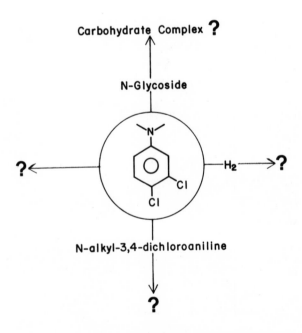

Figure 11.

Figure 12. Nature of 3,4-dichloroaniline bound residues 1975

same as those products observed from the bound residues of rice roots.

When N-ethyl 3,4-dichloroaniline was digested with nitric acid-^{15}N, the ^{15}N was found in both nitro compounds and as one nitrogen atom of the dichlorobenzofurazan (Fig. 11). There was a significant amount of ^{15}N material in the ether and aqueous phases, but these products were not purified or characterized. The speculation that part of the bound residues were 3,4-dichloroaniline-compounds was supported by the isolation and identification of the 5,6-dichlorobenzofurazan derivative from nitric acid-^{15}N treated rice root residues. The similarities of the products from the digestion of the bound residues and the model compound, N-ethyl-3,4-dichloroaniline, support this possibility. We speculate further that the anilium nitrogen may be covalently bound <u>via</u> an unknown alkyl linkage to an insoluble portion of the plant tissue.

Figure 12 summarizes our present knowledge of the nature of the 3,4-dichloroaniline bound residues in rice root tissues. 3,4-Dichloroaniline, from either propanil or some other pesticide precursor, may be translocated and bound by rice plants. At this time, our data appear to support the presence of a transient glycoside and the speculation of an N-alkylated bound product. Animal studies (5), however, have shown that these bound residues are passed through monogastric animals and returned to the soil environment.

Literature Cited

(1) Still, G. G., Science 159, 992 (1968).
(2) Bligh, E. G., and W. J. Dyer, Can. J. Biochem. Physiol. 37, 911 (1959).
(3) Still, G. G., and E. R. Mansager, Pest. Biochem. Physiol. 3, 87 (1973).
(4) Iwan, J. and G. G. Still, private communication.
(5) Dalton, R. L. and H. L. Pease, J.A.O.A.C. 45, 397 (1962).
(6) Sutherland, M. L., Vail Conference (1975).

11

Classification and Analysis of Pesticides Bound to Plant Material

J. WIENEKE

Radioagronomy, Kernforschungsanlage, Julich 517 Julich, Postfach 1913, West Germany

The use of radiotracer technology to follow pesticide metabolism yields a precise accounting of the radiolabel distribution,
but provides very little information about the chemical structure
of the metabolic products. Since most residue analysis methods
are designed to assay the parent compound, the more polar metabolites require additional steps for their isolation and purification. After removal of the nonpolar and polar pesticide and
pesticide metabolites the remaining plant residue is thought to
consist of insoluble material. However, during the course of a
study on the metabolism of the insecticide, azinphos in bean
plants, the nature of the bond residues became questionable.
However, the results of the following preliminary investigations
with ^{14}C-azinphos will contribute to a better understanding of
the extraction and classification of bound pesticide residues
in plant material.
Methods: The first trifoliate leaves of bean plants were
sprayed with the ^{14}C-azinphos. At the time of sampling, the
treated leaves were stripped with benzene (2X) and frozen at
-18°. The frozen plant material was crushed and homogenized in
an "Ultra Turrax" blender with the following solvents: 1. acetone,
2. acetone:water (3 : 1v/v) and 3. two volumes of chloroform.
The volume of each extracting solvent was 5 - 6 ml/gr fresh
weight. Each solvent was removed from the homogenate by vacuum
filtration. After chloroform extraction, the residue was washed
3 times with acetone. Further Soxhlet extraction with acetone
for 28-48 hrs yielded negligible radioactivity (0.02-0.05% of
the applied radioactivity or 1-2% of the radiolabel present in
the extracted residue). Radioactivity in the unextractable residue fraction increased with time. Four weeks after application
of the radiolabeled insecticide, 5-11% of the radioactivity was
in the bound residue fraction.
The nature of the azinphos-^{14}C bound residues was questioned.
Was the parent molecule incorporated or was the bound ^{14}C-residue
a fragment of the original insecticide which was somehow incorporated into the plant polymeric structure? To answer this

AZINPHOSMETHYL (GUTHION)

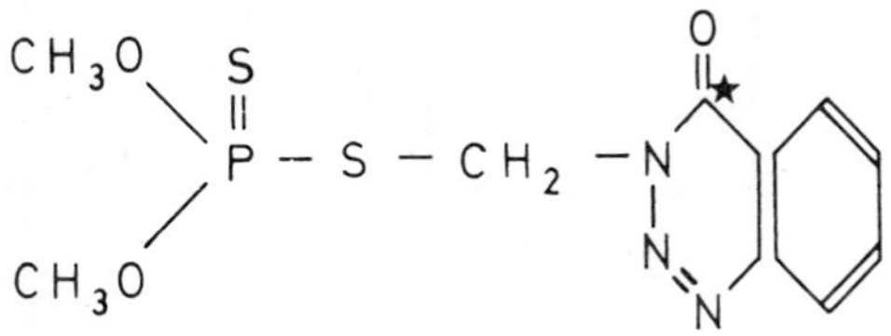

S - (3,4 - dihydro - 4 - oxo - benzo
[d] - [1,2,3] - triazin - 3 - ylmethyl)
0,0 - dimethyl phosphorodithioate

*Figure 1.   Chemical structure and name of azinphos [14]C-
labelled in the carbonyl position*

Table 1:   Distribution of radioactivity in different fractions
after subsequent elution of the unextractable residue
of bean leaves treated with ^{14}C-azinphos, with various
solvents at room temperature

ELUTION SOLVENT	VOLUME ELUTED ML	COLUMN I 749013 DPM/ 304.9 MG DR.M. %	COLUMN II 919169 DPM/ 374.2 MG DR.M. %
Acetone	28 – 38	0.5	0.4
Ethylacetate	27 – 37	0.2	0.1
Methanol	31 – 32	5.2	4.5
Methanol/H_2O (4 : 1)	33 – 44	22.2	23.9
Acetonitril/Ethyl-Acetate/Methanol H_2O (65/15/10/10)	32 – 33	1.7	0.8
Butanol/Acetic Acid/H_2O (4/1/1)	33 – 36	24.3	38.9
6 N Acetic Acid	16	18.9	7.2
H_2O	29 – 30	0.2	0.1
2 N NH_4 OH	4 – 5	6.1	4.4
H_2O	12 – 20	3.2	3.2
Total Soluble		82.5	83.5

question, aliquots of the unextracted residue fraction were treated as follows: 1. Hydrolysis with 25% formic acid (known to cleave hemicellulose and oligosaccharides etc.), 2. dissolution of cellulose in the remaining residue (procedure after Waksmann, 1931). Radioactive material was distributed in the cationic, anionic and neutral fractions, in the cellulose fraction and also in the fraction corresponding to lignin. There was no indication that the parent compound or a major portion of the parent compound was present in these fractions.

A second approach to the problem of the extraction of bound ^{14}C-azinphos residues was to fill a chromatography column with the unextractable residue fraction and elute the material with different solvents and solvent mixtures. The results of these studies (Table 1) demonstrate that greater than 80% of the radiolabel is extracted by this method. Thin layer chromatographic analysis of the various eluted fractions revealed that the radiolabel was distributed in several compounds.

The significance of bound pesticide residues is primarily a consideration of the amount and the character of the substances analyzed as bound residues. The results reported in table 1 suggest that the removal of bound materials from plant tissue is dependent upon the solvent extraction procedure. These results challenge us to examine critically the various commonly used extraction methods for the isolation of fractions reported as "bound pesticide residues". The question arises as to whether all the substances that are not dissolved by a specific solvent method may be classified as bound residues, or should only those portions that remain after extraction with either acid or alkaline solvent mixtures be considered bound pesticide residues? Furthermore it seems questionable to analyze for bound residues in plant materials if the concentration is low, such as with azinphos which was only 5-10% of the total ^{14}C found in treated bean leaves. Even when calculated as azinphos equivalents the total concentration in the unextractable residue is far below the residue tolerance levels for the parent product.

If, however, the concentration of the unextractable residue radioactivity is high, the nature of the radioactivity associated with this fraction should be determined. If these residues can be dissolved, then the solubilized radioactivity should be characterized to determine whether it is the parent compound, or a metabolic fragment.

Waksman, S.A. and Tenney, F.G.: Soil Sci. <u>24</u>, 275 (1931) cited in: Bodenkundliches Praktikum, edited by E. Schlichting and H.P. Blume, Parey-Verlag, Hamburg, Berlin.

12

Bound Residues of Nitrofen in Cereal Grain and Straw

R. C. HONEYCUTT, J. P. WARGO, and I. L. ADLER

Rohm and Haas Co., Research Laboratories, Spring House, Pa. 19477

Nitrofen is a selective herbicide marketed by Rohm and Haas Company and used to control annual grasses and broad-leafed weeds. The structural formula for nitrofen is shown below.

2,4-dichloro-1-(4-nitrophenoxy) benzene
*Site of ^{14}C Labeling

Investigations into the environmental fate of nitrofen-^{14}C applied pre- and postemergence to wheat and rice led to some interesting discoveries. We found that only 50% of the radioactive residue in the straw or grain of these 110-147 day old plants could be extracted by conventional organic solvent Soxhlet extraction. Thus, about 50% of the residue was considered "bound residue."

During the course of this investigation, it was supposed that these "bound residues" may be naturally occurring plant structural or storage molecules. Thus, we set out to isolate starch from the grain of ^{14}C nitrofen treated wheat or rice plants and lignin from the straw to determine if the ^{14}C of nitrofen could be incorporated into these compounds.

Figures 1 and 2 show the methods by which we isolated the starch and lignin from rice or wheat plants treated at 3-4 lb/acre with ^{14}C nitrofen. Wheat or rice grain or straw was harvested at 110-147 days from treatment and worked into homogenous samples. Starch was extracted from wheat or rice grain with DMSO. The isolated starch was then hydrolyzed and the resulting ^{14}C-glucose derivatized to the glucosazone with phenylhydrazine. The resulting osazone was recrystallized to constant specific radioactivity.

Wheat or rice straw was Soxhlet extracted 18 hours with an ethanol-benzene mixture. Cellulose and lignin were then isolated from the remaining solids using hot sodium hydroxide under

pressure. The lignin was then purified to a constant specific
radioactivity.

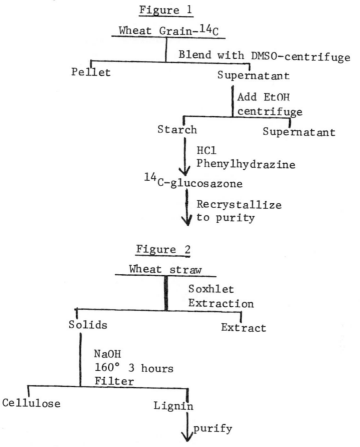

Figure 1

Wheat Grain-^{14}C

Blend with DMSO-centrifuge

Pellet   Supernatant

Add EtOH
centrifuge

Starch   Supernatant

HCl
Phenylhydrazine

^{14}C-glucosazone

Recrystallize
to purity

Figure 2

Wheat straw

Soxhlet
Extraction

Solids   Extract

NaOH
160° 3 hours
Filter

Cellulose   Lignin

purify

Figure 3 shows the specific radioactivity of glucosazones
and lignin isolated from ^{14}C-nitrofen treated wheat grain and
straw respectively and taken through 3 purification steps during
a typical experiment.

We can easily see that constant specific radioactivity can
be obtained through repeated purification of both the ^{14}C-gluco-
sazones from wheat grain and the ^{14}C-lignins from wheat straw.

One can determine what percentage of the total ^{14}C-nitrofen
residue at harvest is in starch or lignin by dividing the
specific radioactivity of the starch or lignin by the specific
radioactivity of the original wheat straw and multiplying by
the percent of starch in grain or the percent of lignin in straw.

Such experiments and such calculations of the data obtained
were carried out on a variety of samples of grain and straw from
wheat and rice plants treated both preemergence and postemergence

with ^{14}C-nitrofen. Figure 4 shows a summary of such work.

These data show that a substantial amount of the radioactive residue found in rice and wheat straw at harvest is in lignin. Application preemergence or postemergence appears to make no difference in the amount of ^{14}C incorporated into lignin. Whether the ^{14}C in lignin is a lignin-nitrofen conjugate or whether the ^{14}C of nitrofen has been incorporated into the carbon skeleton of lignin remains to be elucidated.

It also can be seen from this figure that a great deal of the radioactive residue is in the form of starch. By derivatization of the starch to the glucosazone it has been proven that the ^{14}C of nitrofen was reincorporated through metabolic processes into glucose and finally into starch.

Figure 3

Purification of Glucosazones

^{14}C-Glucosazones	Specific Radioactivity dpm/g
1st Recrystallization	122
2nd Recrystallization	147
3rd Recrystallization	133

Purification of Lignin

^{14}C-Lignin	Specific Radioactivity dpm/g
Crude Lignin	3,129
1st Reprecipitation	3,834
2nd Reprecipitation	3,670
3rd Reprecipitation	3,432

Figure 4

Sample	Treatment	% of Radioactive Residue at Harvest in Lignin	% of Radioactive Residue at Harvest in Starch
Rice Straw	Preemergence	33	----
Rice Straw	Postemergence	25	----
Wheat Straw	Preemergence	30	----
Rice Grain	Preemergence	----	64
Wheat Grain	Preemergence	----	70

Use of Radiotracer Studies in the Estimation of Conjugated and Bound Metabolites of Dichlobenil in Field Crops

A. VERLOOP

Research Laboratories, Philips-Duphar B.V., The Netherlands

For several pesticides the metabolic pathways in plants and in animals are (partly) identical. A common metabolic pathway of pesticides with an aromatic ring structure is hydroxylation followed by conjugation, i.e.

$$P \rightarrow POH \rightarrow POR,$$

where P is a pesticide with an aromatic ring structure, POH is its corresponding phenol and POR is the conjugated form. In animals R is often glucuronic acid or sulphate so that POR is very water soluble and easily excreted in the urine. However in plants, additional to water soluble glucosides, an important and often major part of POR is present in the form of water insoluble glycosides where the phenol is bound to the polymeric part of the plant structure.

One example of this kind is the metabolism of the herbicide dichlobenil (2,6-dichlorobenzonitrile) and of its major soil metabolite BAM (2,6-dichlorobenzamide) in plants and animals [1].

The degradation and transfer routes of dichlobenil in soils and crops are illustrated in Fig. 1, where the main routes are marked with the heavier arrows. The results indicated are based upon laboratory and greenhouse studies with ^{14}C-labeled dichlobenil. In extensive residue studies of field crops grown in soil treated with dichlobenil, under different seasonal and regional conditions, it was found that most of the crops contained no detectable residues of the parent herbicide. But in these field trials BAM residues were present in crops of high water content such as grapes and apples although they were absent in crops of low water content such as rice, wheat and olives[1,2]. It was concluded that the soil metabolite BAM is the chief source of residues in crops, probably caused by its higher rate of transfer in plants and lack of volatility, compared with the parent herbicide.

In some field trials in apple orchards after soil treatment with ^{14}C labeled dichlobenil, additional to BAM, the main metabolite 3-hydroxy-2,6-dichlorobenzamide (HOBAM) was found in even greater amounts in the leaves at harvest (Fig. 2). This metabolite

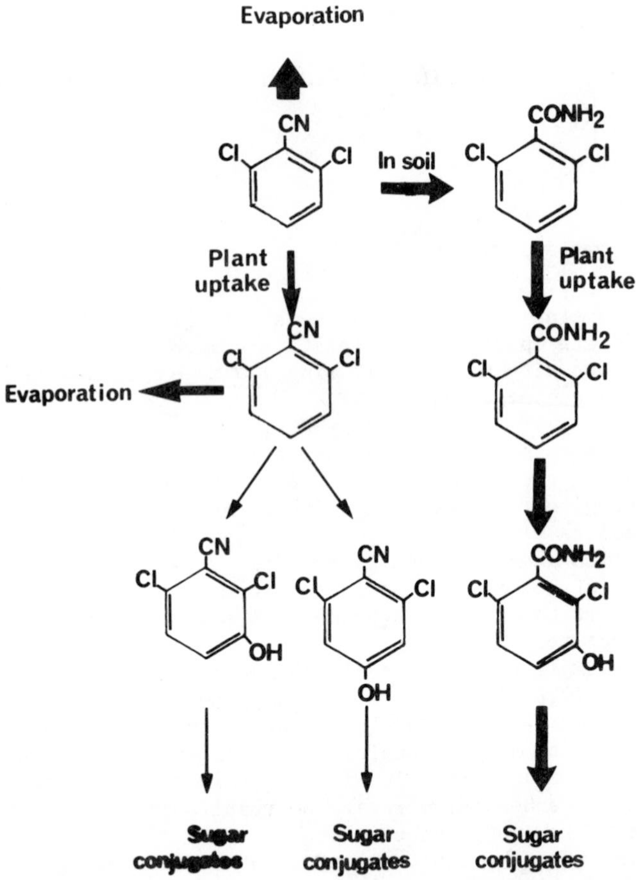

Residue Reviews

Figure 1.   Degradation and transfer routes of dichlobenil in soils
        and in plants given in treated soils (1, 2)

Residues calculated as ppm dichlobenil in

Source	Leaf				Fruit			
	Unextractable	Extractable			Unextractable	Extractable		
		Total	BAM	HOBAM		Total	BAM	HOBAM
Orchard, 1967	3.0	34.0	5.5	25.5	--	--	--	--
Orchard, 1970	1.3	23.0	3.1	18.0	<0.01	0.08	0.08	<0.01
Greenhouse, 1972	1.0	24.0	5.0	15.0	<0.01	0.12	0.10	<0.007

Figure 2. Mean residues of breakdown products in apple leaves and fruits at harvest after soil treatment with ^{14}C-dichlobenil granules (6 kg/ha a.i.)

is mainly present in the conjugated and bound forms; the satisfactory recovery indicated in Fig. 2 could only be obtained
after rather drastic hydrolysis of the plant material with 2N
HCl in ethanol-water at 80° C.

It was found that HOBAM is a main metabolite of BAM in
animals too(3), so it was most appropriate to propose a combined
tolerance for BAM + HOBAM, based upon the toxicity data of BAM
on the one hand and upon the combined residue data of the two
compounds, on the other. For the analysis of residues of BAM
adequate methods could be developed and many residue data under
different seasonal and regional conditions have been obtained
(2). As a result a tolerance of 0.3 ppm BAM in apples has been
obtained in several countries. But for possible residues such
as HOBAM, where a major part is bound to polymeric plant constituents it is often difficult to develop a reliable residue
method. Recovery studies carried out shortly after application
of the metabolite to the commodity do not necessarily give a
realistic picture of the recovery several months after application of the parent pesticide. It may be tried to modify the
method in such a way that the bound residue is liberated and
recovered. But in the case of HOBAM it was found with radiotracer studies that a sufficient result could only be obtained
with rather drastic methods of chemical hydrolysis, which resulted in extracts giving great difficulties in the development
of clean-up procedures. A reliable residue method for HOBAM
is, therefore, as yet not available. Further work might at most
lead to a very time consuming method, so that its application
to the analysis of many samples seems virtually impossible.

Radiotracer studies under normal or simulated field conditions might, however, provide an alternative method to determine the maximum level of HOBAM in crops. This is illustrated in Fig. 2 where the residues of breakdown products in
apple leaves and fruits are given at harvest after soil treatment with ^{14}C-dichlobenil in the granular formulation at the
normal rate of 6kg/ha a.i. The experiment of 1972 was carried
out in a special greenhouse with gauze walls so that the climatic conditions were rather similar to those outside. It can
be concluded that appreciable amounts of HOBAM are present in
apple leaves, but in the fruit no HOBAM was found up to the
detection limit of 0.01 ppm. From these results it might be
concluded that in these experiments the maximum amount of HOBAM
present in the fruits is about 20% of the BAM residue level.

Of course the disadvantage of this approach is the very
limited seasonal variation in the residue data of HOBAM, while
the regional variation is even completely absent. But degradation processes of xenobiotics in plants are known to be
generally (pseudo) first order. This has been found also for
the closely related hydroxylation of dichlobenil in laboratory
studies with bean seedlings (4). If it is assumed, that the
degradation of BAM to HOBAM is also first order, then the maximum

possible HOBAM residues in apple calculated as a fixed percentage
of the BAM residue level would have a more general validity. In
this way seasonal and regional variations in the maximum HOBAM
residues would also be accounted for via the variations in the
BAM levels, although these HOBAM levels were only determined in
a few trials. This reasoning leads to a proposed combined tolerance in apple for BAM + HOBAM of 120% of the BAM tolerance. In
comparable experiments with grapes small residues of HOBAM were
indeed found in grapes with the radiotracer method.

Literature Cited

(1) A. Verloop, Residue Reviews, 43, 55, (1972).
(2) K.I. Beynon and A.N. Wright, Residue Reviews, 43, 23 (1972).
(3) J.G. Wit, unpublished results.
(4) A. Verloop and W.B. Nimmo, Weed Research, 9, 357, (1969).

14

Metabolite Fate of *p*-Toluoyl Chloride Phenylhydrazone (TCPH) in Sheep. The Nature of Bound Residues in Erythrocytes

PREM S. JAGLAN, RONALD E. GOSLINE, and A. WILLIAM NEFF

Agricultural Division, The Upjohn Co., Kalamazoo, Mich. 49001

The fate of p-toluoyl chloride phenylhydrazone (TCPH), an anthelmintic, efficacious against gastrointestinal nematodes and cestodes of ovines was studied in a number of sheep following a single oral dose of 50 mg/kg. TCPH was ^{14}C labeled either as the phenylhydrazine uniformly ring labeled (TCPH-I) or carboxyl labeled (TCPH-II).

Both TCPH-I or II cleared the gastrointestinal tract of sheep over a ten-day period following treatment. The radioactivity observed in the feces was 3-4 times that seen in urine. The overall excretion pattern for the two labeled forms was similar, however, TCPH-II produced a somewhat larger proportion of ^{14}C in the urine than TCPH-I.

Tissues taken at selected intervals following treatment with TCPH-I indicated high residues in blood and blood rich organs such as, liver, lung, kidney, and spleen and slow depletion of radioactivity from these tissues with time. Comparison of the residues from a TCPH-I treated sheep with a TCPH-II treated one showed a distinct difference in residue levels, particularly in the blood, 9.3 ppm for TCPH-I and 1.3 ppm for TCPH-II. Although plasma levels were about equal from both labels, TCPH-I residues were much higher in erythrocytes. With the passage of time, ^{14}C declined uniformly in both plasma and erythrocytes from TCPH-II treated sheep and reached about 1 ppm. On the other hand, ^{14}C levels in erythrocytes of TCPH-I treated sheep are 15 times higher than TCPH-II treated sheep and radioactivity persisted much longer. These data suggested that molecular cleavage of TCPH had taken place and only the phenylhydrazine portion of the molecule was responsible for the high blood residues.

Fractionation of blood from TCPH-I treated sheep showed that the major part of the radioactivity was associated with proteins (66%). The rest of the radioactivity was diffused into several components, viz., RNA, 4%; carbohydrates, 6%; glucosaminglycon, 3% and lipids, 19%. The concentration of the radioactivity in erythrocytes was ten times that of plasma. Both heme and globin

were labeled. Although 49% of blood radioactivity was present in
the globin as compared to 20% in heme, concentration of radioactivity was 700 ppm relative to heme in contrast to 72 ppm in
globin.
Very little ^{14}C-activity could be extracted from hemoglobin
with organic solvents under different pH conditions. The radioactivity in hemoglobin was not dialyzable. It eluted with hemoglobin from Sephadex G-25. Electrophoresis data indicated that
the radioactivity was associated with hemoglobin.

After hydrolysis of globin with pronase, trypsin or pepsin,
the radioactivity remained associated with the aqueous phase and
very little radioactivity partitioned into hexane, methylene
chloride and ether. Hydrolysis in 6N HCl, however, generated
20% chloroform extractable radioactivity. Although concentration
of radioactivity in the globin was low making characterization
difficult, TLC of the hydrolyzed globin produced a number of
compounds with TLC characteristics similar to known aromatic
amino acids, which suggested covalent bonding of the TCPH-I
phenyl group to amino acids.

Oxidation of heme by chromate generated a derivative or
fragment of metabolite most of which was extractable in ether
at pH 2. Rf on silica gel TLC of radioactivity extracted in
ether was identical to authentic benzoic acid. The specific
activity of radioactive benzoic acid isolated from the heme
agreed with theory. Derivatization of ether extractable radioactive compounds with diphenyldiazomethane and silica gel TLC
showed that the derivative had the same Rf as authentic diphenyl
methylbenzoate. The identity of ^{14}C-benzoic acid from oxidation
of heme was further confirmed by GLC and GLC/mass spectrometry
of its methyl ester which demonstrated that the phenyl group was
derived from TCPH, whereas the carboxyl group came from the heme
fraction. Oxidation of globin also generated ^{14}C-benzoic acid,
but the specific activity was only 5% of theory. Globin has
endogenous precursors such as, phenyllalanine, which were also
oxidized to benzoic acid and consequently diluted the ^{14}C-benzoic
acid.

15

Organic Matter Reactions Involving Pesticides in Soil

F. J. STEVENSON

Department of Agronomy, University of Illinois, Urbana, Ill. 61801

Adsorption by organic matter has been shown to be a key factor in the behavior of many pesticides in soil. Numerous examples
where bioactivity, persistence, biodegradability, leachability,
and volatility have been shown to bear a direct relationship to
organic matter content can be found in several reviews on the subject (1, 2, 3, 4, 5). It has been well established, for example,
that the rate at which any given adsorbable herbicide must be
applied in order to obtain adequate weed control can vary as much
as 20-fold, depending upon the nature of the soil and the amount
of organic matter it contains. Soils which are black in color
e.g., most Mollisols) have higher organic matter contents than
those which are light colored (e.g., Alfisols), and pesticide
application rates must often be adjusted upward on the darker
soils in order to achieve the desired result.
    Adsorption by organic matter depends to a considerable extent
upon the physical and chemical properties of the pesticide -- each
type has its own special features and must be considered separately. Information as to how pesticides react with soil organic
matter may provide a more rational basis for their effective use,
thereby reducing undesirable side effects due to carry-over, contamination of the environment, and, in the case of herbicides,
phytotoxicity to subsequent crops.
    The organic fraction of the soil also has the potential for
promoting the nonbiological degradation of many pesticides (3, 6,
7), as well as for forming strong chemical linkages with residues
arising from their partial degradation by microorganisms (3).
These aspects of pesticide-soil organic matter interactions de-
serve further study because such processes would play an important
role in detoxification and protection of the environment. Chemical binding of pesticide-derived residues would increase their
persistence in the soil but in forms unharmful to the environment.

Chemical Nature of Soil Organic Matter

    Soil organic matter chemistry is undoubtedly the least understood field of soil science, and in many ways the most perplexing. As Hayes (2) and Stevenson (3) have pointed out, mechanisms of pesticide-organic matter interactions will remain obscure until more is known about the nature and chemical composition of the organic fraction of soils.

Organic matter exists in many forms in soil, including the unmodified remains of plant and animal tissues (plant detritus, roots, bacterial and fungal tissue) and secondary products of microbial metabolism. The latter is commonly referred to as "humus" and is often used synonymously with "soil organic matter". This report will be devoted largely to pesticide reactions involving "humus" although it should be noted that plant residues and the mycellial tissue of actinomycetes and fungi may also be important in pesticide adsorption (5).

Humus, or soil organic matter, can further be classified into two main groups of compounds, nonhumic substances and humic substances (8, 9, 10, 11). The former includes substances belonging to the well-known classes of organic compounds, such as the carbohydrates, proteins, fats, waxes, and resins. The latter group, the so-called humic and fulvic acids, represents chemically and biologically modified substances which bear little if any resemblance to any of the known organic compounds. The soil, being a graveyard for the bodies of micro- and macrofaunal organisms, would be expected to contain most of the biopolymer and biochemical compounds synthesized by living organisms. As one might suspect, many of the biochemicals will occur in exceedingly small quantities. Nevertheless, certain trace biochemicals have the potential for forming conjugates with pesticides, as will be noted later.

The humified material, which represents the most reactive component of humus, consists of a series of highly acidic, yellow- to black-colored, high-molecular-weight polyelectrolytes referred to by such names as humic acid, fulvic acid, etc. The dynamic nature of these substances is due to their high contents of oxygen-containing functional groups, including COOH, phenolic-, aliphatic-, and enolic-OH, and C=O structures of various types. Amino, heterocyclic amino, imino, and sulfhydryl groups may also be present. The current view is that the various humic fractions represent a complex mixture of molecules which vary in a systematic way with regard to such properties as degree of polymerization, molecular weight, exchange acidity, and content of oxygen-containing groups (Figure 1). In classical terminology, humic acid is defined as the material extracted from soil by alkaline solutions and which precipitates upon acidification: fulvic acid is the material remaining in solution. Humic acid can further be divided into brown humic acid (coagulated with electrolyte under alkaline conditions) and gray humic acid (not coagulated with electrolyte). In the older literature, considerable importance was given to "crenic" and "apocrenic" acids, which are light yellow fulvic acid-type substances. Renewed interest has recently

been shown to these rather low-molecular-weight substances because of their ubiquitous occurrence in natural waters.
    The range of oxygen-containing functional groups in humic
and fulvic acids is given below. Total acidities of fulvic acids
(usual range of 900 to 1,400 meq/100 g) are considerably higher
than humic acids (usual range of 500 to 870 meq/100 g). Both
COOH and acidic OH groups (presumed to be phenolic OH) contribute
to the acidic nature of these substances, with COOH being the
most important.

	Total acidity	COOH	Acidic OH*	Weakly acidic- plus alcoholic OH	C=O
			normal range, meq/100 g		
Humic acids	500–870	150–300	250–570	270–350	90–300
Fulvic acids	900–1,400	610–910	270–670	330–490	110–310

*Usually reported as "phenolic OH"

    For reasons outlined above, humic acid (the most extensively
investigated component of soil humus) cannot be regarded as a
single chemical entity capable of being described by a single
structural formula -- no two molecules may have the precise chemical structure (6, 9, 11). A "typical" molecule is believed to
consist of micelles of polymeric nature, the basic structure of
which is an aromatic ring of the di- or trihydroxyphenol type
bridged by -O-, -NH-, N=, and -S- linkages and containing both
free OH groups and quinone linkages. The dark color of humic
acids, and their ability to form adsorption complexes with a
variety of inorganic and organic substances, is consistent with
this concept.
    The "type structure" for humic acid shown in Figure 2 meets
many of the above requirements. While the humic acids in any
given soil will vary widely in composition, most molecules would
be expected to contain the same basic units and the same types of
reactive groups indicated by the model structure. A number of
sites are illustrated which can combine with herbicides, such as
by electrostatic bonding (attraction of a positively charged
organic cation to an ionized COOH or phenolic OH group), H-bonding
(note large numbers of COOH, OH, and C=O groups), and ligand
exchange (formation of a covalent bond with an attached metal ion).
Other type structures proposed for humic acids are given elsewhere
(9, 10, 11).
    Humic substances also contain rather high concentrations of
stable free radicals, possibly of the hydroxyquinone type (12).
These sites may be of considerable importance in the binding of
certain herbicides, particularly those capable of being ionized
or protonated to the cation form.
    Several aspects of organic matter chemistry require further

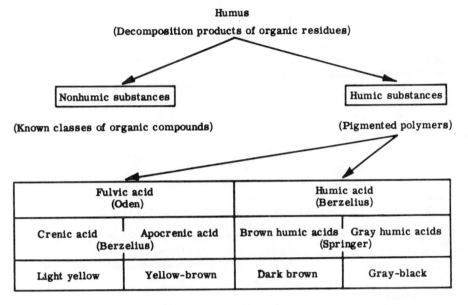

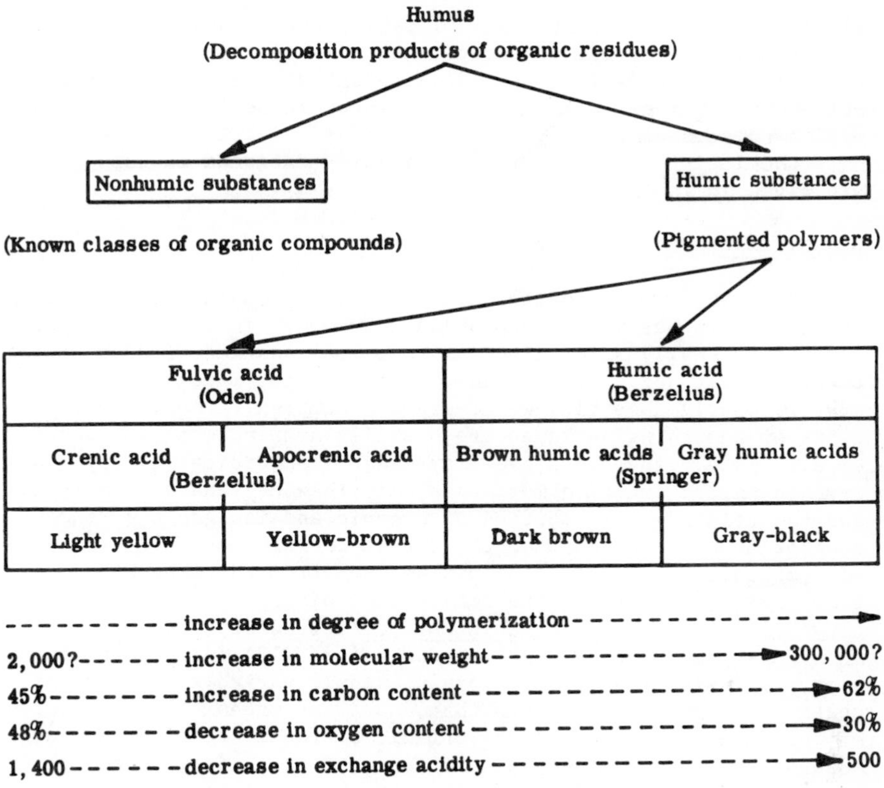

Fulvic acid (Oden)		Humic acid (Berzelius)	
Crenic acid	Apocrenic acid (Berzelius)	Brown humic acids	Gray humic acids (Springer)
Light yellow	Yellow-brown	Dark brown	Gray-black

－－－－－－－－－ increase in degree of polymerization － － － － － － － － － － － ➤

2,000? － － － － － increase in molecular weight － － － － － － － － － － － ➤300,000?

45% － － － － － － increase in carbon content － － － － － － － － － － － － ➤62%

48% － － － － － － decrease in oxygen content － － － － － － － － － － － ➤30%

1,400 － － － － － decrease in exchange acidity － － － － － － － － － － － － ➤500

Figure 1.   Classification and general chemical properties of humic substances (adapted from Ref. 10)

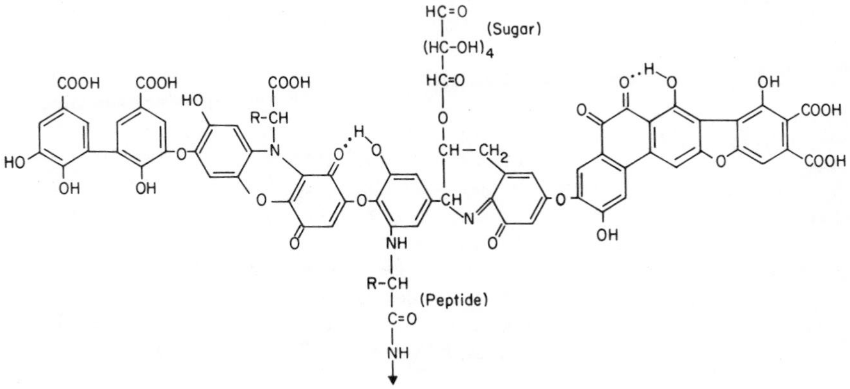

Figure 2.   Type structure for humic acid

elaboration regarding the fate of pesticides in soil, including
(i) organic matter-clay interactions, (ii) quantitative differences in organic matter, and (iii) potential chemical reactions
between pesticides and organic substances in soil. These items
will be discussed briefly in the sections which follow.

Chemical formulas of the pesticides mentioned in this review
are given in Table 1.

Organic Matter Versus Clay as Adsorbent

Clay and organic matter are the soil components most often
implicated in pesticide adsorption. However, individual effects
are not as easily ascertained as sometimes assumed, for the reason
that, in most soils, the organic matter is intimately bound to the
clay, probably as a clay-metal-organic complex. Thus, two major
types of adsorbing surfaces are normally available to the pesticide, namely, clay-humus and clay alone. Accordingly, clay and
organic matter function more as a unit than as separate entities
and the relative contribution of organic and inorganic surfaces to
adsorption will depend upon the extent to which the clay is coated
with organic substances. As can be seen from the schematic diagram shown in Figure 3, the interaction of organic matter with
clay still provides an organic surface for adsorption.

Data published by Walker and Crawford (13) for adsorption of
some s-triazines by 36 soils having widely variable organic matter
contents (Figure 4) suggests that, up to an organic matter content
of about 6%, both mineral and organic surfaces are involved in
adsorption: at higher organic matter contents, adsorption will
occur mostly on organic surfaces. It should be noted, however,
that the amount of organic matter required to coat the clay will
vary from one soil to another and will depend on the kind and
amount of clay that is present. For soils having similar clay and
organic matter contents, the contribution of organic matter will
be highest when the predominant clay mineral is kaolinite and lowest when montmorillonite is the main clay mineral. Bailey et al.
(14) demonstrated that the adsorption capacity of clays for herbicides followed the order montmorillonite > illite > kaolinite.

Comparative studies between known clay minerals and organic
soils suggest that most, but not all, pesticides have a greater
affinity for organic surfaces than for mineral surfaces. Scott
and Weber (15) found that the phytotoxicities of 2,4-D, prometone,
and CIPC to the test plant were reduced to a much greater extent
by addition of an organic soil to the growth media than by addition of montmorillonite or kaolinite. Doherty and Warren's (16)
results show that both fibrous peat and a well-decomposed muck
were more adsorptive than bentonite for pyrazone, linuron, prometone, and simazine. Hance (17) concluded that diuron was a more
effective competitor for water at organic matter surfaces than at
mineral surfaces, and Deli and Warren (18) found that organic

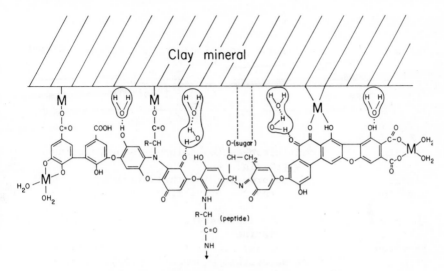

Figure 3.   Clay–metal–organic matter complex

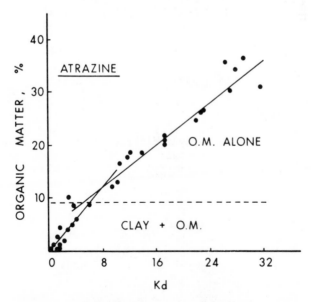

Figure 4.   Relationship between organic matter content
and amount of atrazine adsorbed by 36 soils. Kd = μmoles
adsorbed per g/μmoles per ml equilibrium solution. From
Ref. 3 as adapted from Ref. 13.

Table 1.   Chemical designations of organics mentioned in text.

Common name	Chemical formula
s-Triazines	
Atrazine	2-chloro-4-ethylamino-6-isopropylamino-s-triazine
Simazine	2-chloro-4,6-bis(ethylamino)-s-triazine
Atratone	2-methoxy-4-ethylamino-6-isopropylamino-s-triazine
Ametryn	2-methylthio-4-ethylamino-6-isopropylamino-s-triazine
Prometon	2-methoxy-4,6-bis(isopropylamino)-s-triazine
Prometryn	2-methylthio-4,6-bis(isopropylamino)-s-triazine
Propazine	2-chloro-4,6-bis(isopropylamino)-s-triazine
Substituted ureas	
Diuron	3-(3,4-dichlorophenyl)-1,1-dimethylurea
Monuron	3-(p-chlorophenyl)-1,1-dimethylurea
Fenuron	3-phenyl-1,1-dimethylurea
Linuron	3-(3,4-dichlorophenyl)-1-methoxy-1-methylurea
Neburon	1-butyl-3-(3,4-dichloropheny)-1-methylurea
Phenylcarbamate	
CIPC	isopropyl m-chlorocarbanilate
Bipyridylium quaternary salts	
Diquat	6,7-dihydrodipyrido(1,2-a:2',1'-c)pyrazidinium salt
Paraquat	1,1'-dimethyl-4,4'dipyridinium salt

Table 1 (Cont'd)

Common name	Chemical formula

Others

Amiben	3-amino-2,5-dichlorobenzoic acid
2,4-D	2,4-dichlorophenoxyacetic acid
Picloram	4-amino-3,5,6-trichloropicolinic acid
Dalapon	2,2-dichloropropionic acid
Diphenamid	N,N-dimethyl-2,2-diphenylacetamide
Trifluralin	α,α,α-trifluro-2,6-dinitro-N-N-dipropyl-p-toluidine
DCPA	dimethyl-2,3,5,6-tetrachloroterephthalate
DNPB	4,6-dinitro-o-<u>sec</u>-butylphenol
Amitrole	3-amino-1,2,4-triazol
Pyrazone	5-amino-4-chloro-2-phenyl-3-(2H)-pyridazone
Lindane	1,2,3,4,5,6-hexachlorocyclohexane
DDT	1,1,1-trichloro-2,2-bis(p-chlorophenyl)ethane

matter was more effective in adsorbing diphenamid than clay (bentonite). In other studies, Weber, Perry, and Ibaraki (19) found that, on a weight basis, an organic soil was more effective than montmorillonite in reducing the phytotoxicity of prometone to wheat. For the s-triazines, prometone and prometryne may prefer mineral surfaces (20).

Laboratory studies have, in general, corroborated field observations indicating that organic matter plays a major role in the performance of soil-applied pesticides. This work has generally involved multiple correlation analysis for pesticide adsorption by a series of soils with widely different properties, the usual soil parameters being organic matter content, texture (clay content), clay mineral type, pH, and cation exchange capacity. In a typical study, a given quantity of soil is added to a pesticide solution of known concentration, the mixture is allowed to equilibrate, and the concentration of the pesticide in the solution phase is estimated. The amount of pesticide adsorbed is subsequently calculated from the change in concentration and is usually expressed by such units as μ moles adsorbed per Kg of soil (x/m). By repeating the measurements at several pesticide concentrations, an adsorption isotherm can be obtained by plotting the quantity adsorbed (x/m) vs. the equilibrium concentration (C). In most instances, a straight line is obtained when the data are plotted as log x/m vs. log C, according to the Freundlich adsorption equation,

$$x/m = KC^{1/n}$$

where K and n are constants. The constant K provides a measure of the extent of adsorption and has been used in correlation studies aimed at determining the relative importance of the various soil parameters on adsorption.

Alternately, a distribution coefficient, Kd, can be obtained for a given solution concentration as the ratio of the amount of pesticide adsorbed to the amount remaining in solution

$$Kd = \frac{\text{pesticide adsorbed } (\mu \text{ moles/Kg})}{\text{pesticide in solution } (\mu \text{ moles/liter})}$$

Table 2 gives typical correlations between adsorption of some common herbicides and the soil variables of organic matter content, cation exchange capacity, and pH. It can be seen that, in most cases, the correlation coefficient relating adsorption to organic matter content is considerably higher than for the other soil parameters, including clay content.

Qualitative Differences in the Organic Matter of Natural Soils

The fact that soils differ greatly in their organic matter contents is well known but it is not generally appreciated that

Table 2. Organic matter, clay, and other soil properties
correlated with adsorption parameters.

Compound	No. of soils	Correlation Coefficient				Reference
		Organic matter	Clay	CEC	pH	
s-Triazines						
Ametryn	34	0.41*	0.14	0.19	-0.37*	25
Atrazine	25	0.82**	0.65**	0.63**	-0.28	20
Propazine	25	0.74**	0.71**	0.69**	-0.41*	20
Prometon	25	0.26	0.60**	0.55**	-0.42*	20
Prometryn	25	0.40*	0.68**	0.63**	-0.49	20
Simazine	25	0.83**	0.77**	0.79**	-0.39	20
Simazine	65	0.72**	0.12	0.52**	0.04	21
Simazine	32	0.62**	0.27	0.54**	-0.35	24
Simazine	18	0.82**	0.48**	0.84**	-0.40	26
Substituted Ureas						
Diuron	34	0.73**	0.37*	0.58**	0.10	25
Diuron	32	0.89**	0.28	0.56**	-0.03	24
Linuron	11	0.90**	0.06	0.57*	-0.14	22
Neburon	7	0.76*	-0.37	0.19	0.14	27
Picloram	6	0.90*	0.55	0.65	-	23
Phenyl carbamates						
CIPC	32	0.85**	0.16	0.38*	0.48*	24
Other						
Diphenamid	11	0.91**	0.16	0.60*	0.11	22

*Significant at p = 0.05   **Significant at p = 0.01

major qualitative differences also exist, both with respect to
the known classes of organic compounds (lipids, carbohydrates,
proteins) and with the so-called humic substances (humic acid,
fulvic acid, etc). For example, the percentage of the organic
matter as fats, waxes, and resins ranges from as little as 2% in
some soils to over 20% in others, with the higher value being
typical of forest humus layers and acid peats (28). The percentage of the organic matter as "protein" may vary from 15 to 45%.
A recent review on the subject shows that the carbohydrate content of soil organic matter ranges from 5 to 25% (29).

Humified organic matter may comprise three-fourths of the
total organic matter in some soils but less than one-third in
others. The humic fraction in grassland soils is dominated by
humic acids; that in forest soils is relatively rich in fulvic
acids. The so-called brown humic acids are characteristic of
the humic acids of Alfisols and Ultisols whereas the gray humic
acids are typical of the humic acids in Mollisols.

Generalized diagrams showing the humic acid–fulvic acid
relationships in grassland (Mollisols) and forest soils (e.g.,
Alfisols) are as follows, where FA = fulvic acid, GHA = gray humic
acid, and BHA = brown humic acid.

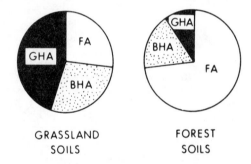

GRASSLAND   FOREST
SOILS   SOILS

Differences in organic matter composition have implications
with respect to correlation studies of pesticide performance and
organic matter content. Hayes, Stacey, and Thompson (30) obtained
results which indicated that fulvic acids were less effective in
adsorbing the s-triazine herbicides than humic acids. More
recently, Dunigan and McIntosh (31) found that the ether- and
alcohol-extractable components of soil organic matter (fats, waxes,
and resins) had a negligible capacity to adsorb atrazine; a hot-
water-extractable component (presumably a polysaccharide) had a
small adsorption capacity. Removal of ether- and alcohol-soluble
material resulted in increased adsorption, apparently due to
uncovering of reactive sites. Experiments conducted with compounds representative of natural soil organic matter showed that
polysaccharide-type constituents had rather low affinities for
atrazine, a protein had an intermediate affinity, and humic acids
and lignins had high affinities (31). Walker and Crawford (13),

in an experiment in which various decomposable organic materials
were incubated with soils low in organic matter, found that both
the type of material being decomposed and its stage of decomposition were important in the adsorption of s-triazines. Addition
of lignin (1%) to a sand culture has been found to be more effective in reducing the phytotoxic effect of atrazine on oats than an
equivalent amount of cellulose (32).

Doherty and Warren (16) found that prometryne, simazine, and
pyrazon were more highly adsorbed by a fibrous peat than by a
muck soil; for linuron, the reverse was true. The conclusion was
reached that the muck soil contained large quantities of an organic constituent (unidentified) which adsorbed linuron but not the
other herbicides. Talbert and Fletchall (20) found that a well-
humified peat adsorbed more simazine and atrazine than peat moss,
while Sherburne and Freed (33) obtained greater retention of a
substituted urea herbicide by a muck soil than by sawdust, straw,
or activated charcoal. In other work, Hance (17) found that
adsorption of diuron was much lower on some "hydrophilic"
materials (cellulose and chitin) than on some "less hydrophilic"
substances (lignin and a methylated soil organic matter prepra-
tion).

Abnormally high retention of herbicides has been observed in
burned-over fields and those containing wind-blown carbon particles (34, 35, 36). In general, activated charcoal tends to
adsorb pesticides, although the amount adsorbed varies greatly
with the different compounds. Weber et al. (19, 37), for example,
found that prometone and 2,4-D were adsorbed to a greater extent
than diquat and paraquat. The order of adsorption for the eight
herbicides examined by Coffee and Warren (38) was : CIPC >
trifluralin > 2,4-D > diphenamid > DCPA > DNPB > amiben > paraquat (no adsorption). The most readily desorbed herbicide was
2,4-D; CIPC and DNPB showed little or no desorption.

The possibility of using activated charcoal to detoxicate
herbicide treated soil has been discussed by Ahrens (39) and
Coffee and Warren (38).

Special Role of Fulvic Acids

Because of their low molecular weights and high acidities.
fulvic acids are more soluble than humic acids, and they may have
special functions with regard to herbicide transformations. First
they may act as transporting agents for pesticides in soils and
natural waters. Ogner and Schnitzer (40), and Schnitzer and
Ogner (41), suggested that fulvic acids act as carriers of alkanes
and other normally water-insoluble organic substances in aquatic
environments, and it is possible that these constituents also
function as vehicles for the transport of pesticides. According
to Ballard (42), the downward movement of the insecticide DDT in
the organic layers of forest soils is due to water-soluble, humic-
like substances.

Second, fulvic acids by virtue of their high acidities, may

catalyze the chemical decomposition of certain pesticides. The
suggestion has been made, for example, that these constituents
might catalyze the hydroxylation of the chloro-s-triazines (30).
For additional information regarding fulvic acids and their reactions,the reader is referred to the recent book of Schnitzer
and Khan (43).

Potential Chemical Reactions Involving Pesticides and Organic
Substances in Soil

There seems little doubt but that the organic fraction of the
soil has the potential for promoting the nonbiological degradation
of many pesticides. Organic compounds containing nucleophilic
reactive groups of the types believed to occur in humic and fulvic
acids (e.g., COOH, phenolic-, enolic-, heterocyclic-, and ali-
phatic-OH, amino, heterocyclic amino, imino, semiquinones, and
others) are known to produce chemical changes in a wide variety
of pesticides (7, 44). Of additional interest is that humic substances are rather strong reducing agents and have the capability
of bringing about a variety of reductions and associated reactions,
as discussed by Crosby (7). The occurrence of stable free radicals in humic and fulvic acids further implicates organic matter
in chemical transformations of pesticides. For example, the
heterocyclic ring of amitrole is known to be highly susceptible to
attack by free radicals (45, 46).

Basic amino acids and similar compounds have the potential
for catalyzing the hydrolysis of organophosphorus esters (47), as
well as the dehydrochlorination of DDT and lindane (48). Miskus
et al. (49) demonstrated that certain chlorophyll degradation
products (reduced porphyrins) can convert DDT to DDD. Substances
in soil organic matter which contain hydroxyl and amino groups,
such as humic and fulvic acids, are potentially capable of being
alkylated by the action of chlorinated aliphatic acids (e.g.,
chloroacetic, dichloropropionic), as shown below (50).

$$R-NH_2 + Cl-CH_2-(CH_2)_n-COOH \rightarrow R-NH-(CH_2)_{n+1}{}^{-COOH}$$
$$R-OH + Cl-CH_2-(CH_2)_n-COOH \rightarrow R-O-(CH_2)_{n+1}{}^{-COOH}$$

Specific examples of nonbiological transformations brought
about by the organic fraction of the soil includes hydroxylation
of the chloro-s-triazines (51-56) and decomposition of amitrole
(45,46). With regard to the former, Armstrong and Chesters (51)
concluded that hydrolysis of atrazine resulted from the sequence
of events shown in Figure 5. Adsorption was believed to take place
between a ring nitrogen atom and a protonated COOH group of the
organic matter. Hydrogen bonding of the ring nitrogen was believed to cause the withdrawal of electrons from the electron deficient carbon atom bonded to the chloride; thereby enabling water
to replace the chloride atom. Nearpass (55)found that propazine
hydrolysis was enhanced in the presence of organic matter irrespec-
tive of the pH of the system and was related in some way to

adsorption. In other work, Hance (53) was unable to establish a
relationship between rate of atrazine decomposition and extent of
adsorption.

The review of Crosby (7) should be consulted for other
examples of nonbiological degradation of pesticides by reaction
with organic substances.

Chemical Binding of Pesticides and Their Decomposition Products

   Substantial evidence exists to indicate that pesticide-
derived residues can form stable chemical linkages with organic
substances and that such binding greatly increases the persistence of the pesticide residue in the soil (57-61). Two main
mechanisms can be envisioned: (i) direct chemical attachment of
the residues to reactive sites on colloidal organic surfaces and
(ii) incorporation into the structures of newly formed humic and
fulvic acids during the humification process (3).

   A key to the fate of pesticides and their intermediate decomposition products may be provided by consideration of the process whereby humic and fulvic acids are formed. The lignin-protein theory in its original form is now believed by many investigators to be obsolete, and the modern view is that humic substance
are formed by a multiple stage process which includes: (i) decomposition of all plant components, including lignin, into simpler
monomers, (ii) metabolism of the monomers with an accompanying
increase in the soil biomass, (iii) repeated cycling of the biomass carbon with synthesis of new cells, and (iv) concurrent polymerization of reactive monomers into high-molecular-weight polymers
(8-11). The general consensus is that polyphenols (quinones)
synthesized by microorganisms, together with those liberated from
lignin, polymerize alone or in the presence of amino compounds
(amino acids, etc.) to form brown colored polymers. An alternate
pathway is by condensation of amino acids and related substances
with reducing sugars, according to the Maillard reactions.

   The reaction between polyphenols and amino compounds involves
simultaneous oxidation of the polyphenol to the quinone form, such
as by polyphenol oxidase enzymes. The addition product readily
polymerizes to form brown nitrogenous polymers according to the
general sequence shown in Figure 6.

   In the case of the Maillard reaction, the initial step
involves addition of the amine to the C=O group of the sugar, with
the formation of an aldosylamine (Figure 7). This is followed by
the Amadori rearrangement to form the N-substituted keto derivative, which subsequently undergoes dehydration and fragmentation
to yield a variety of unsaturated intermediates (62, 63). In the
final stages of browning, the intermediates polymerize into brown
polymers and copolymers. The rate of the reaction increases with
temperature, pH, and the basicity of the amine. Under laboratory
conditions, brown polymers can readily be synthesized from amino
acid-sugar mixtures within hours in aqueous solution at 50° C.

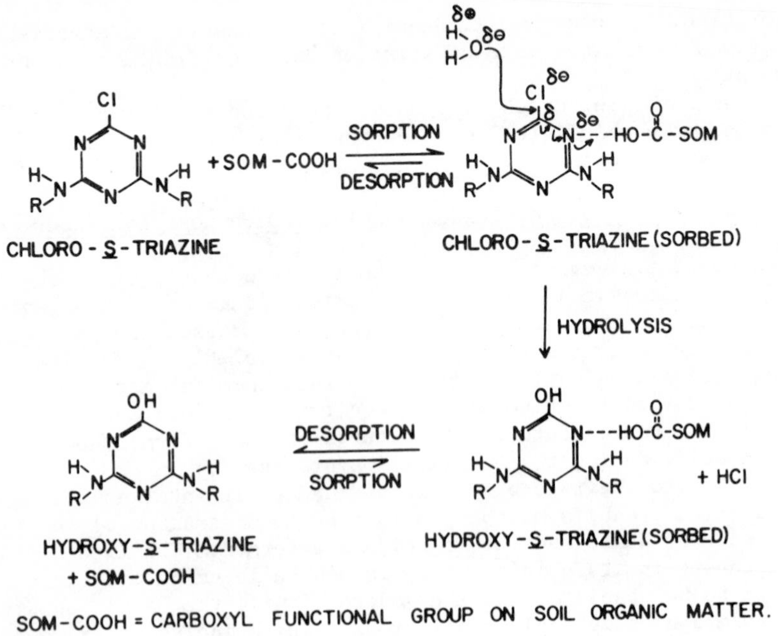

Pesticides in Soil and Water

Figure 5.   Proposed model for the sorption-catalyzed hydrolysis of chloro-s-triazines by soil organic matter (6)

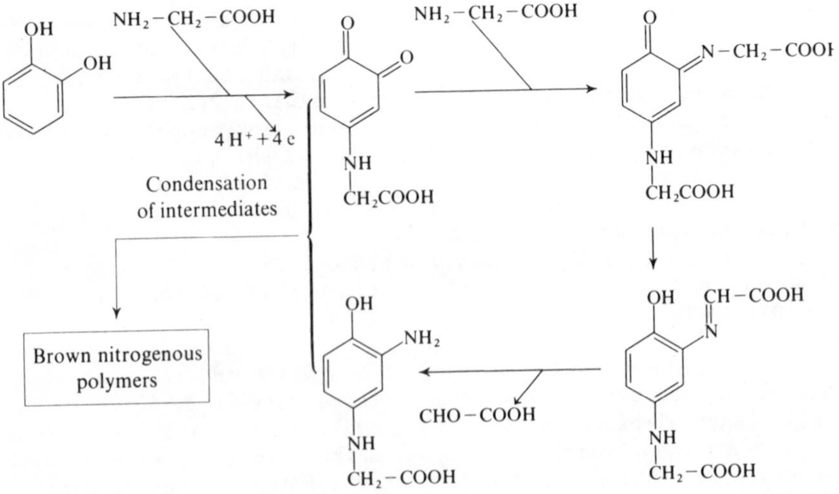

Figure 6.   General scheme for the formation of brown nitrogenous polymers by condensation of polyphenols and amino acids as exemplified by the reaction between catechol and glycine

Condensation reactions between pesticides and their degradation products with organic substances in soil would be enhanced by such processes as freezing and thawing, wetting and drying, and the intermixing of reactants with mineral matter having catalytic properties.

It is rather evident that reactions similar to these shown in Figures 6 and 7 could be involved in pesticide transformations in soil. Many weakly basic compounds, including amino acids, pyrrols, amides, amines, and imines, are known to have the ability to combine chemically with an array of carbonyl-containing substances, including reducing sugars, reductones, the common aldehydes and ketones, and furfural. Many of the common pesticides fall into one of these categories. Those pesticides which are basic in character, have the potential for forming a chemical linkage with C=O constituents of soil organic matter; those containing a C=O group are theoretically capable of reacting with amino constituents. Condensation and conjugate reactions of pesticides with metabolic products have been postulated to constitute a form of pesticide transformations by microorganisms and high plants.

Another factor to consider is that the partial degradation of many pesticides by microorganisms leads to the formation of chemically reactive intermediates which can combine with amino- or C=O containing compounds, as illustrated in Figure 8. Thus, loss of the side chain from the phenoxyalkanoic acids by enzymatic action leads to the formation of phenolic constituents which can either be oxidized further via the enzymatic route or undergo condensation (probably as quinones) with amino compounds to form "humic-like" substances. On the other hand, amines (or chloroamines) produced by biological decomposition of such herbicides as the acylanilides, phenylcarbamates and phenylureas may react with C=O constituents occurring naturally in soil. Entry into the carbon cycle by this mechanism may constitute a form of natural detoxification.

Thus, it must be concluded that some pesticides or their decomposition products can become part of the pool of precursor molecules for humus synthesis, and, in so doing, lose their identity.

Bartha (57), Bartha and Pramer (58), and Chisaka and Kearney (59) concluded that the bulk of chloroanilines liberated by partial degradation of the phenylamide herbicides (acylanilides, phenylcarbamates, and phenylureas) becomes immobilized in soil by chemical bonding to organic matter. The chemically bound residues could not be recovered by extraction with organic solvents or inorganic salts; partial release was possible by acid or base hydrolysis (60). According to Hsu and Bartha (60), binding occurs when the amino group of the anilines react with C=O and COOH groups appropriately positioned on the humic acid core with formation of a heterocyclic ring. The soil-bound chloroaniline residues resist attack by microorganisms (57, 61).

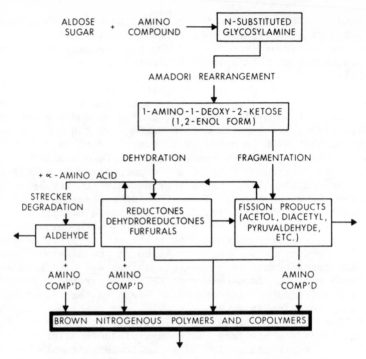

Figure 7. Formation of brown nitrogenous polymers according to the Maillard reaction

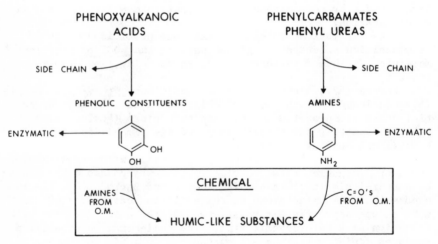

Environmental Quality

Figure 8. Chemical reactions involving intermediate products of herbicide decomposition and constituents of soil organic matter (3)

Fate of Organics in Sediments

The role of organic matter in chemical transformations de-
serves serious attention in determining the long-time fate of
persistent pesticides in the environment. Indirect information
on this subject is provided by the many biogeochemical studies
dealing with the fate of naturally occurring organics in sediments
and sedimentary rocks (64). Since this subject is beyond the
scope of the present review, the discussion which follows will be
confined to a consideration of the diagenesis of amino acids in
sediments, as outlined elsewhere (65).

The net effect of nonbiological reactions involving amino
acids (see Figures 6 and 7) is incorporation of nitrogen into the
structures of humic and fulvic acids. Thus, whereas amino acids
constitute 80% or more of the organic nitrogen of microbial tissue (biomass), they account for only about one-third of the
nitrogen in soil and marine humus. Following burial in sediments,
further changes occur, with transfer of amino acid-N into the
humified remains.

The significance of chemical transformations of amino acids
following burial can be illustrated by consideration of data
obtained for the forms of nitrogen in sediments of the Argentine
Basin (66) and the Experimental Mohole (67). In the case of the
Argentine Basin sediments (estimated maximum age of 125,000 years),
data were reported (65) for total amino acids and organic carbon
in 72 individual samples from two cores (V-15-141 and V-15-142).
As indicated in Figure 9, there was a progressive decrease in the
percentage of the organic carbon as amino acids with increasing
age. A similar result was obtained for the somewhat older Experimental Mohole sediments, for which quantitative data were available for eight samples ranging in age from 3 to 14 million years.
These data are shown in Figure 10.

For both basin sediments, the disappearance of amino acids
from the sedimentary organic matter, as estimated from composi-
tional studies, was considerably greater than would be anticipated
from kinetic studies of amino acids in aqueous solution. This
observation lends support to the conclusion that, during diagenesis,
amino acids are transformed to other products as a consequence of
chemical reactions with other organics, presumably by reactions
of the type shown in Figures 6 and 7.

Diagenetic changes in the amino acids of sediments have also
been observed for Saanich Inlet, British Columbia, where it was
found that losses of amino acids with depth exceeded that for
organic carbon (68). Abelson (69) earlier postulated that nonbiological processes involving complex heteropolymers were
responsible for the disappearance of amino acids from sediments.
Coupling of amino acids with porphyrins may be of geochemical
significance (70). The high stability of amino acids in petroleum
brines has been attributed to their linkage with phenol- and
quinone-containing substances (71).

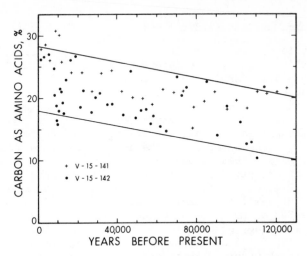

Advances in Organic Geochemistry

Figure 9. Relationship between estimated age of Argentine Basin sediments and the percentage of organic carbon which occurred as amino acids. Regression equation: $\hat{Y} = 22.92 - 0.05\ x$ (r $= 0.47$, significant at p $= 0.01$) (65).

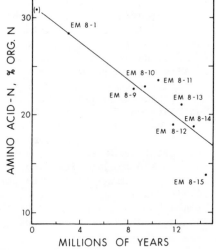

Figure 10. Relationship between age of sedimentary material in Experimental Mohole sediments and the percentage of organic nitrogen which occurred as amino acids. Value shown in brackets in the upper left hand corner represents an average for the younger Argentine Basin sediments. Regression equation: $\hat{Y} = 32.03 - 0.001\ x$ (r $= 0.95$, significant at p $= 0.01$).

Using the analogy given above, one would expect that natural organics would exert an appreciable kinetic influence on the disappearance of certain persistant pesticides (or their intermediate decomposition products). Rate of loss would, of course, depend upon the nature of the compound and environmental conditions existing in the sediment, such as pH and temperature.

Adsorption Mechanisms

Bonding mechanisms for the retention of pesticides by organic substances in soil include ion exchange, protonation, H-bonding, van der Waal's forces, and coordination through an attached metal ion (ligand exchange). In addition, nonpolar molecules may be partitioned onto hydrophobic surfaces through "hydrophobic bonding." For some pesticides, adsorption is apparently not completely reversible (18, 20, 32, 38, 72), a factor which is of considerable importance in determining the environmental impact of pesticides in soil and water.

Ion Exchange and Protonation.

Adsorption through ion exchange is restricted to those pesticides which either exist as cations (diquat and paraquat) or which become positively charged through protonation (s-triazines; amitrole).

Diquat and paraquat, being divalent cations, have the potential for reacting with more than one negatively charged site on soil humic colloids, such as through two COO$^-$ ions (illustrated below for diquat), a COO$^-$ ion plus a phenolate ion combination, or a COO$^-$ ion (or phenolate ion) plus a free radical site.

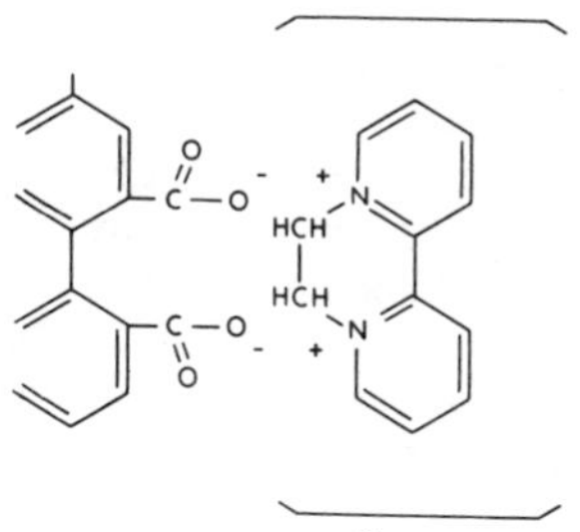

Diquat

On the basis of infrared studies, Khan (73, 74) suggested that bipyridylium herbicides form charge-transfer complexes with humic substances. This could not be confirmed by Burns et al, (75), who subjected some paraquat-humic acid complexes to ultraviolet analysis.

Paraquat has been found to be complexed in greater amounts by humic and fulvic acids (73), and by an organo-clay complex (74), than diquat. A histosol and its humic and fulvic fractions has also been observed to show selective preference for paraquat (76).

Factors which influence the availability of exchange sites for adsorption include the presence of competing metal cations and pH. Soil pH has a direct bearing on the relative importance of organic matter and clay in retaining organic cations. Unlike clay, organic colloids have a strongly pH-dependent charge. Therefore, the relative contribution of organic matter to cation exchange capacity, and subsequently retention of cations, will be higher in neutral and slightly alkaline soils than in acidic ones. For each unit change in pH, the change in cation exchange capacity for organic matter is several times greater than for clay.

Less basic compounds, such as the s-triazines, may become cationic through protonation. Whether or not protonation occurs will depend upon: (i) the nature of the compound in question, as reflected by its pK_a, and (ii) the proton-supplying power of the humic colloids. Reactions leading to adsorption of the s-triazines, as postulated by Weber et al. (77), are shown by the following equations:

$$T + H_2O \rightleftharpoons HT^+ + OH^- \tag{1}$$

$$RCOOH + H_2O \rightleftharpoons R\text{-}COO^- + H_3O^+ \tag{2}$$

$$RCOO^- + HT^+ \rightleftharpoons R\text{-}COO\text{-}HT \tag{3}$$

$$RCOOH + T \rightleftharpoons R\text{-}COO\text{-}HT \tag{4}$$

where R is the organic colloid, T the s-triazine molecule, HT^+ the protonated molecule, and H_3O^+ the hydronium ion.

Equation [1] represents pH-dependent adsorption through protonation in the soil solution while equation [2] represents ionization of the colloid COOH group. Ionic adsorption of the cationic s-triazine molecule, formed by reaction [1], is shown by equation [3]. Adsorption through direct protonation on the surface of the organic colloid is shown by reaction [4].

Amitrole is another example of a weak base that can be adsorbed to soil organic matter through ion exchange (78).

Soil pH has a profound effect on adsorption of s-triazines and other weakly basic pesticides by organic matter. Soil reaction governs not only the ionization of acidic groups on the organic colloids but the relative quantity of the pesticide which occurs in cationic form, in accordance with equation [1]. The pK_a of acidic groups in humic acids (COOH plus phenolic- and/or enolic-OH) is of the order of 4.8 to 5.2. Thus, it would appear that ion exchange would not be an important mechanism for adsorption of weakly basic pesticides with pK_a's much lower than 3.0. It should be pointed out however, that the pH at the surface of soil organic colloids may be as much as two pH units lower than that of the liquid environment. The adsorption capacities of soil organic matter preparations for the s-triazines has been

found to follow the order expected on the basis of pK_a values for
the herbicides, with maximum adsorption occurring at pH values
near the pK_a of the respective compound (78).

Ion exchange is but one of several mechanisms for adsorption
of the s-triazines to organic colloids, as illustrated below:

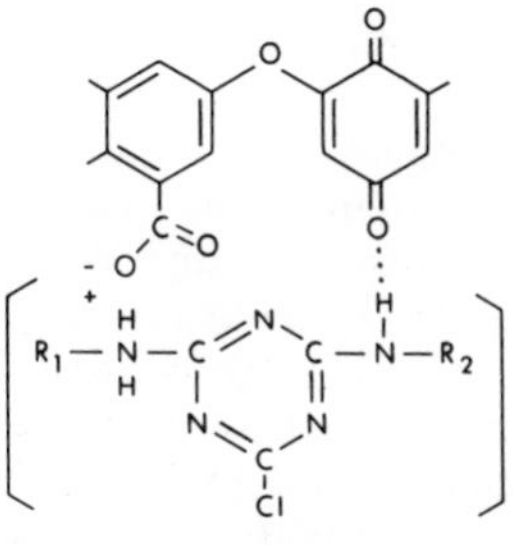

s-Triazine

On the basis of an infrared study of some s-triazine-humic
acid complexes, Sullivan and Felbeck (79) concluded that one
secondary amino group was bound to either a C=O or quinone group
of the humic acid through a hydrogen bond whereas the other
secondary amino group became protonated and was bound by ion
exchange to a COO⁻ group.

It should be noted that the mechanism described above is
somewhat different from that shown earlier in Figure 5. Walker
and Crawford (13) suggested that adsorption of the s-triazines by
organic matter could best be regarded as partitioning out of solution onto hydrophobic surfaces (discussed later).

For anionic pesticides, such as the phenoxyalkanoic acids,
repulsion by the predominantly negatively charged surface of
organic matter may occur. Positive adsorption of anionic herbicides at pH values below their pK_a can be attributed to adsorption
of the unionized form of the herbicide to organic surfaces, such
as by H-bonding between the COOH group and C=O or NH_2 groups of
organic matter.

H-Bonding, van der Waals Forces, and Coordination.

Adsorption mechanisms for retention of nonionic polar pesticides, such as the phenylcarbamates and substituted ureas, are
illustrated in Figure 11. The great importance of H-bonding in
retention is suggested, with multiple sites being available on
both pesticide and organic matter surface. Other adsorption
mechanisms include van der Waals forces (physical adsorption),
ligand exchange (-M²⁺ ..... O=C), and, for pesticides containing
an ionizable COOH group, a salt linkage through a divalent cation
on the organic exchange site. For chlorinated phenoxyalkanoic
acids, such as 2,4-D, physical adsorption to aromatic constituents

of organic matter may be involved; H-bonding will be limited to acid conditions where COOH groups are unionized.

Considerable variation can be expected in the adsorption capacity of organic matter for nonionic polar herbicides, depending upon steric effects and the number and kinds of electronegative atoms in the molecule.

Hydrophobic Bonding

Partitioning on hydrophobic surfaces has been proposed as a mechanism for retention of nonpolar organic pesticides by soil organic matter. Active surfaces include the fats, waxes, and resins, as well as possible aliphatic side chains on humic and fulvic acids. Weber and Weed (80) pointed out that "humus," by virtue of its aromatic framework and presence of polar groups, may contain both hydrophobic and hydrophylic adsorption sites.

Pierce et al. (81, 82) suggested that chlorinated hydrocarbons (such as DDT) would have greater affinity for hydrophobic sites on organic substances than for clay and that scavenging by organic particulates provided a means whereby these persistent pollutants are transported through the water column and concentrated in bottom sediments. For DDT, it would be noted that chlorine atoms on the ethyl group may impart a slight negative charge to the molecule (83); consequently, part of the adsorption attributed to hydrophobic bonding may be due to attraction to positively charged sites such as to an amino group.

Considerable emphasis has been given to hydrophobic bonding as a mechanism for adsorption of the s-triazines (13) and the phenylureas (17) by soil organic matter. These claims require confirmation; for these pesticides, the bulk of the evidence favors the idea that specific adsorption sites (functional groups) are involved (see earlier discussion).

Relative Affinities of Pesticides for Organic Matter

The deliberations of the previous section serve to emphasize that the various pesticides differ greatly in their relative affinities for soil organic colloids. The approximate order for some common herbicides are given in Figure 12. Thus, the cationic herbicides (diquat and paraquat) would be expected to be the most strongly bound, followed by those weakly basic types capable of being protonated under moderately acidic conditions. For the s-triazines, differences in adsorbability can be accounted for by variations in pK_a, with the more basic compounds (high pK_a) being adsorbed the strongest. Herbicides in the next order of adsorption are those having very low pK_a values but which contain one or more polar groups suitable for H-bonding. Anionic pesticides may or may not be adsorbed, depending upon soil pH.

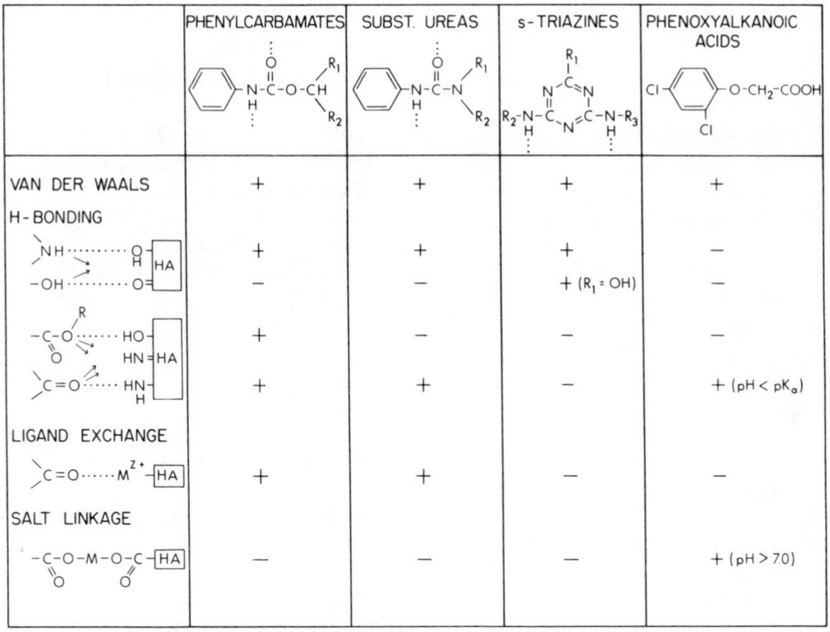

	PHENYLCARBAMATES	SUBST. UREAS	s - TRIAZINES	PHENOXYALKANOIC ACIDS
VAN DER WAALS	+	+	+	+
H - BONDING				
$\diagdown NH \cdots O \diagup$ ... $\diagdown$ HA / $-OH \cdots O=$ HA	+ / −	+ / −	+ / + (R_1 = OH)	− / −
$-C-O \cdots HO-$ / $-C=O \cdots HN-$ HA	+ / +	− / +	− / −	− / + (pH < pK_a)
LIGAND EXCHANGE				
$\diagdown C=O \cdots M^{z+}$ HA	+	+	−	−
SALT LINKAGE				
$-C-O-M-O-C-$ HA	−	−	−	+ (pH > 70)

Figure 11.   Typical bonding mechanisms for adsorption of some common herbicides by soil organic matter (3)

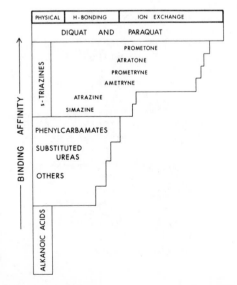

Figure 12.   Relative affinities of herbicides for soil organic matter surfaces (3)

Literature Cited

1. Bailey, G. W., and White, J. L., J. Agr. Food Chem. (1964) 12, 324-333.
2. Hayes, M. H. B., Residue Rev. (1970) 32, 131-174.
3. Stevenson, F. J., J. Environ. Quality (1972) 1, 333-343.
4. Weed, S. B., and Weber, J. B., In Guenzi, W. D. (ed.), "Pesticides in Soil and Water," pp. 39-66. American Society of Agronomy, Madison, Wisc. (1974).
5. Wolcott, A. R., In "Proceedings Int. Symposium on Pesticides in Soil." pp. 128-138. Michigan State University, East Lansing. (1970).
6. Armstrong, D. E., and Konrad, J. G., In Guenzi, W. D. (ed.), "Pesticides in Soil and Water," pp. 123-131. American Society of Agronomy, Madison, Wisc. (1974).
7. Crosby, D. G., In "Proceedings Int. Symposium on Pesticides in Soil," pp. 86-94. Michigan State University, East Lansing. (1970).
8. Dubach, P., and Mehta, N. C., Soils Fert. (1963) 26, 293-300.
9. Kononova, M. M., "Soil Organic Matter," pp. 544. Pergamon Press, Inc., New York. (1966).
10. Scheffer, F., and Ulrich, B., "Humus and Humusdüngung. Bd I," pp. 266. Ferdinand Enke, Stuttgart, Germany. (1960).
11. Stevenson, F. J., and Butler, J. H. A., In Englinton, G., and Murphy, M. T. J., (eds.), "Organic Geochemistry," pp. 534-557. Springer-Verlag, Berlin. (1966).
12. Steelink, C., and Tollin, G., In McLaren, A. D., and Peterson, G. H. (eds.). "Soil Biochemistry," pp. 147-169. Marcel Dekker, Inc., New York. (1967).
13. Walker, A., and Crawford, D. V., In "Isotopes and Radiation in Soil Organic Matter Studies," pp. 91-108. Atomic Energy Agency, Vienna, Austria. (1968).
14. Bailey, G. W., and White, J. L., Residue Rev. (1970) 32, 29-92.
15. Scott, D. C., and Weber, J. B., Soil Sci. (1967) 104, 151-158.
16. Doherty, P. J., and Warren, G. F., Weed Res. (1969) 9, 20-26.
17. Hance, R. J., Weed Res. (1965) 5, 108-114.
18. Deli, J., and Warren, G. F., Weed Sci. (1971) 19, 67-69.
19. Weber, J. B., Perry, P. W., and Ibarki, K., Weed Sci. (1968) 16, 134-136.
20. Talbert, R. E., and Fletchall, O. H., Weeds (1965) 13, 47-52.
21. Day, B. F., Jordan, L. S., and Jolliffe, V. A., Weed Sci. (1968) 16, 209-213.
22. Dubey, H. D., Sigafus, R. E., and Freeman, J. F., Agron. J. (1966) 58, 228-231.
23. Grover, R., Weed Res. (1968) 8, 226-232.
24. Harris, C. I., and Sheets, T. J., Weeds (1965) 13, 215-219.
25. Liu, L. C., Cibes-Viade, H., and Koo, F. K. S., Weed Sci. 18, 470-474.
26. Nearpass, D. C., Weeds (1965) 13, 341-346.

27. Obien, S. R., Suehisa, R. H., and Younge, O. R., Weeds (1966) 14, 105-109.
28. Stevenson, F. J., J. Amer. Oil Chem. Soc. (1966) 43, 203-210.
29. Swincer, G. D., Oades, J. M., and Greenland, D. J., Adv. Agron. (1969) 21, 195-325.
30. Hayes, M. H. B., Stacey, M., and Thompson, J. M. In "Isotopes and Radiation in Soil Organic Matter Studies," pp. 75-90. Atomic Energy Agency, Vienna, Austria (1968).
31. Dunigan, E. P., and McIntosh, T. H., Weed Sci. (1971) 19, 279-282.
32. Tompkins, G. A., McIntosh, T. H., and Dunigan, E. P., Soil Sci. Soc. Amer. Proc. (1968) 32, 373-377.
33. Sherburne, H. R., and Freed, V. H., J. Agr. Food Chem. (1954) 2, 937-939.
34. Lichtenstein, E. P., Fuhremann, T. W., and Schulz, K. R., J. Agr. Food Chem. (1968) 16, 348-355.
35. Hilton, H. W., and Yuen, Q. H., J. Agr. Food Chem. (1963) 11, 230-234.
36. Yuen, Q. H., and Hilton, H. W., J. Agr. Food Chem. (1962) 10, 386-392.
37. Weber, J. B., Perry, P. W., and Upchurch, R. P., Soil Sci. Soc. Amer. Proc. (1965) 29, 678-688.
38. Coffee, D. L., and Warren, G. F., Weed Sci. (1969) 17, 16-19.
39. Ahrens, J. F., Proc. N. E. Weed Control Conf. (1965) 19, 364.
40. Ogner, G., and Schnitzer, M., Geochim et Cosmochim. Acta (1970) 34, 921-928.
41. Schnitzer, M., and Ogner, G., Israel J. Chem. (1970) 8, 505-512.
42. Ballard, T. M., Soil Sci. Soc. Amer. Proc. (1971) 35, 145-147.
43. Schnitzer, M., and Khan, S. U.,"Humic Substances in the Environment," pp. 327. Marcel Dekker Inc., New York. (1972).
44. Burchfield, H. P., and Schechtman, J., Contrib. Boyce Thompson Inst. (1958) 19, 411-416.
45. Kaufman, D. D., Plimmer, J. R., Kearney, P.C., Blake, J., and Guardia, F. S., Weed Sci. (1968) 16, 266-272.
46. Plimmer, J. R., Kearney, P. C., Kaufman, D. D., and Guardia, F. S. , J. Agr. Food Chem. (1967) 15, 996-999.
47. Gatterdam, P. E., Casida, J. E., and Stoutamire, D. W., J. Econ. Entomol. (1959) 52, 270-276.
48. Lord, K. A., J. Chem. Soc. (London) (1948), 1657-1661.
49. Miskus, R. P., Blair, D. P., and Casida, J. E., J. Agr. Food Chem. (1965) 13, 481-483.
50. Kearney, P. C., Harris, C. I., Kaufman, D. D., and Sheets, T. J., Adv. Pest Control Res. (1965) 6, 1-30.
51. Armstrong, D. E., and Chesters, G., Environ. Sci. Technol. (1968) 2, 683-689.
52. Armstrong, D. E., Chesters, G., and Harris, R. F., Soil Sci. Soc. Amer. Proc. (1967) 31, 61-66.
53. Hance, R. J., Soil Biol. Biochem. (1974) 6, 39-42.

54. Harris, C. I., J. Agr. Food Chem. (1967) 15, 157-162.
55. Nearpass, D. C., Soil Sci. Soc. Amer. Proc. (1972) 36, 606-610.
56. Li, G. C., and Felbeck, G. T., Jr., Soil Sci. (1972) 114, 201-209.
57. Bartha, R., J. Agr. Food Chem. (1971) 19, 385-387.
58. Bartha, R., and Pramer, D., Adv. Appl. Microbiol. (1970) 13, 317-341.
59. Chiska, H., and Kearney, P. C., J. Agr. Food Chem. (1970) 18, 854-858.
60. Hsu, T-S., and Bartha, R., Soil Sci. (1974) 116, 444-452.
61. Hsu, T-S., and Bartha, R., Soil Sci. (1974) 118, 213-220.
62. Ellis, G. P., Adv. Carbohydrate Chem. (1959) 14, 63-134.
63. Hodge, J. E., J. Agr. Food Chem. (1953) 1, 928-943.
64. Englinton, G., and Murphy, M. T. J., "Organic Geochemistry" pp. 828. Springer-Verlag, Berlin. (1969).
65. Stevenson, F. J., In Tissot, B., and Bienner, F. (eds.). "Advances in Organic Geochemistry," pp. 701-714. Editions Technip, Paris. (1974).
66. Stevenson, F. J., and Cheng, C-N., Geochim. et Cosmochim. Acta (1972) 36, 653-671.
67. Stevenson, F. J., and Tilo, S. N., In Hobson, G. D., and Speers, G. C. (eds.) "Advances in Organic Geochemistry" pp. 237-263. Pergamon Press, London. (1970).
68. Brown, F. S., Baedecker, M. J., Nissenbaum, A., and Kaplan, I. R., Geochim et Cosmochim. Acta (1972) 36, 1185-1203.
69. Abelson, P. H., Carnegie Inst. of Washington Yearbook (1958) 58, 181-185.
70. Hodgson, G. W., Holmes, M. A., and Halpern, B., Geochim. et Cosmochim. Acta (1970) 34, 1107-1119.
71. Degens, E. T., Hunt, J. M., Reuter, J. H., and Reed, W. E., Sedimentology (1964) 3, 199-225.
72. Grover, R., Weed Sci. (1974) 22, 405-408.
73. Khan, S. U., Can. J. Soil Sci. (1973) 53, 199-204.
74. Khan, S. U., J. Soil Sci. (1973) 24, 244-248.
75. Burns, E. G., Hayes, M. H. B., and Stacey, M., Pesticide Sci. (1973) 4, 201-209.
76. Best, J. A., Weber, J. B., and Weed, S. B., Soil Sci. (1972) 114, 444-450.
77. Weber, J. B., Weed, S. B., and Ward, T. M., Weed Sci. (1969) 17, 417-421.
78. Nearpass, D. C., Soil Sci. Soc. Amer. Proc. (1969) 33, 524-528.
79. Sullivan, J. D., Jr., and Felbeck, G. T., Jr., Soil Sci. (1968) 106, 42-52.
80. Weber, J. B., and Weed, S. B., In "Guenzi, W. D. (ed.) "Pesticides in Soil and Water," pp. 223-256. American Society of Agronomy, Madison, Wisc. (1974).
81. Pierce, R. H., Jr., Olney, C. E., and Felbeck, G. T., Jr., Environ. Lett. (1971) 1, 157-172.

82. Pierce, R. H., Jr., Olney, C. E., and Felbeck, G. T.,Jr.,
    Geochim. et Cosmochim. Acta (1974) 38, 1061-1073.

83. Champion, D. F., and Olsen, S. R., Soil Sci. Soc. Amer.
    Proc. (1971) 35. 887-891.

16

Clay—Pesticide Interactions

JOE L. WHITE

Department of Agronomy, Purdue University, West Lafayette, Ind. 47907

This paper is concerned with an evaluation of the current status of information on clay-pesticide interactions as related to the problem of "bound" or unavailable residues. Attempts will be made to point out deficiencies in our understanding of soil-pesticide interactions which need to be overcome in order to provide a better basis for prediction and understanding of the behavior of pesticides in soils on a long-term basis. In addition, some examples of experimental approaches useful in such studies will be given.

Whether or not a pesticide may pose a problem as a "bound" or unavailable residue depends largely on the nature and extent of the interactions of the pesticide with the soil constituents having high surface areas. Since high specific surface is usually associated with small particle size, the colloidal fraction of the soil will be the controlling factor in interactions between pesticide molecules and the soil.

Several comprehensive reviews of the factors which influence adsorption, desorption and movement of pesticides in soils have been made recently ($\underline{1}$, $\underline{2}$, $\underline{3}$, $\underline{4}$, $\underline{5}$, $\underline{6}$). Mortland ($\underline{7}$) and Theng ($\underline{8}$) have provided detailed treatments of clay-organic complexes and interactions.

Structure and Properties of Clay Minerals

This contribution is concerned with the nature of the interactions between the colloidal mineral fraction of soils and pesticides. The minerals in the colloidal fraction form very stable complexes with components of the soil organic matter. The behavior of this natural clay-organic complex with respect to the adsorption of pesticides has received little attention. The role of organic matter in interactions between pesticides and soils has been discussed by Stevenson ($\underline{9}$).

The clay fraction ($<2\mu$) is composed of crystalline clay minerals, quartz, amorphous silica, and crystalline and amorphous oxides and hydroxides of iron and aluminum. The cation exchange capacity and surface area values for some of the main

inorganic soil constituents are shown in Table I.
Clay minerals also influence the behavior and fate of
pesticides in their role as constituents of sediments (3).
The residues reported in surveys of rivers, lakes, etc. are
usually of insecticides carried on particulate matter suspended
in the water. Keith and Hunt (10) demonstrated that insecticides quickly became partitioned between the water, suspended
material and bottom sediments (Table II).
Although the structure and properties of clay minerals
which occur in soils have been given previously (1, 2, 11), it
is pertinent here to briefly review some general features of
clay mineral structures and to focus attention on certain properties of the clay–water–cation system that are related to the
problem of "bound" or unavailable residues.
The layer lattice silicates may be divided into two main
structural groups on the basis of the ratio of tetrahedral
sheets to octahedral sheets in the unit layer. The kaolin
group is an example of a 1:1 structure, i.e., it consists of
one sheet of tetrahedrally coordinated cations (Si- or Al-O
tetrahedra) with one sheet of octahedrally coordinated cations
(Al^{3+}, Mg^{2+}, Fe^{2+}, etc.). The thickness of a single 1:1 layer
is about 7.2 A. The 1:1 layer silicate group includes kaolinite, dickite, serpentine and halloysite. In general, the 1:1
type layer silicates are electrically neutral or possess a very
low negative charge. With the exception of halloysite, the 1:1
minerals do not swell under normal conditions; the active surface area is therefore limited to external surfaces only. Because of the small surface area and small negative charge possessed by the minerals in this group they interact with pesticides to a very limited extent.
The other type of structure, which is the basic layer of
micas, chlorites, pyrophyllite, talc, vermiculite and the montmorillonite minerals (smectites), is the 2:1 type. This group
consists of two sheets of tetrahedrally coordinated cations
(Si-O or Al-O tetrahedra), with one sheet of octahedrally coordinated cations (Al^{3+}, Mg^{2+}, Fe^{2+}, etc.). The 2:1 layer has
a thickness of slightly less than 10 A. The 2:1 layers often
carry a negative charge due to isomorphous substitutions in
which Si^{4+} in tetrahedral positions is replaced by Al^{3+}, or
Mg^{2+} replaces Al^{3+} in octahedral sites. These negative charges
are balanced by positively-charged ions or groups of atoms
which occur between successive 2:1 layers. The presence of
interlayer cations is associated with the swelling properties
of the 2:1 clay minerals. The expanding 2:1 minerals, such as
montmorillonite and vermiculite, have a high cation exchange
capacity and a high surface area; the interlayer regions can
accommodate water molecules as well as inorganic and organic
cations. Organic molecules may also be adsorbed on the internal surfaces of these minerals.

Table I.   Cation exchange capacity and surface area
          values for clay minerals.

Mineral	Cation Exchange Capacity, meq/100 g	Surface Area, m^2/g
Vermiculite	100 to 150	600 to 800
Montmorillonite	80 to 150	600 to 800
Dioct. Vermiculite	10 to 150	50 to 800
Illite	10 to 40	65 to 100
Chlorite	10 to 40	25 to 40
Kaolinite	3 to 15	7 to 30
Oxides & Hydroxides	2 to 6	100 to 800

(after Bailey and White (1))

Table II. Distribution of pesticides in a lake.

Constituent	No. of Samples	Average Residues (Ranges in parentheses) DDT and Related Compounds
Water $(pp10^{12})(ng/1)$	82	0.62 (0 - 22.0)
Particles (ppm)(µg/g)	33	14.74 (1.8 - 78.0)
Bottom Sediment (ppm)(µg/g)	39	4.44 (0.01 - 94.0)

(from Keith and Hunt (10)).

Many of the 2:1 minerals in soils and sediments are the result of weathering actions on micas such as muscovite and illite. Under moderately intensive weathering conditions these micas may be altered to vermiculite, montmorillonite, or to a chlorite-like mineral with hydroxy aluminum interlayers.

Mention should also be made of attapulgite (palygorskite) and sepiolite. These minerals consist of chains or ribbons of the 2:1 structure joined through oxygen ions and with open channels running parallel to the chains. These minerals cannot expand nor can the channels accommodate pesticide molecules, but because of their corrugated external surfaces they have very high surface areas. They have been used as carriers for pesticides in powder and granule formulations.

Crystalline and amorphous oxides and hydroxides of silica, iron, and aluminum occur in soils as separate phases as well as coatings on surfaces of layer silicates. It appears that surface coatings of silica, aluminum and iron hydrous gels on mineral particles are more common than generally realized. Some of the crystalline materials may have very low surface areas, whereas some of the amorphous materials such as allophane may have large surface areas (500 m^2/gram) and be positively charged ($\underline{12}$).

The swelling clay minerals such as montmorillonite and vermiculite provide very large surface areas for interaction with pesticide molecules. The clay mineral surfaces and associated cations have a very strong attraction for water molecules and the competition between pesticide molecules and water molecules for adsorption sites on the mineral surfaces may determine the degree of adsorption, volatility, biological activity, etc. In addition, the interlayer cations which neutralize the negative charges on the planar surfaces of the clay interact very strongly with the water molecules. Multivalent cations such as Ca^{2+}, Mg^{2+}, Al^{3+}, etc. strongly polarize the water molecules associated with them; this results in a greater degree of dissociation of water in the vicinity of the clay surface than in the bulk water. Protons are thus more accessible at the clay surface than would be indicated by the glass electrode pH measurement on the suspension ($\underline{13}$). This ability of clay minerals to donate protons characterizes them as Bronsted acids. This "surface acidity" or proton donating capacity is a function of moisture content ($\underline{14}$). The range of acidities of some clay minerals measured by Solomon et al. ($\underline{15}$) using Hammett indicators is shown in Table III. The "surface acidity" of the clay minerals may be important in adsorption of basic pesticides as well as in the degradation and detoxification of some of the pesticides.

Adsorption

Adsorption Mechanisms. There are several mechanisms by

Table III. Acid strength of some clay minerals as
           measured using Hammett indicators.

Mineral	H_o (pK_a of strongest sites)
Kaolinite (untreated)	<-8.2
Kaolinite (Calgon-treated)	$+1.5$ to -3.0
Montmorillonite	-5.6 to -8.2
Attapulgite	-8.2
Talc	$+4.0$ to $+3.3$

(from Solomon et al. (15); measured on dry samples
H_2O content $\approx 0\%$).

which organic pesticides may be adsorbed at clay mineral surfaces. These mechanisms may be grouped under the headings of
physical adsorption and chemisorption. Physical adsorption
would include van der Waals forces and hydrogen bonding. Chemisorption would include coordination complexes, ion exchange,
and protonation or charge-transfer. The extent of adsorption
of a pesticide compound depends upon the nature and properties
of the compound itself, the kind of clay mineral present, and
the environment provided. Once adsorbed at a clay mineral surface, a compound may be easily displaced, it may be released
with difficulty, or not at all. The latter two cases are of
special interest from the standpoint of "bound" or unavailable
residues. Since the clay minerals are negatively charged, one
would expect the positively-charged pesticide molecules to be
most strongly retained. This group includes cationic compounds
such as paraquat and diquat as well as basic compounds such as
amitrole and the s-triazines. The degree of retention will
influence persistence, biological availability, and ease of
extraction. For example, the combination of the strong electrostatic forces in addition to van der Waals forces in the
case of adsorption of the cationic herbicides paraquat and
diquat by montmorillonite results in their complete inactivation when present in amounts less than cation exchange capacity of the montmorillonite. This represents the extreme case
of a "bound" residue uniquely attributable to the mineral constituents of a soil.

Biological Activity of Adsorbed Organic Compounds. Adsorption of organic compounds such as proteins (16), organic
phosphorus compounds (17), and antibiotics (18) by montmorillonite resulted in retardation of their chemical and biological
decomposition. In the case of strongly basic antibiotics, adsorption by montmorillonite made the antibiotic biologically
unavailable.
    The biological activity of an adsorbed pesticide is the
ultimate criterion for characterization of the adsorbed compound
as a "bound" or unavailable residue. Weber and Weed (5) have
recently summarized the effects of soil constituents on the
biological activity of pesticides in a very comprehensive manner.
They assumed that the biological activity of pesticides adsorbed
on soil colloids was related to the mechanism by which the compounds are bound and discussed the relationships on the basis
of the chemical properties of the pesticides. They reported
that diquat and paraquat, when adsorbed by montmorillonite in
amounts less than the cation exchange capacity, are biologically unavailable to both microorganisms and plant roots. X-ray
diffraction studies showed the compounds to be adsorbed in the
interlayer region of the montmorillonite. They cited unpublished work of Weber which indicated that ^{14}C-diquat adsorbed
on the internal surfaces of montmorillonite clay in aqueous

soil-nutrient suspensions was not degradable by microorganisms over a 1-year period and the herbicide was extracted in its original form at the end of the experiment. When diquat and paraquat were adsorbed on kaolinite clay having only external surfaces, cucumber seedlings growing in systems containing the herbicide-kaolinite mixture were injured because the bioactivity of the herbicide had not been eliminated.

Re-examination of Adsorption Parameters. Most of the information concerning adsorption and leaching behavior of pesticides has been obtained in experiments of relatively short duration (a few hours or days) and for systems that have water/soil ratios much higher than those possible under field conditions. For the average mineral soil about half of the total volume is pore space; thus, the maximum water content of the soil would be 50% by volume. In addition to the low water/soil ratio which occurs under natural conditions, the surface soil is subjected to diurnal variations in temperature of 20° C. or more as well as fluctuations in moisture content and relative humidity. This suggests that pesticides in the surface one or two cm. of soil are subjected to almost daily wetting and drying cycles. Freezing results in partial dehydration of soils.
       One of the most commonly used techniques for predicting the behavior of pesticides in soils is that of establishing adsorption isotherms. Such adsorption studies usually involve dilute suspensions of soils or colloids and short equilibration times. The amount of pesticide adsorbed is most often based on analysis of the solution phase concentration before and after adsorption. The soil or colloid phase is separated from the solution phase by centrifugation or filtration. The short time period generally used for equilibration is dictated by the need to eliminate or minimize complications arising from hydrolysis, oxidation, microbial decomposition, volatilization, etc., if the adsorption experiment is continued for weeks or months.
       Hamaker and Thompson (19) have previously pointed out the lack of data on long-term adsorption. The work of McGlamery and Slife (20) and Obien (21) points to the existence of a further slow adsorption occurring over weeks and months. Table IV from Obien (21) shows the effect of time on the amount of atrazine adsorbed by soils having various pH values.
       From the standpoint of long-term behavior of pesticides in soils desorption studies are of critical importance. Hamaker and Thompson (19) state that desorption appears to differ from adsorption in being slower and in that a portion of the material is very difficult to remove. Graham-Bryce (22) found for disulfoton that if the soil was dried and rewetted, the desorption isotherm represented a considerably higher proportion of the material retained in the soil. This desorption behavior is consistent with the general experience that it is more difficult to extract an "aged" residue from soil than

Table IV. Changes in percent of atrazine adsorbed as
a function of time, pH and soil type.

Soil	pH	Days	Atrazine Adsorbed %
Kapaa	4.9	0	34
		5	46
		20	48
		60	60
Lualualel	7.9	0	6
		5	7
		20	8
		60	6
Molokai	6.3	0	26
		5	34
		20	37
		60	41
Kaipoipoi	5.4	0	60
		5	70
		20	65
		60	81

(from Obien (21).

freshly incorporated material.

It is obvious that a better understanding of the effects
of time and cyclic variations in temperature and moisture content on adsorption and desorption of pesticides by soils and
soil constituents must be developed before the problem of
"bound" or unavailable residues can be adequately treated.

Surface Chemistry of Adsorbed Pesticides

As indicated above, normal adsorption studies involve
analysis of the equilibrium solution for the pesticide which
has not reacted with the adsorbent. For a better understanding
of the nature of "bound" or unavailable residues we need to
examine that portion of the pesticide which has reacted with
the adsorbent surface. Information on changes in the pesticide
molecule resulting from the interaction with the adsorbent can
provide a basis for predicting ease of protonation, hydrolysis,
degradation, desorption, etc. Such studies might include interactions between plant and animal conjugates and the soil mineral surfaces.

The surface chemistry of the interactions between clay
minerals and pesticides can be studied by infrared spectroscopy for certain favorable cases. Applications of infrared
and related spectroscopic techniques to the study of the surface
chemistry of soil-pesticide interactions include those on EPTC
by Mortland and Meggitt (23), amitrole by Russell et al. (24),
s-triazines by Cruz et al. (25), and malathion by Bowman et al.
(26). Russell et al. (24) demonstrated that amitrole was protonated as a result of the "surface acidity" arising from the
dissociation of water bound to exchangeable cations on montmorillonite. Cruz et al. (25, 27) were able to provide direct
infrared evidence for the state of adsorbed s-triazine species.
In addition, their studies showed that dissociation of adsorbed
water on the clay surface plays a major role in the protonation
and hydrolysis of the s-triazine herbicides. Protonation and
deprotonation with concomitant adsorption and desorption was
demonstrated for propazine by gas phase titrations with NH_3
and HCl in a vacuum infrared cell.

White (28) has recently described a simple infrared
technique for determining the relative ease of protonation and
hydrolysis of s-triazines when adsorbed on calcium-saturated
montmorillonite. Air dry films of calcium-saturated montmorillonite were exposed to chloroform solutions of s-triazines for
14 days. s-Triazines susceptible to protonation were transformed into organic cations and adsorbed into the interlamellar
region of the montmorillonite; protonation was indicated by
a relative decrease in the intensity of bands in the 1530-
1550 cm^{-1} region and a relative increase and slight shift in
the band near 1620 cm^{-1}. Hydrolysis of the s-triazines resulted in the appearance of a band at 1750 cm^{-1} due to the carbonyl

group in the keto form of the hydroxytriazine. The relative
degree of protonation and hydrolysis was measured for twenty-
seven chloro-, bromo-, methoxy-, and methylthio-s-triazines.
Ten of the twenty-seven were not hydrolyzed to a significant
degree; desorption of these s-triazines would be expected to
result in their showing biological activity.

The above example illustrates the kind of information that
can be obtained by infrared techniques concerning the state of
the adsorbed pesticide--the interlamellar position of the adsorbed s-triazine can be established and the biological activity
of the desorbed triazine molecule can be predicted. This type
of information is of considerable significance in terms of
potential problems posed by desorption of strongly held residues.

Bailey and Karickhoff (29) have recently shown that u.v.
spectroscopy makes possible studies of protonation of pesticides in the concentration range characteristic of field applications. They were able to detect trace amounts of indicator in clay (less than 0.1 per cent of CEC for hectorite)
in aqueous systems. The u.v. sensitivity makes it possible
to work within the water solubility limits of many of these
compounds.

Raman spectroscopy offers the potential for obtaining
information on clay-pesticide interactions complementary
to that available from infrared and u.v. spectroscopy.

Abstract

The structure and properties of clay minerals are discussed and related to the retention of pesticides in "bound"
or unavailable forms. A re-examination of adsorption parameters suggests that effects of time and cyclic variations in
temperature and moisture content of soils on adsorption and
desorption reactions must be evaluated before the "bound"
residue problem can be resolved. It is shown that surface
chemistry studies using infrared and u.v. techniques can provide useful information on protonation, hydrolysis, degradation
and biological activity of adsorbed pesticides.

Literature Cited

1. Bailey, G. W., and White, J. L., Residue Rev. (1970) 32, 29.
2. Weber, J. B., Residue Rev. (1970) 32, 93.
3. Weber, J. B., Advan. Chem. Ser. (1972) 111, 55.
4. White, J. L., and Mortland, M. M., Proc. Symp. Pesticides
     Soil: Ecology, Degradation, Movement, 95, Mich. State
     Univ., Feb. 25-27, 1970.
5. Weber, J. B., and Weed, S. B., "Pesticides in Soil and
     Water", 223, Soil Sci. Soc. Amer. Inc., Madison, Wis., 1974.

6. Green, R. E., "Pesticides in Soil and Water", 3 , Soil Sci.
   Soc. Amer. Inc., Madison, Wis., 1974.
7. Mortland, M. M., Advan. Agron. (1970), 22, 75.
8. Theng, B. K. G., "The Chemistry of Clay-Organic Reactions",
   136, Adam Hilger, London, 1974.
9. Stevenson, F. J., Advan. Chem. Ser. (1975)(this volume)
   Amer. Chem. Soc., Washington.
10. Keith, J. O., and Hunt, E. G., Trans. 31st North Amer. Wildl.
    Res. Conf. (1966) 150.
11. Grim, R. E., "Clay Mineralogy", McGraw-Hill, New York, 1968.
12. Aomine, S., and Otsuka, H. Trans. 9th Int. Cong. Soil Sci.
    (1968) 1, 731.
13. Bailey, G. W., White, J. L., and Rothberg, T.,Soil Sci. Soc.
    Amer. Proc. (1968) 32, 222.
14. Mortland, M. M., Tran. 9th Int. Cong. Soil Sci. (1968) 1, 691.
15. Solomon, D. H., Swift, J. D., and Murphy, A. J., J. Macromol.
    Sci. Chem. (1971) A5, 587.
16. Ensminger, L. E., and Gieseking, J. E., Soil Sci. (1942)
    53, 205.
17. Mortland, M. M., and Gieseking, J. E., Soil Sci. Soc. Amer.
    Proc. (1952) 16, 10.
18. Pinck, L. A., Clays Clay Miner. (1960) 9, 520.
19. Hamaker, J. W., and Thompson, J. M., "Organic Chemicals in
    the Soil Environment", 1, 49, C. I. Goring and J. W.
    Hamaker, ed., Marcel Dekker, Inc., New York, 1972.
20. McGlamery, M. D., and Slife, F. W., Weeds (1966) 14, 237.
21. Obien, S. R., Ph.D. Thesis, Univ. of Hawaii, Honolulu, 1970.
22. Graham-Bryce, I. J., J. Sci. Food Agr. (1967) 18, 72.
23. Mortland, M. M., and Megitt, W. F., J. Agr. Food Chem.
    (1966) 14, 126.
24. Russell, J. D., Cruz, M., and White, J. L., J. Agr. Food
    Chem. (1968) 16, 21.
25. Cruz, M., White, J. L., and Russell, J. D., Israel J. Chem.
    (1968) 6, 315.
26. Bowman, B. T., Adams, R. S., Jr., and Fenton, S. W., J. Agr.
    Food Chem. (1970) 18, 723.
27. Cruz, M., and White, J. L., Environ. Qual. Safety (1972)
    1, 221.
28. White, J. L., Arch. Environ. Contam. Toxicol. (1975)
    (in press).
29. Bailey, G. W., and Karickhoff, S. W., Clays Clay Miner.
    (1973) 21, 471.

Turnover of Pesticide Residues in Soil

JOHN W. HAMAKER and CLEVE A. I. GORING

Dow Chemical U.S.A., Ag-Organics Department, Walnut Creek, Calif.

Turnover of material is a matter of rates, whether in a soil, a living tissue, a lake or even the entire earth. The significant quantities such as the amount of material, residence time or rate of change of the material depend upon rates at which materials enter and leave the system. "Leaving," of course, includes decomposition.

In the case of pesticides added to soil, it is useful to distinguish between material that is available and that which is not immediately available to plants, animals and microorganisms in the soil. This will include the bound residue according to the definition suggested for this conference: residue remaining after exhaustive extraction. It will, however, also include other material which can be extracted but will not be immediately available to plants, soil animals or microorganisms. This material may be held as an unchanged molecule or have reacted with soil organic matter and become attached by a chemical bond, but the bonding is considered to be slowly reversible under the appropriate soil conditions.

The unavailable chemical will tend to resist extraction to varying degrees, not only by soil water but also by other solvents, depending on the efficiency of the solvent and conditions of extraction. For this reason, any definition of "bound" residue based on a defined solvent extraction system is arbitrary and not necessarily well correlated with lack of availability to plants, animals, and microorganisms. The situation is conceptually similar to the problem of quantifying "bound" and available phosphate residues in soil, except that losses of phosphate from the reservoir rarely occur, whereas pesticides and/or their metabolites are lost by leaching, volatilization, and conversion to CO_2, H_2O and mineral elements. These considerations lead

to a model which is graphically represented in Figure 1.
In this model, the pesticide in the soil solution as
well as that adsorbed on the soil surface are considered part of a single pool of mobile and labile
chemical. Material in this pool is subject to chemical
transformations and to movement, i.e., leaching and
volatilization. On the other hand, pesticide in the
unavailable condition is considered to be immobile and
non-labile. It is held in the soil in such a way that
it is protected from chemical degradation. Movement in
and out of this compartment is slow compared to the
rapid exchange between dissolved and adsorbed material
in the labile compartment.

Decomposition is related to the labile pool as a
whole rather than to the adsorbed or to the dissolved
material. It is probable that decomposition takes
place at catalytic sorption sites such as portions of
the mineral soil surfaces, enzymes deposited on the
soil, or enzymes in microorganisms. However, because
of the rapid exchange between adsorbed and dissolved
material, all the material in the pool is available for
chemical degradation, whether it is in the solution or
is on the solid surface. Even material that is
adsorbed on non-reactive portions of the soil will
reach active sites through desorption and diffusion.
The rate of degradation for the labile pool is, therefore, considered as if it were a single reaction, even
though it is internally complex.

The state of the chemical in the unavailable
condition is not characterized beyond being immobile,
resistant to chemical attack, and slowly exchanged with
chemical in the labile condition. The detailed mechanism of binding, i.e., type and strength of the bonds,
is not specified because this information is not
generally available for soil. Soil is a heterogeneous
medium and must have a variety of sorption sites of
different types and bond strength for any given
chemical, but we cannot specify the specific nature of
the soil adsorptive surface with the present state of
knowledge in this area. Nor do we understand the ways
in which pesticides may temporarily become integrated
into the soil organic matter.

This model assumes that the bound material can be
adequately represented for mathematical purposes as a
single compartment with average behavior for the
unavailable material. It is recognized that portions
of the material will vary from the average availability
to the labile pool, depending on the way they are bound.
In some cases, part of the material could be sufficiently more tightly bound as to require a second
compartment to describe it. The model shown in Figure 1
is only the simplest member of a family of models.

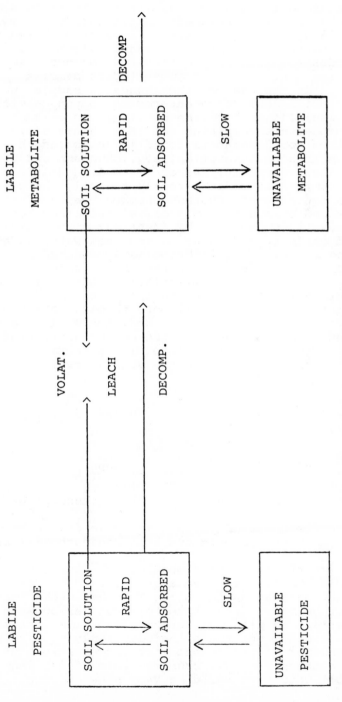

Figure 1.   Compartment model for soil decomposition

This proposed model can qualitatively explain many of the phenomena related to soil residues. For example, Saha, et al ([1],[2]) has shown that dieldrin is easily extracted from soil immediately after addition, but more drastic methods are required to remove it from soil samples that have been incubated in a moist condition. In the case of the incubated sample, part of the material would have entered the bound chemical pool and would, therefore, be difficult to extract.

The model also explains the fact that a different distribution coefficient is found if approached by desorption than be adsorption, e.g., Swanson and Dutt ([3]), Saltzman, et al ([4]) and others. If a dilute solution of chemical is added directly to soil, the soil initially contains no adsorbed chemical and equilibrium is approached by adsorption. If, on the other hand, the soil is first "loaded" by treatment with a small volume of concentrated solution and then water added to dilute the solution, equilibrium is approached by desorption from the soil. In the latter case, more chemical remains on the soil relative to that in the solution than if equilibrium is approached by adsorption. In terms of the model, more chemical has been transferred to the unavailable condition while the soil was loaded with chemical from the more concentrated solution. Thus, the ratio of total concentration in the soil to concentration in the soil water is also greater.

$$(C_{Bound})_{Desorption} > (C_{Bound})_{Adsorption}$$

Therefore:

$$\left(\frac{C_{Ads} + C_{Bound}}{C_{Solution}}\right)_{Desorption} > \left(\frac{C_{Ads} + C_{Bound}}{C_{Solution}}\right)_{Adsorption}$$

This effect is observed quite quickly (a few hours or days) so the bound compartment must begin to fill right away.

The importance of this model to the question of turnover is that it can give a quantitative understanding of the degradation in soil. For this, some assumptions must be made of the rates of transfer between the compartments. We plan to concentrate on the behavior of the pesticide, but the principles apply equally well to the behavior of its metabolites and even to bound residue, although in this case the rate of leaving the unavailable condition would be very slow.

In applying this model to the rate of disappearance of chemicals from soil, it is assumed that the rates of decomposition and of transfer are first-order. Thus k, k_1 and k_{-1} in the following diagram are first-order rate constants applying to the concentrations of labile and unavailable material: c_1 and c_2. Also shown are the pair of simultaneous differential equations derived from the model and the first-order rate constants. Solution of these equations is given in Appendix 1.

Rate Model for Soil Decomposition

$$c_1 \quad \boxed{\text{Labile}} \quad \xrightarrow{\quad k \quad}$$

$$k_{-1} \Big\uparrow \Big\downarrow k_1$$

$$c_2 \quad \boxed{\text{Unavailable}}$$

$$dc_1/dt = -(k + k_1)c_1 + k_{-1}c_2$$
$$dc_2/dt = k_1 c_1 - k_{-1}c_2$$

The idea that soil decomposition should be a first-order reaction was early proposed by Freed and others. Their reasoning was that, since the soil was present in large excess, the rate would depend only upon the concentration of the chemical. This should be true even if the reaction were biological or chemical in nature or if it were catalyzed on the soil surface, since all of these cases approach first-order for very low concentrations. Unfortunately, for this idea, however, many of the observed soil decompositions were either clearly not first-order, or only approximately so, and then only for limited time periods. In most cases, the decomposition reaction behaves as if its rate depended upon some higher power of the concentration of the pesticide, i.e., greater than first-order. Many examples could be cited, but the extremes are illustrated by Figures 2 and 3 from the work of Zimdahl, et al and Wolfe, et al. In these graphs, it is important to realize that they are semi-logarithmic--that is, the pesticide concentration is plotted as a logarithm. On such a graph, a first-order degradation will be a straight line, while higher orders will give curves that are concave upward. In the cases of atrazine and simazine, there is a suggestion of an upward concavity, but with parathion there is no doubt that rate of degradation deviates from first-order. It is probably quite significant that parathion is much more strongly adsorbed by soil than either atrazine or simazine.

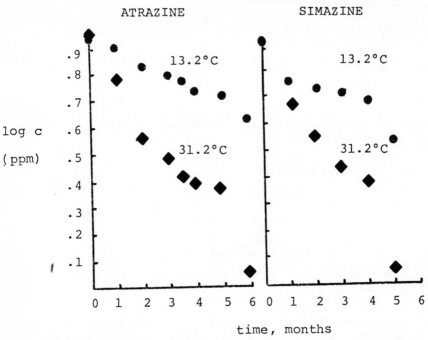

Figure 2. Rate of degradation of atrazine and simazine applied to soil at 8 ppmw (Zimdahl, Freed, Montgomery, and Fertick, Weed Research (1970) 10, 18–26)

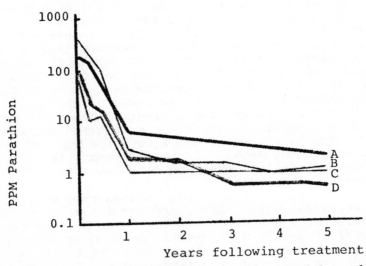

Figure 3. Disappearance of parathion from top 1 in. of sandy loam soil following topical applications of two concentrations using two formulations (Wolfe, Staiff, Armstrong, and Comer, Bull. Environ. Contam. and Toxicol. (1973) 10, 1–9)

The present model for bound material explains the behavior very nicely as due to the slow formation of a "pool" of bound material that does not participate in the degradation process. Thus, when the chemical is first added to the soil, it is all available for degradation, which, in many cases, is shown by a rapid initial degradation. However, as time passes, chemical is also passing into the bound residue compartment where it is not available for degradation. Because of this, the rate of degradation will be less than would be expected from the total chemical left in the soil. Eventually, however, the system will reach a steady-state condition and the decomposition will proceed at a rate that is determined by the reservoir of mobile material maintained by the reservoir of bound material.

The precise shape of the disappearance curve will depend upon the rate laws and rate constants that control the entrance of material into the unavailable residue compartment and its departure from the compartment. Mathematically, there is, of course, an inexhaustible range of choices for these rate laws and rate constants, but in this treatment, the first-order rate law has been chosen to represent transfer to and from the unavailable residue compartment. To some extent, the choice was arbitrary and motivated by the simplicity of the mathematics associated with first-order kinetics. On the other hand, the low pesticide concentration and large excess of soil make it probable that these processes can be represented by first-order kinetics.

One approach to the best kinetics for the model would be through understanding the mechanism by which chemical becomes unavailable for degradation. Many possibilities exist. One can picture, for example, the presence of adsorption sites on the soil surface that are partially blocked so that molecules have to over-come a potential barrier to reach one such site. Once in the site, the molecule would have even more diffi-culty leaving, since now it must also break loose from the adsorption bonds. The other possibility is that the bound material may have become "lost" in cracks, crevices or gels in the soil. In this case, the rate of entry or exit would essentially be controlled by diffusion, but the distribution might also depend upon bonding within the solid phase, i.e., solubility. Analysis of these two possibilities for the limiting case of low pesticide concentrations indicates that in both cases, transfer can be represented by first-order kinetics so they are not distinguishable in this way.

A least squares fit of the quantitative model was made to disappearance curves of the herbicidal chemical, triclopyr (2,3,5-trichloro-2-pyridyloxyacetic acid), in two soils (5). In judging the fit of these curves, shown in Figures 4 and 5, allowance should be made for the logarithmic vertical scale which makes differences appear larger for lower concentrations. A clearer idea can be obtained from the comparison of observed and calculated values in Tables 1 and 2. Inspection will show that percentage differences are more nearly the same than absolute differences. The percentage differences are approximately what is thought to be the experimental error for the California soil but a little higher for the Illinois soil. The reason for the larger difference in the 250-day point is not known.

Of particular interest is the fourth column of the tables, which shows the calculated concentration in the labile and bound pools. This indicates a larger bound residue relative to the labile pool for the Illinois soil than for the California soil. This may be related to the higher organic carbon content of the Illinois soil: 4.2% versus 0.8%.

The three parameters, k, k_1 and k_{-1}, have physical implications, but these should not be pursued without some idea of the uncertainty for values for the k's. Quite often in non-linear systems, one or more of the parameters will be very poorly determined by the data. However, the discussion in Appendix 2 suggests that all three constants have sufficient precision to permit qualitative discussion, at least. The constant, k, is determined by all the data, while k_1 and k_{-1} are largely determined by experimental points in the steady-state portion of the curve, i.e., the second straight-line portion. The constant, k, which represents the rate of degradation of labile material, is larger for the Illinois soil, and triclopyr disappears more rapidly at first in that soil. However, for the Illinois soil, the constant, k_{-1} (rate of leaving the bound state), is smaller than k_1 (rate of entering the bound state), so more material will accumulate in the bound reservoir and not be available for decomposition. In spite of the more rapid decomposition rate, decomposition in the Illinois soil is eventually slower than in the California soil. The value of k_{-1} is larger than k_1 in the California soil, and it might at first be thought that the reservoir would empty thereby. Actually, the ratio, k_1/k_{-1}, determines the relative sizes of the labile and bound reservoirs: If k_{-1} is larger, the labile reservoir would be smaller than the bound reservoir if there were equilibrium.

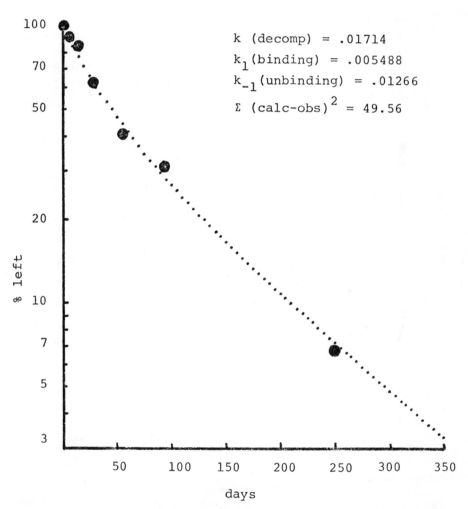

Figure 4. Best fit for disappearance of 1 ppmw triclopyr from a California soil at 35°C, 1/3 bar moisture

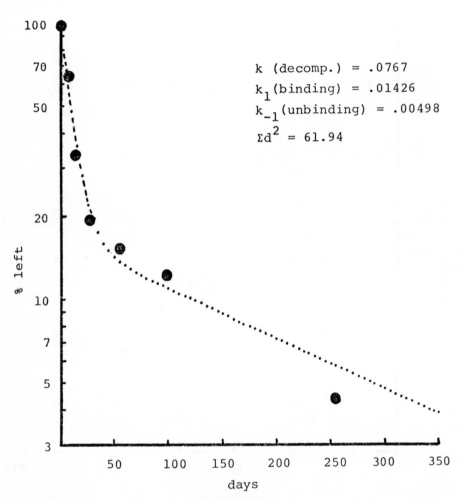

Figure 5. *Best fit for disappearance of 1 ppmw triclopyr from an Illinois soil at 35°C and 1/3 bar moisture*

TABLE I - Calculated Versus Observed Values for
          Triclopyr in a California Soil[1] (Davis)
          at 35°C and 100% of 1/3 Bar Moisture

Time, Days	Observed Conc. (% of Appl.)	Calc. Conc. (% of Appl.)	Calc. Labile + Conc. (% of Appl.)	Calc. Bound Conc. (% of Appl.)	% Difference (Calc.-Obs.) Obs.
0	99.2	99.2	(99.2 +	0)	--
7	88.9	88.9	(84.8 +	3.4)	-0.78
14	84.4	78.8	(72.8 +	6.0)	-6.63
28	62.2	63.6	(54.2 +	9.4)	2.25
56	40.5	43.6	(31.7 +	11.9)	7.65
93	30.9	28.6	(17.4 +	11.2)	-7.44
249	6.8	7.3	(3.4 +	3.8)	7.35
Average					0.4
Average					5.35%[2]

1/ A Yolo County loam containing 0.8% organic carbon
   and pH 6.5.

2/ Average of the absolute values.

TABLE II - Calculated Versus Observed Concentration
for Triclopyr in an Illinois[1] Soil at
35°C and 100% 1/3 Bar Moisture

Time, Days	Observed Conc. (% of Appl.)	Calc. Conc. (% of Appl.)	Calc. Labile Conc. (% of Appl.) +	Calc. Bound Conc. (% of Appl.)	% Difference (Calc-Obs) / Obs.
0	97.2	97.2	(97.2 +	0)	--
7	63.6	58.6	(51.5 +	7.0)	-7.86
14	33.2	38.0	(27.5 +	10.5)	-14.7
28	19.3	21.0	(8.2 +	12.9)	9.32
56	16.3	13.7	(1.28 +	12.4)	-16.0
100	12.3	11.0	(0.61 +	10.4)	-11.8
255	4.4	5.78	(0.31 +	5.47)	24.1
Average					2.08
Average[2]					14.0

1/ Flanagan Silty Clay Loam, 4.2% organic matter and pH 5.8.

2/ Average of the absolute values.

Actually, there is not equilibrium, since decomposition
keeps the labile reservoir small, and for the California soil, the two reservoirs end up about the same
size at the steady state.

A more complete picture of the features of this
model are shown in Figure 6, which shows a family of
curves based on a single first-order constant:
$k = 0.0152$ or half-life = 45.6 days. The straight line
on the graph represents the degradation from this source
only with no bound residue--$k_1 = k_{-1} = 0$. As finite
values are assigned to the constants, k_1 and k_{-1}, the
curves are seen to deviate from the first-order pattern.
The extent of deviation increases as the ratio, k_1/k_{-1},
increases, which is entirely reasonable since the more
slowly the chemical leaves the bound condition (k_{-1}
decreasing) relative to the speed at which it enters
(controlled by k_1), the larger the bound residue there
would be relative to the labile pool. The size of k
compared to k_1 will determine the proportion of the
material in the labile pool that decomposes to that
which passes into the bound condition. In any case,
it would appear that a range of degradation curves from
the slight curvature of the triazines to the sharp
change in direction with parathion can be fitted by the
appropriate selection of values for the three constants
of the system: k, k_1 and k_{-1}.

For pesticides, an important special case of turnover is the accumulation of residues from repeated
addition, often annually. This model can be used to
estimate this residue by repeated calculations. Results
of such a calculation are shown in Figures 7 and 8 for
the two cases of triclopyr. They indicate no significant accumulation of residue and show a rapid disappearance of chemical once the annual addition ceases.

A more complete picture of accumulation behavior
can be seen in Tables 3 and 4. In both tables, the
constants are adjusted so decomposition in 1 year is
90% (essentially no significant accumulation if unavailable residues do not occur). In Table 3, a value of
0.006308 is assigned to k_{-1}, which controls the rate at
which material leaves the unavailable condition. This
is between the values for k_{-1} in the two triclopyr
cases. The size of the available residue is varied by
changing the binding rate constant, k_1, and the
condition of 10 percent left after the first year, met
by adjusting the decomposition rate constant, k. The
interesting feature of this table is that the steady-
state accumulation is mainly controlled by the size of
k_{-1}. In Table 4 where k_{-1} is varied, it becomes
severely limiting as in the third case, where the

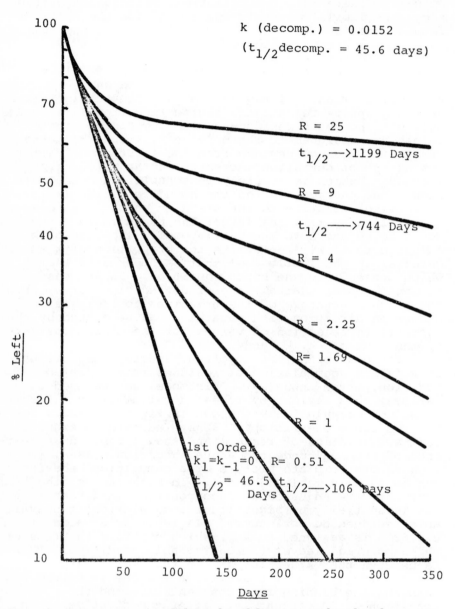

Figure 6.   Disappearance of total chemical for different sizes of bound residue reservoir, i.e., k_1 (binding)/k_{-1} (unbinding) = R

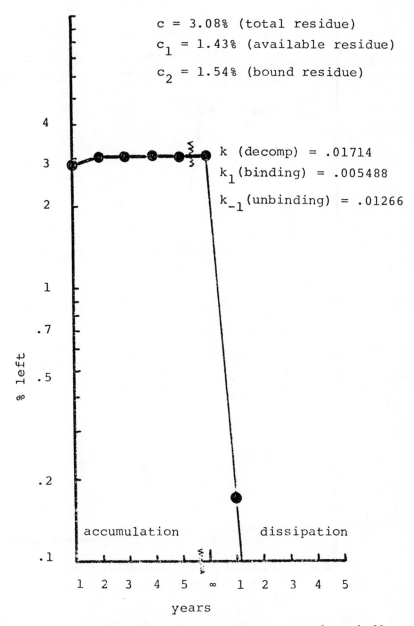

Figure 7.   *Accumulation and dissipation of residue from repeated annual addition of 1 ppmw triclopyr to a California soil at 35°C and 1/3 bar moisture*

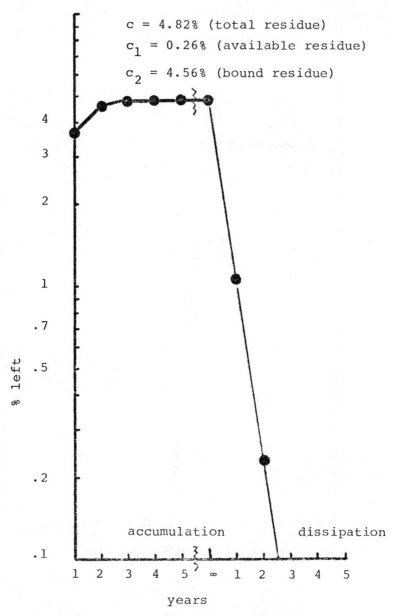

Figure 8. Accumulation and dissipation of residue from repeated annual addition of 1 ppmw triclopyr to an Illinois soil at 35°C and 1/3 bar moisture

TABLE III - Accumulation and Dissipation of Residues from Annual Addition of Chemical. Constant Conditions (1) 10% Remaining After First Year; (2) k_{-1} = 0.006308.

k	$0.006308^{a/}$	$0.007260^{b/}$	$0.0152^{c/}$	$0.0818^{d/}$
k_1	0	0.000789	0.006308	0.0505
k_{-1}	0	0.006308	0.006308	0.006308
k_1/k_{-1}	--	1/8	1	8
1 Yr	10.0	10.0	10.0	10.0
2 Yr	11.0	11.5	12.3	12.5
∞	11.11	11.8 (8.8,3.0)*	13.0 (3.5,9.5)*	13.3 (.6,12.7)*

Clearing:

1 Yr	1.11	1.83	3.0	3.3
2 Yr	0.11	0.31	0.69	0.81
3 Yr	0.011	0.05	0.15	0.20

* Percent remaining (% available; % unavailable).

$\underline{a}/(t_{1/2}=110d)$; $\underline{b}/(t_{1/2}= 95d)$; $\underline{c}/(t_{1/2}= 45d)$; $\underline{d}/(t_{1/2}=8.5d)$

TABLE IV - Accumulation and Dissipation of Residues
from Annual Additions of Chemical to Soil:
Ten Percent Remaining after First Year and
k_1 ("Binding" Rate Constant) = 0.006308

k 0.007156 ($t_{1/2}$=96d)	0.0152 ($t_{1/2}$=46d)	0.0442 ($t_{1/2}$=15.7d)
k_1 0.006308	0.006308	0.006308
k_{-1} 0.05046	0.006308	0.000789
k_1/k_{-1} 1/8	1	8
1 Yr 10.0	10.0	10.0
2 Yr 11.0	12.3	17.8
3 Yr 11.1	12.8	23.8
∞ *11.12(1.77,1.38)	*12.99(3.46,9.53)*	45.0(0.70,44.30)
1 Yr 1.13	2.98	21.2
2 Yr 0.12	0.68	16.5
3 Yr 0.012	0.016	12.8
		<1.0% in 16 Yrs
		<0.10% in 25 Yrs

*Total Percent (% available; % unavailable)

accumulated residue rises slowly to the relatively
large value of 45 percent. Nevertheless, even this
extreme does not represent an excessive accumulation.
The range of values for k_1/k_{-1} (1/8 to 8/1) exceeds
the values observed in the two triclopyr cases (1/2.3
to 2.9). In Table 5, the effect of increasing the
amount left after a year is explored. As would be
expected, the accumulation is greater, as for example
with 25 percent left after the year, the estimated
accumulated residue at the end of the season, 1.4 times
the annual dose. In the case of 80 percent left after
1 year, a level of 16 times the annual dose is predicted. Obviously, pesticides that degrade slowly will
give more accumulation, but the model suggests that
excessive accumulation is not likely to be a problem
for less persistent pesticides. It is recognized that
these predictions are extrapolations and need to be
confirmed experimentally.

The biological effects of these unavailable
residues (including the bound portion) is of primary
concern for pesticides.

If these and other cases are examined, it will be
seen that where annual additions produce an accumulation much greater than the residue remaining after 1
year, most of it is in the bound compartment and very
little immediately available for degradation, leaching
or plant uptake. Moreover, the degradation rate is
relatively large to compensate for the small fraction
of the material available for degradation. For
example, in the third case in Table 2, over 98 percent
of the steady-state residue is in the unavailable
condition, and the degradation rate constant, k, or
0.0442 represents a half-life of 15.7 days. Thus, we
would predict the following for the unavailable
residue reservoir:
1. This residue is released gradually to the labile
compartment where it will be available to plant roots
or soil animals.
2. Any abnormally large release of unavailable
material will be decomposed rapidly so that it would
have only a relatively transient effect.
From these considerations, it does not seem likely that
residues from repeat applications will present the
problem suggested by untrammeled imagination.

It would be appropriate to comment on the treatment of metabolites. Assuming that this same model
applies to both the parent material and the metabolites,
a more complex set of differential equations would
result, but it would still be manageable with the use
of an adequate computer. The residue of metabolites

TABLE V - Accumulation of Residues from Annual Additions of Chemical to Soil: k_1 (Binding Rate Constant) = 0.006308 and 25% or 80% Left after First Year

	0.0075 ($t_{1/2}$≅92d)	0.01542 ($t_{1/2}$≅45d)	0.001030 ($t_{1/2}$≅673d)	0.00152 ($t_{1/2}$≅456 da)
k				
k_1	0.006308	0.006308	0.006308	0.006308
k_{-1}	0.006308	0.0007885	0.006308	0.0007885
k_1/k_{-1}	1	8	1	8
1 Yr	25.0 [a]	25.0	80.0 [a]	80.0
2 Yr	34.3	45.4	146.8	155.3
3 Yr	37.7	62.1	197.6	201.4
∞	*39.8 (14.5, 25.2)	*136.5 (4.5, 131.6)	*484.7 (232.9, 251.8)	1,575 (151, 1424)
∞/1 Yr	1.58	5.46	6.05	19.7

[a] 25% after 1 yr eq. to first-order half-life of 0.5 yr.
80% after 1 yr eq. to first-order half-life of 3.1 yr.
*Total percent (percent available, percent unavailable)

will depend upon the relative rate constants, but since they are not added all at once, metabolites will be distributed between the bound and unbound condition more nearly at the steady-state proportion.

Literature Cited

1. Saha, J. G., B. Bhavaraju and Y. W. Lee. Validity of Using Soil Fortification with Dieldrin to Measure Solvent Extraction Efficiency. J. Agr. Food Chem. (1969) 17:874-876.
2. Saha, J. G., B. Bhavaraju, Y. W. Lee and R. L. Randell. Factors Affecting Extraction of Dieldrin-^{14}C from Soil. J. Agr. Food Chem. (1969). 17:877-882.
3. Swanson, R. A. and G. R. Dutt. Chemical and Physical Processes That Affect Atrazine and Distribution in Soil Systems. Soil Sci. Soc. Amer. Proc. (1973) 37:872-876.
4. Salzman, S., L. Kliger and B. Yaron. Adsorption-Desorption of Parathion as Affected by Soil Organic Matter. J. Agr. Food Chem. (1972) 20:1224-1226.
5. Regoli, A. J. and D. A. Laskowski. Unpublished data of The Dow Chemical Company, Walnut Creek, California and Midland, Michigan (1974).

Abstract

A model is proposed for decomposition and movement of pesticides in soil. For this model, the soil residue is partitioned into two reservoirs: a mobile and labile fraction and a fraction that is immobile and resists degradation. The degradation of chemical depends, therefore, upon the fraction of the chemical in the labile condition. The model was made quantitative by assigning first-order rate constants, k, k_1 and k_{-1}, respectively, to the rates of degradation, of entry into the bound condition and of leaving the bound condition. The resulting system of differential equations was solved and degradation curves calculated for a range of values for the three rate constants. The shapes of these calculated curves corresponded well with the experimental curves very often observed. Degradation curves for the herbicide, triclopyr, were fitted to the model by least squares with satisfactory precision.

The accumulation of soil residues from repeated annual additions was investigated mathematically for 90 percent degradation in a year and different relative

sizes of the mobile and immobile reservoirs. It was
concluded that the accumulation would not be excessive
(<2X first-year residue) and that any accumulation
would be compensated for by a large degradation rate
which would quickly degrade any material released from
the unavailable state.

APPENDIX 1

A Solution of the Differential Equations

This system of differential equations,

$$dc_1/dt = -(k_1 + k)c_1 + k_{-1}c_2,$$

can be solved by substituting

$$c_1 = Ae^{mt} \text{ and } c_2 = Be^{mt} (dc_1/dt = mAe^{mt} \text{ and } dc_2/dt = mBe^{mt}).$$

This gives the simultaneous equations,

$$(m + k_1 + k)A = k_{-1}B$$
$$k_1A = (m + k_{-1})B$$

from which A and B can be eliminated. The resulting
quadratic yields two values for m:

$$m_1 = (-(k + k_1 + k_{-1}) + \sqrt{X})/2$$
$$m_2 = (-(k + k_1 + k_{-1}) - \sqrt{X})/2$$

where $X = k^2 + k_1^2 + k_{-1}^2 + 2kk_1 = 2kk_{-1} + 2k_1k_{-1}.$

Solving for A and B by using these two values for m
gives the following general solution:

$$c_1 = A_1e^{m_1t} + A_2e^{m_2t}$$
$$c_2 = A_1Z_1e^{m_1t} + A_2Z_2e^{m_2t}$$

where: $$Z_1 = (k + k_1 - k_{-1} + \sqrt{X})/2k_{-1},$$
$$Z_2 = (k + k_1 - k_{-1} - \sqrt{X})/2k_{-1}$$

and since for t = 0, $(c_1)_o = A_1 + A_2$ and $(c_2)_o = A_1Z_1 + A_2Z_2,$

$$A_1 = ((c_1)_o(-(k + k_1 - k_{-1}) + \sqrt{X}) + 2(c_2)_ok_{-1})/2\sqrt{X}$$
$$A_2 = -((c_1)_o(-(k + k_1 - k_{-1}) - \sqrt{X}) - 2(c_2)_ok_{-1})/2\sqrt{X}$$

Non-linear curve-fitting methods are used to find the
values of k, k_1 and k_{-1} which best fit a set of data
for disappearance of a chemical. These are essentially
repetitious trial-and-error procedures based on initial

estimates. It is possible, however, to estimate k, k_1 and k_{-1} directly from the curve using three measurements: initial slope, steady-state slope, and steady-state intercept. These are indicated on the accompanying sketch:

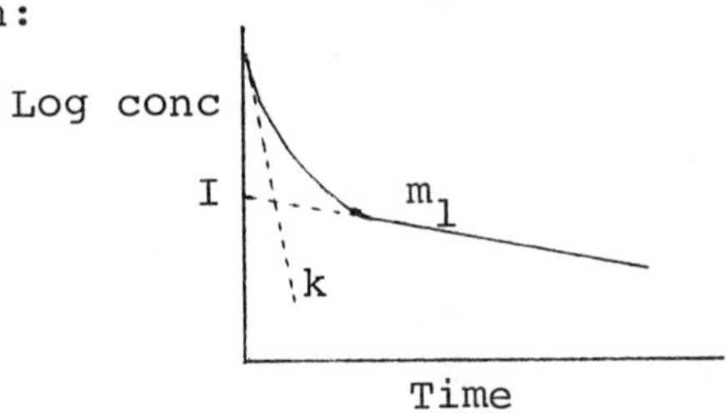

$$k = \ln (c_o/c)/t = \text{initial slope for the semi-log plot;}$$
$$m_1 = \ln (c_{final}/c_{initial})/(t_{final} - t_{initial}) = \text{slope}$$
of the steady-state line; I = intercept of the steady-state line (as conc.).

The quantities can be read off the curve with reasonable accuracy for good data. That the initial slope on the semi-log plots is, in fact, the rate constant, k, can be seen by adding the rate equations:

$$dc/dt = dc_1/dt + dc_2/dt = kc_1,$$

where c_1 is the labile chemical. However, when the chemical is first added to the soil, it is all in the labile or available condition, so c_2, the unavailable chemical, is zero and $c_1 = c$. This can be represented mathematically as follows:

$$(dc/dt)_{initial} = k(c)_o$$

The quantities, m_1 and I, are related to the general solution as follows: If m_1 and m_2 are examined, it will be noted that m_2 is always less than m_1 (because the $\sqrt{x}$ is subtracted). This means that as t increases, the second term becomes smaller in relation to the first and for sufficiently long times the equations become:

$$(c_1)_{t_{large}} = A_1 e^{m_1 t}$$
$$(c_2)_{t_{large}} = A_1 z_1 e^{m_1 t}$$
$$(c)_{t_{large}} = (c_1 + c_2)_{t_{large}} = A_1 (1 + z_1) e^{m_1 t}$$

And taking logarithms:
$$\ln (c)_{t_{large}} = \ln A_1 (1 + z_1) + m_1 t.$$

The last equation describes the steady-state line (i.e., the large times) for the semi-log plot (log c vs. time) where m_1 is the slope and A_1 $(1 + Z_1)$ is the intercept, I.

Since m_1, A_1 and Z_1 are defined in terms of k, k_1 and k_{-1}, the k's can be solved for by straightforward though rather tedious algebraic manipulation with the following results:

$$k_1 = -m_1 - k - (1 + k/m_1)k_{-1} \text{ and}$$
$$k_{-1} = (-b \overset{+}{-} \sqrt{b^2 - 4ac})/2a$$

where $a = (R^2-1)k_2^2/m_1^2$; $b = -4k_2^2/m - 2(R^2 + 1)k$;
$c = (R^2-1)m_1^2 - 4m_1k - 4k^2$; $R = 2I/(c_1)_o - 1$.

These relationships were applied to the triclopyr curves in Figures 3 and 4 with the following results:

1.   Figure 3 - Davis Soil

m_1 = ln (3.3/55)/350 = -0.008038; R = 2X55/99.2 - 1
      = -0.10887
I = 55.0; k_2= 0.017137; $(c_1)_o$ = 99.2
a = $(R^2-1)k_2^2/m_1^2$ = -4.49119
b = $4k^2/m - 2(R^2+1)k$ = 0.111458209
c = $(R^2 -1)m_1^2 - 4m_1k - 4k^2$ = -0.000687545
k_{-1}= $(-b + \sqrt{b^2- 4ac})/2a$ = 0.01147 (optimum k_{-1}=0.01266)
      $(-b - \sqrt{b^2- 4ac})/2a$ discarded)
k_1= $m_1 - k - (1 + k/m_1)k_{-1}$ = 0.003882 (optimum k_1 = 0.005488)

2.   Figure 4 - Illinois Soil

m_1= ln(3.9/17)/350 = -0.004206; R_1 = -0.650206
I= 17 k_{-1} = $(-b + \sqrt{b^2 - 4ac})/2a$
k= 0.0767   = 0.005049 (optimum k_{-1}= 0.00498)
$(c_1)_o$=97.2 k_1 = 0.01452 (optimum k_1=0.01426)
a = -191.9208; b = 5.3760; c = -0.022251

These comparisons are somewhat biased because the optimum value of k was used rather than a visual estimate, and the calculated curve was used to determine I and m_1. Greater difference could be expected for curves fitted to the data by inspection.

This procedure for estimating values of k, k_1 and k_{-1} from the disappearance curve not only are useful to give initial estimates for non-linear curve-fitting, but would also help the investigator for whom non-linear curve-fitting is not accessible. It should be

kept in mind, however, that (1) the results are no
better than the estimates of m, I and k and (2) the
quadratic solution may involve small differences from
large numbers and, therefore, be inherently inaccurate.
The latter expressed the fact that from some data k_1
and k_{-1} is poorly determined.

18

Microbial Synthesis of Humic Materials

K. HAIDER

Institut für Biochemie des Bodens der Forschungsanstalt für Landwirtschaft,
Braunschweig-Völkenrode, West Germany

Humus is primarily derived from higher plants during
the microbial decomposition of the original plant constituents, and from new substances synthesized by the
soil microorganisms. Exact concepts about the chemical
structure of humic compounds and how the essential constituents are linked, are not yet available. However,
it seems reasonable to say, that during humus formation reactive compounds occur which are linked or complexed together, either through microbial activity or
chemical reactions. These reactive constituents are
abundant and may be derived from many different sources. But generally, they are either formed by transformation of plant constituents, such as simple or more
condensed phenols, or are synthesized by microorganisms. The first type involves the transformation of
lignin and other plant phenolic constituents. The second type of reaction involves intracellular microbial
transformation of carbohydrates and other aliphatic
organic substanes. These compounds are rapidly used as
a source of energy and the synthesis of cell tissues,
but parts of it are transformed by secondary metabolism into phenols, quinones, or other aromatic substances. These compounds then react oxidatively with
peptides or other nitrogen containing cell constituents to form dark colored melanins inside or outside
the cells. These melanins are sometime soluble in dilute NaOH and are similar to soil humic acids. Sometimes melanins are incorporated into cells or spores
and protect them by being attacked and degraded by
other microorganisms (1).

Theories were forwarded that soil humus is mostly
composed of the so called lignin-protein or lignin-
ammonia complexes which should be responsible for the
biological stability and other properties of soil
organic matter (2, 3). But attempts to isolate lignin-

protein complexes from soil have shown that these complexes, if present in soil, occur in very small quantity (4, 5). Moreover there is evidence that soil organic matter does not contain very much material which can be directly traced back to lignin. Nitrobenzene oxidation delivers only trace amounts of phenolic aldehydes as compared to lignin (6, 7, 8). Also other degradative procedures, such as permanganate oxidation (9) or reductive Na-amalgam degradation (10, 11, 12) deliver only small amounts of typical lignin-derived compounds. Therefore the structure of lignin is essentially altered if it is converted into humic materials.

Only attempts can now be made to describe the microbial conversion of lignin into humus (13). The microbes able to degrade lignin are found within the basidiomycetes and a number of fungi, grouped together as the so called "soft rot fungi". The wood destroying basidiomycetes are divided into white and brown rot fungi, whether they use mainly lignin or cellulose as a carbon source. They mostly live in wood and their occurrence in soil seems to be more or less restricted to forest soils. Soft rot fungi are a diversified group of fungi imperfecti and ascomycetes and can be isolated in appreciable numbers from arable soils (14). They belong to the early colonizers of plant materials which are brought into the soil. The former belief that they attack mainly the cellulose part of plant material and only to a minor degree the lignin (15, 16) seems to have changed, since some of them were found to have appreciable lignolytic activity (17, 18, 19). With respect to humus formation, some of these fungi are of great interest, since they form dark colored, high molecular weight polymers even when cultured on glucose as the only carbon source. If they grow on lignified plant material, the melanins contained lignin derived phenols in addition to the typical phenols of microbial biosynthesis (20). Electron micrographs made by Kilbertus et al. (21) show the microbial transformation of plant cell wall material into humic compounds.

Some experiments were started (22) with specifically ^{14}C-labeled lignin like polymers which were prepared from coniferyl alcohol or other lignin alcohols, labeled either in the methoxy groups, the side chain carbons, or in the carbons of the aromatic nuclei. The differently labeled polymers were added to cultures of melanin forming soft rot fungi or to white-rot fungi with appreciable lignin degrading capacity.

According to Freudenberg (23) the lignins of

different plants are polymers of the lignin alcohols.
These units are mainly linked by ß-arylether linkages,
followed by α-arylether and C-C-linkages. A lignin
labeled in the ß-position was prepared by polymerisation of a coniferyl alcohol labeled in the 2'-position.
$^{14}CO_2$ release from this position should indicate a depolymerization of the lignin by disconnection of the
ß-arylether linkages. Furthermore by polymerisation
of a methyl-labeled coniferyl alcohol, a methyl-labeled lignin was prepared. The $^{14}CO_2$ from these labeled methyl groups in several experiments should indicate the degree of demethylation during the microbial
degradation. Polymerisation of ring-labeled coniferyl
alcohol resulted in a ring-labeled lignin. The release
of $^{14}CO_2$ from these ring carbon structures indicated
a more or less complete assimilation of the lignin as
a carbon and energy source. The time course of the
$^{14}CO_2$-release from these several labeled groups by
white-rot fungi showed in the first few days a release
of CO_2 from the ß-^{14}C-atoms of the side chain. This
was followed later by an increasing release of $^{14}CO_2$
from the methyl groups which, after several days, exceeded that of the side-chain carbons. The release of
$^{14}CO_2$ from the ring carbons showed some interesting
differences dependent upon the lignin degrading
capacity of white-rot fungi as compared to soft-rot
fungi. With white-rot fungi, the release of $^{14}CO_2$ from
ring carbon was sometimes equivalent to that evolved
from the ^{14}C-methyl groups. With soft-rot fungi, especially with melanin forming species, the release of
$^{14}CO_2$ from the ß-carbons and the methyl groups was
higher than that of the ring carbons. These experiments show that with melanin forming fungi, side-chains
and methoxyl groups were degraded more rapidly than
the ring structures and these became incorporated into
the melanins and were partly found in the humic acid
like material (24).
    During the formation of these melanins, lignin
degradation products seem to react together with units
of the microbial metabolism and become stabilized
against further microbial degradation. Ring structures
of lignin derived material seem to be more favoured
than methoxyl or the side chain carbons in these
reactions.
    Other plant phenolic constituents, such as fla-
vonoids, lignans or naphthalenic compounds are sometimes present in relatively high concentrations in
roots or leaves. During the decomposition of these
plant residues, degradation products of these phenolic
constituents, or even the constituents themselves

may become integrated into the soil humus. This was
shown recently by Metche et al. (25, 26) with juglone
(5-hydroxynaphthaquinone) during the humification of
walnut leaves.

During the decomposition of plant residues a number of phenolic compounds can be extracted from the
soil especially after weak hydrolysis treatment (27,
28, 29, 30). Phenols such as p-hydroxybenzoic, vanillic,
protocatechuic, syringic, p-hydroxycinnamic, ferulic
and other aromatic acids were isolated from soils by
extraction with dilute alkaline solvents or with
ether. Vanillic, syringic or ferulic acids were un-
doubtely derived from lignin, whereas other phenolic
compounds could also have been synthesized by microorganisms. McCalla (31) reported concentrations of
about 15 ppm of p-coumaric acid in a subtilled soil,
which corresponds to 0.04 per cent of the soil organic
matter. Smaller amounts were isolated from plowed soil.

The viewpoint has been expressed that phenolic
substances are stabilized in soil against biodegradation through physical adsorption or cofixation on
humic colloids or clays (30). By following degradation
of uniformly labeled vanillic, p-coumaric, and ferulic
acids, Batistic and Mayaudon (32) suggested a stabilisation through complex formation or polymerisation.
Numerous soil organisms, however, readily degrade most
aromatic compounds in pure or mixed cultures and an
extensive literature exists on mechanisms of the degradation processes (33, 34, 35). It would appear
likely that simple phenolic substances would not
accumulate as such in the soil and that when applied
to the soil they would be readily utilized by the soil
population, unless they have been linked or complexed
into the humic complexes or fixed into the interlamellar spaces of clays.

Studies were made (36) of the decomposition of a
number of specifically labeled benzoic and cinnamic
acids in soil. The use of these ^{14}C-phenolic substances
facilitated following the decomposition or metabolism
of different carbons, including ring, side-chain and
side-group carbons of the phenols. It was also facilitated determinations of how the decomposition
rates of the phenols were influenced by other organic
amendments such as plant materials or humic acids. The
loss of the labeled carbon was followed over a 12-weeks
incubation period in soil. The labeled phenols were
added with and without addition of 0.5 % finely ground
peach wood or 2 % freeze dried humic acid. Under the
experimental conditions used, most of the phenolic
compounds in concentrations up to 1000 ppm were quickly

utilized as a source of carbon and energy by the soil
population. When the hydroxyl groups of the phenols
were methylated, the loss of ring, methoxyl, carboxyl
and C_3-side chain carbons was almost always 70 - 90 %.
The added plant material or the humic acid exerted
little or no influence on this decomposition. Therefore it seems likely that most of these phenols are
not highly stabilized if they are synthesized or
relased from the decaying plant material by microbial
action. However, differences in decomposition rates
were observed with phenols obtaining methylated or
free hydroxyl groups. Phenolic compounds such as
veratrol and veratric acid were rapidly deomposed and
the organic amendments exerted no significant influence on their decomposition rates. The corresponding
compounds with free hydroxyl groups, such as catechol
or protocatechuic acid, however, were degraded to a
smaller extent, and this was decreased in the presence of organic amendments. These oberservations show
some interesting influence of soil organic matter on
the degradability of phenols. Most of the added phenols are degraded through protocatechuic acid or
catechol before they undergo ring splitting and final
metabolism. Veratric acid, for example is taken up
by microbes, demethylated to protocatechuic acid, and
then degraded by ring splitting. Therefore, protocatechuic acid itself should be somewhat more rapidly
degraded than its methylated derivative. This was
observed in pure cultures of pseudomonads (37). However, in soil it seems likely that the more reactive
phenols with free hydroxyl groups are complexed or
linked into soil organic matter and are therefore
more resistant than the methylated compounds.
     These phenomena could also effectively influence
the degradation of pesticide residues in soil. Pure
culture studies with Hendersonula toruloidea have
indicated that substantial percentages of the ring
carbons from 2,4-D (38) are linked into the humic
acid-type polymers of this organism. These linked
materials are much more resistant against decomposition in soil than the nonlinked compounds. The ring
carbons of the linked 2,4-D are much more stabilized
as compared to the side-chain carbons and the resistance was greater if the molecule was incorporated
into the humic acid-like polymers.
     The introductory remarks indicated that an essen-
tial part of the soil humus is formed microbiologic-
ally from carbohydrates through secondary pathways.
Also, this part is formed through reactive constituents which are essentially phenols or quinones.

Many fungi and actinomycetes synthesize by secondary metabolism, either simple or more condensed structures. The great number of antibiotics and other fermentation products, the many kinds of aromatic molecules and the dark polymers of fungi and other microorganisms are formed by secondary metabolism. The structures of most of the dark colored polymers formed by fungi are not yet well known, but they most probably originate through polymerisation of aromatic compounds largely formed through the acetate - malonate pathway, and to a minor extent through the shikimate pathway (39, 47).

A great number of fungi belonging to the imperfecti group synthesize phenols from nonaromatic precursors and transform them into dark colored polymers. Two of the most prominent phenols are formed from these by introduction of further methyl and hydroxyl groups, oxidation of methyl into carboxyl groups, and decarboxylation. Some of the intermediate phenols are easily oxidized to radicals and quinones and polymerize together with other phenols or amino acid compounds to form the dark colored humic acid-like polymers (40, 41). Some of the resulting phenols, like 2,3,5-trihydroxytoluene, are oxidized by air even at weakly acid or neutral pH, and react with other phenols. Other types of phenols such as 4-methylresorcinol are synthesized by microbial introduction of methyl groups at different positions. Numerous other fungi synthesize 6-methylsalicylic acid and transform it into easily oxidizable phenols such as gentisic acid or 2-methylhydroquinone through decarboxylation and introduction of a second hydroxyl group.

Most of the fungi so far studied also form small to moderate amounts of p-hydroxybenzoic and p-hydroxycinnamic acids. These compounds are synthesized through the shikimate pathway. Both compounds are also transformed by introduction of additional OH-groups and by degradation of the side chain of the cinnamic acid derivative into several other very easily oxidized phenols such as gallic acid, pyrogallol and 2,3,4-trihydroxybenzoic acid.

Similar phenols, e.g. ferulic, vanillic or p-hydroxycinnamic acids, were also isolated during lignin degradation. If these acids were added to cultures of melanin forming fungi, they were found to be altered in a similar way as observed with p-hydroxycinnamic acid synthesized by fungi (20). Therefore it seems possible that lignin derived phenols or chunks of lignin could be incorporated into the melanin of soil fungi.

Many of the reactions involved in the formation of
melanin by fungi appear to be autoxidative processes.
It was demonstrated that phenols such as 2,3,5-,
2,4,5-, 2,3,6- or 3,4,5-trihydroxytoluene or 2,3,4-
trihydroxybenzoic acid react even under weakly acid
or neutral conditions with oxygen of the air and form
reactive quinones or radicals. These compounds react
with other phenols present in the mixture to form
polymers. It also appears likely that some fungi or
bacteria can form melanins through oxidation of phenols by phenolases.

A great variety of quinone structures such as
naphtha- or anthraquinone derivatives are also formed
by many soil fungi. These compounds are of interest in
view of quinones which are directly isolated from soil
by extraction procedures (42, 43). Some humic acid
fractions appear to be derived or associated with more
highly condensed quinone structures. The so called
P-type humic acids, isolated first by Kumada and Sato
(44, 45) are probably related to perylene quinone of
fungal origin (46). Similar structures were isolated
from the black fruiting bodies of Daldinia concentrica
and of Aspergillus sp. (47, 48), and are formed by
condensation of 1,8-dinaphthol units with intermediate
accumulation of tetrahydroxy-dinaphthyl and dihydroxy-
perilene quinone. Other fungi which also form dark
colored polymers, synthesize several anthraquinones
in relatively high amounts during growth on glucose-
asparagine- or -sodium nitrate media. These anthraquinones are generally synthesized in addition to
several phenols and are incorporated together into
the humic acid like pigments. Saiz et al. (49) have
found that the soil fungus Eurotium echinulatum
synthesizes both phenols and anthraquinones. The
phenols were mostly those already found to be synthesized by other fungi. The anthraquinones, however,
are mostly derived from endocrocin and emodin which
are formed by condensation of acetate-malonate units.
The transformation of these anthraquinones occurs by
introduction of hydroxy groups, methyl groups, and
by oxidation of methyl into carboxyl groups. By
Introduction of additional OH-groups, catenarin,
dermoglaucin, dermocybin and others were formed.
Stepwise oxidation of methyl into carboxyl groups also
occured. During the period of rapid polymer formation
the phenols and anthraquinones largely disappeared
from   the medium. They were found to form constituent
units of the polymers.

Nitrogen containing compounds can be released by
acid hydrolysis from humic and fulvic acids as well as

from fungal melanins. The N in the humic acids and
in the fungal polymers is not readily available to
microorganisms. Up to 50 % or more of the polymer N
is released in the form of amino acids, amino sugars
or ammonia by hydrolysis with 6 N HCl (50, 51). Peptides have been released from soil and fungal melanins
by treatment with proteases (52, 53). Also, small
amounts of protein were isolated by treatment of soil
humic acids with phenol (54, 55). Mayaudon (56) reported some stabilization of labeled protein from
spinach leaves by flocculation with soil humic acid
residues. Several enzymes such as proteases (57) or
urease (58) were found to be stabilized in soil by
complexation with soil organic matter. In contrast, a
high degree of stabilization of amino acids, peptide
and amino sugars against microbial degradation has
been noted when these compounds were oxidatively poly-
merized with mixtures of severals phenols (59, 60, 61).
The stabilization of amino compounds which occurred
during the formation of fungal melanins, appears to be
the result of the nucleophilic addition of free amino
groups to quinones formed from phenols during oxidation, and was explicated for the addition of amino
acids or peptides to trihydroxytoluenes during oxidation (40, 62). Significant differences in the reactivity of various phenols were found (62) when they
were oxidized together with amino acids or peptides.
For example, 2,3,5-trihydroxytoluene largely binds
amino acids and peptides, whereas 2,4,5-trihydroxy-
toluene actively deaminates the amino acids and peptides.
    Studies of the linkage (61) of ^{14}C-labeled
glucosamine and chitosan from Mucor rouxii into model
phenolic polymers with mixtures of phenols commonly
synthesized by soil fungi showed that both the glucosamine and the chitosan were significantly stabilized
against microbial degradation in the soil by incorporation into polymers. More than 70 % of the glucosamine was shown to be decomposed when this amino
sugar was added to the soil. Upon linkage of glucosamine into phenolic polymers the decomposition rate
was only 20 %. Chitosan was stabilized, to a smaller
extent than glucosamine. Similar experiments were
made (63) on the decomposition rate of labeled amino
acids, peptides and proteins. Included were experiments to determine the effect of humic acid, when
mixed with protein and amino acids, on the decomposition of these N-containing compounds. Furthermore
the influence of an intimate association through
lyophilization of a solution of the amino acids or

proteins together with model or natural humic acid-
type polymers on decomposition rates in soil was
determined. The protein added to soil decomposed
quickly, and after 12 weeks 84 % of the added ^{14}C
had evolved as $^{14}CO_2$. Mixing with humic acid reduced
decomposition to about 50 % and more intimate association of the protein with humic polymers reduced
decomposition to 35 %. Linkage with the polymers
during phenolase oxidative polymerisation reduced
decomposition to 11 %. Mixing or association of the
amino acids from the hydrolized protein with humic
acid did not greatly effect the decomposition rates,
but linkage into the polymers reduced total $^{14}CO_2$-re-
lease from 80 to 17 %. These experiments indicate
that proteins are stabilized to a marked extent
against microbial degradation by close association
with humic acid-type polymers. This close association
could prevent the microbial enzymes from approaching
the proper position to carry out their degradation
reactions.Basaraba and Starkey (64) noted similar
protective effects of tannins on the biodegradability
of proteins. From electrophoretic measurements,
Mayaudon (56) concluded that proteins could be adsorbed to the humic material through H-bonding and
electrostatic forces. Davies (65) suggested that the
tanning of leaf proteins during senescens is caused
by free phenolic hydroxy groups that are linked
through hydrogen bonds to the peptide bonds. Similar
viewpoints for the reaction of plant phenols with
proteins were recently published by van Sumere (66).

Humic compounds are produced from many different
sources and their chemical structure may vary greatly
with units available in any soil microenvironment.
Their structure and amount is furthermore strongly
influenced by climatic factors and by the parent
mineral structure of the soil. However, humic compounds
have similar properties which could be related to the
numerous functional groups, primarily the COOH and
phenolic OH groups.

Humic compounds are considered as a system, where
reactive compounds of various origin are linked or
complexed together. The possibility therefore, that
compounds not actually present in the natural soil
environment may also become incorporated in the system
ist quite easy to imagine. Furthermore, as in the
case of pesticides and their partially metabolized
residues the chemical structures are largely similar
to those of the natural constituents. A great number
of pesticides contain aromatic structures which become
hydroxylated and may easily interact with humic

compounds during their formation. Others contain
heterocyclic nitrogen or amino groups in aromatic
structures.

In spite of an appreciable mean residence time
of humic compounds in soil (67, 68, 69) which is
known to be several hundred to thousands of years, the
viewpoint has been expressed that a rather small fraction of the soil humus is rapidly formed and decomposed (70, 71). Furthermore, it was found (12, 72, 73)
that upon soil incubation with labeled glucose or
cellulose most of the remaining carbon was combined
with hydrolyzable N-containing structures, but a
small part was also found in melanin-like structures
(12, 73, 74). This fraction might also be involved in
the adsorption of actual available pesticide residues
(75). Their availablility might be increased by
accumulation of pesticide residues in cells of microorganisms or plants as it was found by Ko and Lockwood (76) and Lockwood (77) for chlorinated hydrocarbons.

On the other hand surface structures of humic
compounds might interact with pesticide residues
through hydrogen bonding, ion exchange or chemisorp-
tive processes. These kinds of adsorptive processes
seems to be of great importance and have been extensively discussed as regards the adsorption of triazine
herbicides in soil by Hayes (78).

Literature Cited

1)   Webley, D.M. and Jones, D., "Soil Biochemistry",
          Vol.2, and J.Skujins (eds.), M.Dekker,
          Inc., New York, pp.446-485, 1971
2)   Waksman, S.A. and Iyer, K.R.N., Soil Sci.(1932)
          34, 54
3)   Hobson, R.P. and Page, H.J., J. Agric.Sci. (1932)
          34, 43
4)   Jenkinson, D.S. and Tinsley, J., J.Soil Sci.
          (1959) 10, 245
5)   Jenkinson, D.S. and Tinsley, J., Sci.Proc.Roy.
          Dublin Soc. (1960) A 1, 141
6)   Bremner, J.M., J.Agr.Sci. (1955) 46, 247
7)   Morrison, R.I., J.Soil Sci. (1958) 9, 130
8)   Morrison, R.I., J.Soil Sci. (1963) 14, 201
9)   Schnitzer, M. and Khan, S.U., "Humic Substances
          in the Environment", M.Dekker, Inc.,
          New York, 327 pp., 1972
10)   Burges, W.A., Hurst, H.M. and Walkden, Geochim.
          Cosmochim.Acta (1964) 28, 1547
11)   Piper, T.J. and Posner, A.M., Soil Biol.Biochem.
          (1972) 4, 513

12) Martin, J.P., Haider, K. and Saiz-Jimenez, C.,
        Soil Sci.Soc.Amer.Proc. (1974) 38, 760
13) Christman, R.F. and Oglesby, R.T. "Lignins", K.V.
        Sarkanen and C.H. Ludwig (eds.) Wiley-
        Interscience, New York, pp.769-795 (1971)
14) Domsch, K.H. and Gams, W., "Pilze aus Agrarböden",
        G. Fischer, Stuttgart, 222 pp, 1970
15) Savory, J.G. and Pinion, L.C., Holzforsch. (1958)
        12, 99
16) Seifert, K., Holz, Roh- und Werkstoff (1966) 24,
        185
17) Levi, M.P. and Preston, R.D., Holzforsch. (1965)
        19, 183
18) Haider, K. and Domsch, K.H., Arch. Mikrobiol.
        (1969) 64, 338
19) Mangenot, F. and Reisinger, O., Symp. Biol. Hung.
        (1972) 11, 147
20) Martin, J.P. and Haider, K., Soil Sci. (1969)
        107, 260
21) Kilbertus, G., Mangenot, F. and Reisinger, O.,
        Rech.Coop.Programme du C.N.R.S., 40,
        Univers.de Nancy
22) Haider, K. and Trojanowski, J., Arch. Microbiol.
        (1975 in press)
23) Freudenberg, K., "Constitution and Biosynthesis
        of Lignin", K. Freudenberg and A.C.
        Neish (eds.), Springer Verl., Berlin,
        pp. 45-122, 1968
24) Haider, K. and Martin, J.P., "Isotopes and Radia-
        tion in Soil Organic Matter Studies",
        IAEA, Vienna, pp. 189-196, 1968
25) Metche, M., Mangenot, F. and Jaquin, F., Soil
        Biol. Biochem. (1970) 2, 81
26) Metche, M. and Andreux, F., C.R. Acad. Sci.
        Paris (1974) 279 (Ser.D) 1035
27) Guenzi, W.D. and McCalla, T.M., Soil Sci. Soc.
        Amer.Proc. (1966) 30, 214
28) Whitehead, D.C., Soil Fert. (1963) 26, 217
29) Bruckert, S., Jaquin, F. and Metche, M., Bull.
        Ecol. Natl. Sup. Agron. (Nancy) (1967)
        9, 73
30) Wang, T.S.C., Yeh, K.-L.- Cheng, S.-Y. and Yang,
        T.-K., "Biochemical Interactions among
        Plants", Nat.Acad.Sci., Washington D.C.,
        pp.113-120, 1971
31) McCalla, T.M., "Biochemical Interactions among
        Plants", Natl.Acad. Sci., Washington
        D.C., pp. 39-45, 1971
32) Basistic, L. and Mayaudon, J., Ann.Inst. Pasteur
        (1960) 118, 199

33) Dagley, S., "Advances in Microbial Physiology"
    Vol. 6, A.H. Rose and J.F. Wilkinson
    (eds.) Acad.Press, New York, pp 1-42
    1971
34) Evans, W.C., "Fermentation Advances", D. Perlman
    (ed.), Acad.Press, New York, pp. 649-
    687, 1969
35) Stanier, R.Y. and Ornston, L.N., "Advances Micro-
    bial Physiology", A.H. Rose and D.W.
    Tempest (eds.), Acad.Press, London,
    New York, pp. 89-151 1973
36) Haider, K. and Martin, J.P., Soil Sci.Soc.Amer.
    Proc. (1975 in press)
37) Reber, A., Arch.Microbiol. (1973) 89, 305
38) Martin, J.P. and Focht, D.D., "Biological Proper-
    ties of Soils", University of Califor-
    nia, Riverside, 148 pp, 1975
39) Swan, G.A., "Structure, Chemistry and Biosynthe-
    sis of the Melanins", Fortschr.Chemie
    organ.Naturst., W.Herz, H.Grisebach, and
    G.W. Kirby (eds.),Springer Verl., Wien,
    pp.521-581, 1973
40) Martin, J.P. and Haider, K., Soil Sci. (1971)
    111, 54
41) Haider, K., Martin, J.P. and Filip, Z., "Soil
    Biochemistry;, Vol. 4, E.A. Paul and
    A.D. McLaren (eds.), M.Dekker, Inc.,
    New York, pp. 195-244, 1975
42) McGrath, D., Chem. Ind. (Lond.) (1970) 42, 1353
43) McGrath, D., Geoderma (1972) 7, 167
44) Kumada, K. and Sato, O., Soil Sci. Plant Nutr.
    (1962) 8, 31
45) Kumada, K. and Sato, O., "Humus et Planta IV",
    B.Novàk and V. Rypácek (eds.), Prague,
    pp. 131-133, 1967
46) Kumada, K. and Hurst, H.M., Nature (London) (1967)
    214, 631
47) BuLock, J.D., "Essays in Biosynthesis and Micro-
    bial Development", J.Wiley and Sons,
    New York, 71 pp, 1967
48) Robertson, A. and Whalley, W.B., J.Chem.Soc.
    (1953) 2434
49) Saiz-Jimenez, C., Haider, K., Martin-Martinez, F.
    and Martin, J.P., Soil Sci.Amer.Proc.
    (1975 in press)
50) Bremner, J.M., "Soil Nitrogen", W.V. Bartholomew
    and F.E. Clark (eds.), Agron. No. 10.
    Amer.Soc.Agron., Madison, Wisc., pp.
    93-149, 1965

51) Bremner, J.M., "Soil Biochemistry", Vol. 1, A.D.
     McLaren and G.H. Peterson (eds.), M.
     Dekker, Inc., New York, pp. 19-66, 1967
52) Bremner, J.M., Pontificia Acad.Sci. Roma, Scripta
     varia (1968) 32, 143
53) Ladd, J.N. and Brisbane, P.G., Austr.J.Soil Res.
     (1967) 5, 161
54) Simonart, P., Batistic, L. and Mayaudon, J.,
     Plant Soil (1967) 27, 153
55) Biederbeck, V.O. and Paul, E.A., Soil Sci. (1973)
     115, 357
56) Mayaudon, J., "Isotopes and Radiation in Soil
     Organic Matter Studies", IAEA, Vienna,
     pp. 177-188, 1968
57) Ladd, J.N. and Butler, J.H.A., "Soil Biochemistry"
     Vol. 4, E.A. Paul and A.D. McLaren (eds.)
     M.Dekker, Inc., New York, pp. 143-194,
     1975
58) Bremner, J.M. and Douglas, L.A., Soil Biol. Bio-
     chem. (1971) 3, 297
59) Haider, K., Frederick, L.R. and Flaig, W., Plant
     Soil (1965) 22, 49
60) Ladd, J.N. and Butler, H.J.A., Austr.J.Soil Res.
     (1966) 4, 41
61) Bondietti, E., Martin, J.P. and Haider, K., Soil
     Sci.Soc. Amer. Proc. (1972) 36, 597
62) Flaig, W. and Riemer, H., Liebigs Ann. Chem.
     (1971) 746, 81
63) Verma, L., Martin, J.P. and Haider, K., Soil Sci.
     Soc. Amer. Proc. (1975) 39, 279
64) Basaraba, J. and Starkey, R.L., Soil Sci. (1966)
     101, 17
65) Davies, R.I., Soil Sci. (1971) 111, 80
66) Sumere van, C.F., Albrecht, J., Dedouder, A.,
     de Pooter, H. and Pé, J., "The Chemistry
     and Biochemistry of Plant Proteins",
     J.B. Harborne and C.F. van Sumere (eds.)
     Acad. Press, London, pp. 211-264, 1975
67) Paul, E.A., Campbell, C.A., Rennie, D.A. and
     McCallum, K.J., Trans. Int. Congr. Soil
     Sci., 8th, Roumania, pp. 201-208, 1964
68) Scharpenseel, H.W., Ronzani, C. and Pietig, F.,
     "Isotopes and Radiation in Soil Organic
     Matter Studies), IAEA, Vienna, 67-73,
     1968
69) Scharpenseel, H.W., "Humic Substances", D. Povo-
     ledo and H. Golterman (eds.), Pudoc,
     Wageningen, pp. 281-292, 1975
70) Kononova, M.M., Pontificia Acad. Sci. Romana,
     Scripta varia (1968) 32, 361

71) Sauerbeck, D. and Johnen, B., Landw. Forsch. (1973) 30/II, 137
72) Jenkinson, D.S., Soil Sci. (1971) 111, 64
73) Fustec-Mathon, E., Haider, K. and Martin, J.P., C.R. Acad. Sci. Paris, Ser. D. (1973) 276, 929
74) Wagner, G.H., "Isotopes and Radiation in Soil Organic Matter Studies", IAEA, Vienna, pp. 197-204, 1968
75) Wolcott, A.R., "Pesticides in Soil: Ecology, Degradation and Movement", Michigan State University, pp. 128-138, 1970
76) Ko, W.H. and Lockwood, J.H., Rev. Ecol. Biol. Sol. (1970) 7, 465
77) Lockwood, J.L., "Pesticides in the Soil: Ecology, Degradation and Movement", Michigan State University, pp. 47-50, 1970
78) Hayes, M.H.B., "Residue Reviews", Vol. 31, F.A. Gunther (ed.) Springer Verl., New York, pp. 131-174, 1970

19

Spectroscopic Characterization of Soil Organic Matter

R. BARTHA and T.-S. HSU

Department of Biochemistry and Microbiology, Cook College, Rutgers University, New Brunswick, N. J. 08903

In studies concerning the environmental fate of pesticides and other man-made chemicals, it has been noted repeatedly that some of these compounds or the products of their partial degradation "disappear." They do so in the sense that they become undetectable by the conventional techniques of residue analysis, yet radiotracer evidence negates their mineralization, i.e. their complete conversion to mineral constituents. It is now clear that in some of these cases chemical reactions take place between humic substances and the introduced compounds, leading to complexed and often immobilized residues. Such chemical binding of residues was observed primarily in soil and in sediments, but similar reactions can take place in natural waters containing dissolved humic substances. While absorption by clay minerals and humic substances can also lead to a temporary immobilization of residues, this phenomenon can be distinguished from covalent binding by the fact that it can be overcome by exhaustive solvent extraction or by ion exchange techniques. In contrast, covalently bound residues can be released only by relatively severe treatment, e.g. hydrolysis by strong acid or alkali, and such treatment may irreversibly alter both the residue and its binding site. Some of the bound residues can not be released at all in a chemically recognizable form by currently available procedures. For the above reasons, a nondestructive technique for bound residue analysis and for the study of the nature of the chemical attachment is clearly desirable. It is the aim of this brief review to assess the potential usefulness of various spectrometric techniques for this purpose. Some ongoing studies in our laboratory on the covalent binding of chloroaniline residues to humus are discussed in this context.

Origin, Classification and Composition of Humic Substances.

The soil organic matter derives from remains of plants, animals and microbes. Humic substances are defined as that portion of the soil organic matter than has undergone sufficient

transformation to render the parent material unrecognizable.
Humic materials are present in mineral soils typically at less
than 5% by weight. They are generated not only in soils but
also in aquatic environments; additional humic material may be
added to natural waters by leaching from soils (1,2).

The genesis of humic material is a two stage process involving the predominantly microbial degradation of organic
polymers to monomeric constituents such as phenols, quinones,
amino acids, sugars, etc., and the subsequent polymerization
of these due to spontaneous chemical reactions, autoxidation and
oxidation catalyzed by microbial enzymes such as oxidases, poly-
phenoloxidases and peroxidases (3,4). The humic material is
in a dynamic state of equilibrium, its synthesis being compensated for by gradual mineralization of the existing material.

According to their solubility characteristics, humic substances can be fractionated into fulvic acid (soluble in alkali,
not precipitated by acid) humic acid (soluble only at alkaline
pH) and humin (insoluble in alkali) (5). To none of these
fractions can a definite chemical structure be assigned, all
three being randomly assembled irregular polymers. The main
differences between fulvic and humic acids are the lower molecular weight, higher oxygen to carbon ratio, and higher ratio
of acidic functional groups per weight of the former as compared
to the latter, but the spectrum is continuous and the dividing
line is arbitrary. Molecular weights range from around 700
to 300,000, total acidities from 485 to 1,420 meq/100g. Humin
is regarded as a strongly bound complex of fulvic and humic acids
to mineral material rather than a class of compounds by itself
(6,7). The alcohol-soluble hymatomelanic acid, a minor compound
previously suspected to be an artifact of extraction, now appears
to be a genuine fraction consisting of esterified or methylated
humic acids(8).

Since humic compounds are random polymers, we cannot hope to
learn their exact chemical structure. At best, we can establish
type structures, e.g. representative sequences of interconnected
atoms within the humic acid molecule. The theories on humic
type structures are controversial and subject to constant re-
vision and refinement. The perhaps most accepted current theory
visualizes an aromatic "core" consisting of single and condensed
aromatic, heterocyclic and, perhaps, quinoidal (9) rings, linked
and cross-linked by carbon-carbon, ether, amino, and azo bonds.
The rings bear a variety of functional groups, the more prominent
of which are carboxyl, phenolic hydroxyl and carbonyl groups.
Attached to this core are amino acids, peptides, sugars and
phenols, which form further cross linkages. The result is a
three-dimensional sponge-like structure that readily absorbs
water, ions and organic molecules in an exchangeable manner and,
in addition, may chemically bind suitable compounds to its
reactive functional groups (9,10,11,12,13,14). As a consequence,
virtually all natural organic compounds and apparently also
numerous man-made chemicals can occur in bound or absorbed

form in humic substances; even active enzymes were recently
recovered in humus-bound form (15).

The ring structures that serve as building blocks of the
humic acid core may originate from the microbial degradation of
lignin or may be synthesized by various microorganisms from
other carbon substrates. Directly or after oxidation to quinones
the phenolic compounds condense with amino acids in a process
that can be modelled in vitro. Substances closely resembling
humic acids were synthesized by enzymatic oxidation of phenol
mixtures in the presence of amino acids or peptides (16,17).
In addition, simple mixtures such as methylglyoxal-glycine
and glucose-glycine were reported to form, upon heat activation,
substances resembling humic acids (14).

Spectrometry of Humic Substances.

In an effort to elucidate their chemical structure, spectrometric techniques were applied extensively to both supposedly
"intact" and to intentionally modified (e.g. methylated or
acetylated) humic compounds. These studies are instructive as
to the possibilities and limitations of spectrometric techniques as applied to humic materials. They were recently reviewed by Schnitzer (18) and will be discussed here breifly.
Only a limited body of literature exists at this time on the
primary concern of this discussion i.e. on spectrometric studies
of humus-bound pesticide residues.

Visible and Ultraviolet (UV) Spectra. These spectra reflect transitions between electronic energy levels and are, for
practical purposes, confined to the wavelength range of 200 to
800 nm. The principal chromophores are conjugated double (or
triple) bonds both in aliphatic and in aromatic carbon compounds
(19). The broad and overlapping nature of the absorption bands
does not allow much information to be obtained by this technique on the fine structure of such large and complex molecules
as the humic compounds. The usefulness of visible and UV
spectra is largely confined to the characterization of monomeric degradation products of humic compounds. Overall decreases
in conjugated double bonds occur during oxidation, catalytic
hydrogenation, and other treatments that lead to extensive
depolymerization of humic substances, and these can be detected
by a decreased absorption in the visible and UV range. The
comparative absorption of humic compounds in the visible range
is believed to be correlated to the degree of condensation of
their aromatic nuclei, hence darker color indicates a greater
abundance of condensed nuclei (2).

Infrared (IR) Spectra. The energy levels of most molecular
vibrations fall into the energy range of IR radiation. IR
spectra contain a great deal of rather specific information

about the inner structure and the functional groups of an organic molecule and, consequently, IR spectrometry is one of the
preferred tools of structural identification. The most useful
wavelength range for the chemist lies between 2.5 and 16 µ or,
as more commonly expressed, between the wave numbers 4,000 and
625 cm^{-1} .
    The stretching and bending vibrations of various functional
groups have their characteristic wave number regions. While the
molecular environment of the functinal groups has a marked influence on the exact position of the absorption maxima, it is
frequently possible to assign an absorption band to a specific
functional group even if its molecular environment is not known
with certainty (19,20). It is this particular feature of the
IR spectra that makes them quite useful in the study of humic
substances.
    The main IR absorption bands of interest are summarized by
Stevenson and Butler (6). The low energy or "fingerprint" region
(1500-625 cm^{-1}) is relatively featureless in humic compounds
because of the overlapping inner vibrations of the large and
complex humic molecules. In the higher evergy region, the 3,300
cm^{-1} band is assigned to the O-H stretching of H-bonded OH
groups, the 2,900 cm^{-1} band to the C-H stretching of aliphatic
carbon chains, the 1,720 cm^{-1} band to the C=O stretching of
carboxyl and ketone groups, the 1,610 cm^{-1} band to the C=C
vibrations of aromatic rings with a contribution also from the
C=O stretching of H-bonded ketone groups, and the 1,250 cm^{-1}
band to the C-O, stretching and OH deformation of carboxyl
groups.
    Acetylation introduces a strong new band of 1,375 cm^{-1} due
to the   C-C deformation of the additional acetyl groups
(Figure 1). Methylation of humic substances increases the
intensity of 2,900, 1,720 and 1,250 cm^{-1} bands due to the increase of C-H,C=O and C-O stretchings, respectively. The normally
small band at 1,450 cm^{-1} , due to the C-H deformation of methyl
groups,   also increased. Specific acetylation, methylation
and IR spectrometry of the so modified humic compounds lends
support to the presence of quinones in fulvic and humic acids
(9). The need for brevity has necessitated here the over-
simplification of a highly complex subject, and for qualification
for the above statements additional references should be consulted (21,22,23,24).
    IR spectra were utilized in the study of metal ion inter-
actions with humic substances (25). A decrease of the 1,725 cm^{-1}
band (carboxylic C=O) and increases of the 1,600 and 1,400 cm^{-1}
bands (carboxylate) indicated the formation of an iron-carboxyl-
ate) complex. The decomposition of oxygen-containing functional
groups during the gradual pyrolysis of humic material was also
followed by changes in the IR spectra (26).

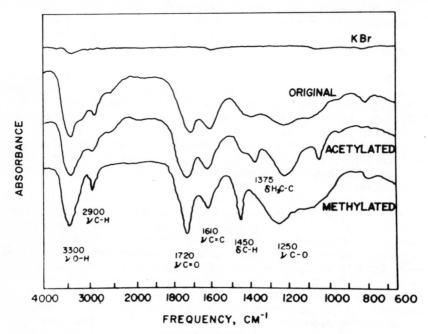

Figure 1. IR spectra of original, acetylated, and methylated humic acids. After Ref. 18

Electron Spin Resonance (ESR) Spectra. Referred to also
as electron paramagnetic resonance (EPR) these spectra reflect
energy absorption due to magnetic spin resonance of unpaired
electrons. The "free radicals" of organic compounds contain
such unpaired electrons. Under the influence of an external
magnetic field, microwave excitation allows transitions to higher
energy levels to be observed, as indicated by absorption of the
microwave radiation. The transition energies are under the influence of the molecular environment of the free radical, and
thus contain useful information about it (27). Typically, free
radicals are highly labile and short-lived intermediates of some
chemical reactions, including some biologically catalyzed oxi-
dations. In certain condensed aromatic systems and quinoidal
structures free radicals may become stabilized. It is a typical
feature of humic substances that they regularly contain stabilized free radicals, presumably in form of semiquinones (28,29,
30,31). The ESR-signal is one of the arguments in favor for
the presence of quinoidal structures in humic substrances. For
a fulvic acid a spin count of 0.58×10^{18} per g was measured,
and from this value it was estimated that one semiquinone radical

is present for every 44,000 carbon atoms (28).

The origin of the free radicals is as intriguing as their persistence, and the most plausible theory is that they were formed during and preserved since the original biological oxidation that lead to the formation of the humic compound. The biological effect of these stabilized free radicals and their potential to initiate polymerization reactions or the binding of pesticide residues to humic materials are virtually virgin areas awaiting experimental work.

Nuclear Magnetic Resonance (NMR) Spectra. The nuclei of certain atoms exhibit a magnetic spin momentum. The most important of these atoms are 7-H,13-C,19-F and 31-P. When placed in a homogeneous magnetic field and excited with radiowaves, energy transitions, as evidenced by radiowave absorption, take place. Again, the molecular environment of the proton influences the resonance energy resulting in a resonance shift relative to the arbitrary reference point of the trimethylsilane (TMS) signal (19). NMR spectrometry is a powerful tool for the exploration of the immediate chemical environment of a proton (32) that has been definitely underutilized in the study of soil organic matter. The limited solubility of the humic compounds in the suitable deuterated solvents presents a problem, but the rapidly increasing sensitivity of the instruments is likely to improve this situation.

To date the NMR-technique was applied only to hydrogen protons in humic acid. Barton and Schnitzer (33) investigating a low molecular weight methylated fulvic acid noted the absence of aromatic and olefinic protons. This surprising result seems to indicate that practically all hydrogens are replaced by substituents on the aromatic core of humic substances. Felbeck (34) applied NMR spectrometry to products of hydrogenolysis of soil organic matter and noted the lack of deuterium-exchangeable protons on nitrogen atoms, and also noted a prevalence of methylene peaks over methyl and methine. The latter finding was indicative of a low degree of branching of the carbon chains, at least in the material modified by hydrogenolysis.

General Limitations of Spectroscopic Characterization of Humic Substances. Spectroscopic methods work best in the characterization of homogeneous substances of small or intermediate molecular size. Under such circumstances spectra are sharp, reasonably simple, and can be interpreted with relative ease. In case of humic substances the individual molecules are large and complex and dissimilar to each other. The overlap of a multitude of absorption bands results in broad areas of absorption rather than in distinct absorption maxima. For this reason, the spectra contain a severely reduced amount of useful information and, in addition, the interpretation of the residual information is fraught with complexity. Solubility of humic

compounds in suitable solvents at high enough concentrations may
present technical problems. Molecular size and a consequent lack
of volatility largely prevents the application of mass spectrometry to intact humic compounds and, therefore, discussion of
mass spectra was omitted here. However, especially in combination with gas chromatography, this technique can be very useful
in characterization of degradation or pyrolysis products.

Spectrometric Studies on Pesticide Residue - Humus Interactions.

The reduced activity of many preemergence herbicides in high
humus soils (35,36) is a general indication of the ability of
humic compounds to bind man-made chemicals by various mechanisms.
Such binding typically leads to increased persistence conbined
with immobilization and decreased biological activity, but there
are marked exceptions to this rule. Dialkyl phthalates (37,38)
and DDT (39) were reported to be mobilized by absorption to or
complexing with water soluble fulvic acids. The absorption
mechanisms of herbicides to humic compounds was recently subject
to a lucid review by Stevenson (7). He lists ion exchange,
H-bonding, van der Vaals forces and coordination through a metal
ion as the prevalent modes of attachment. The molecular structure of herbicides determines the predominant mechanism, but
more than one absorption mechanism may act on the same herbicide.

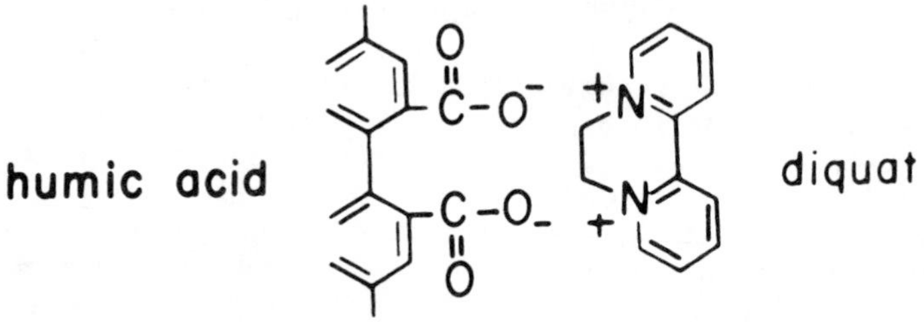

Figure 2. Charge transfer complex of diquat with humic acid. After
Ref. 40

Bipyridylium Herbicides. Herbicides of cationic nature,
such as the two bipyridylium herbicides paraquat and diquat are
bound by ion exchange reactions. This type of binding was
successfully investigated by Khan (40,41) using IR spectrometry.
Based on the shift of C-H out-of-plane bending vibrations
(from 815 cm^{-1} to 825 cm^{-1} for paraquat, from 792 cm^{-1} to
765 cm^{-1} for diquat, respectively) he deduced the formation of a
charge transfer complex (Figure 2). Less basic herbicides may
undergo similar reactions due to protonation (40).

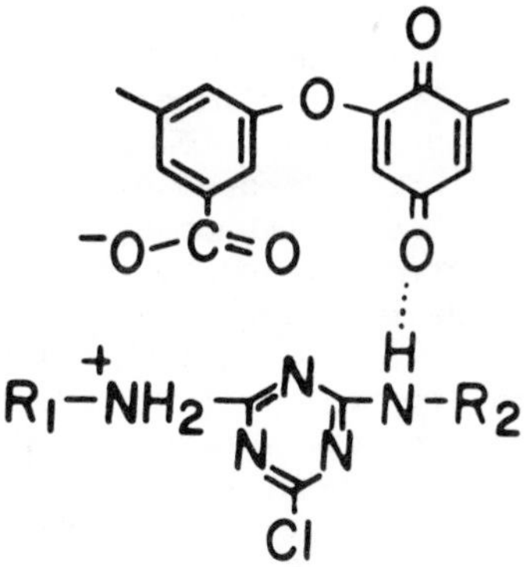

Figure 3. *Attachment of s-triazines to humic acid by charge transfer and hydrogen bonding mechanisms. After Ref. 42*

s-Triazines. H-bonding may take place between C=O groups of the humic compounds and the secondary amino groups of s-triazines (Figure 3). Evidence for this type of bonding was obtained from IR-spectra by Sullivan and Felbeck (42). These workers reacted ethanolic solutions of humic acids with various triazine herbicides. Carbonyl absorption (1720 cm^{-1}) was reduced in every case and in addition, the bands at 2,900 cm^{-1} (C-H) and 3,300 cm^{-1} (-OH) were reduced to varying degrees. New absorption bands appeared at 1,625 cm^{-1} (COO^{-} of humic acid and/or C=N of the s-triazines) and at 1,390 cm^{-1} (indicative of a salt of a carboxylic acid). From these data it was concluded that the primary binding sites of the humic acid are carboxyl groups with a contribution from phenolic hydroxyls. Since in some cases the reduction in carbonyl groups (1,720 cm^{-1}) was not accompanied by an increase of the COO^{-} (1,625 cm^{-1}) and carboxyl salt (1,390 cm^{-1}) bands, it was suggested that C=O from quinones may have also served as a binding site. As there was no evidence for involvement of any other portion of the herbicide molecule, the amino group was proposed as the active binding site of the s-triazines. Additional work on the binding of s-triazines by other techniques (14,43) tended to support the above conclusions.

Potential Applications. Aside of the reviewed IR work, we are not aware of studies on herbicide absorption by humic material which used spectrometric techniques as their primary research tool. In the light of widespread speculations that free radicals may play a role in herbicide residue binding, the lack of studies on quenching of humus ESR signals by pesticide residues is rather surprising. Another spectrometric technique that has great potential in bound pesticide residue research is the 13-C NMR (44).

13-C emits a relatively weak NMR signal, but modern instrumentation using the Fourier transformation principle and computer-
processed multiple scannings to filter out random noise, have
enormously increased the sensitivity and time-efficiency of the
NMR instruments, and made their application for routine structural investigations possible. The natural abundance of 13-C is
low (1.1%) but artificially enriched 13-C compounds are becoming
available for research. NMR-spectrometry of humic compounds
with 13-C enriched bound pesticide residues would give extremely
useful information not otherwise obtainable, since it would
elucidate the actual binding environment. In case of 14-C
labeling, this is to be deduced indirectly from degradation
studies, greatly increasing the danger of artifacts, side reactions and, consequently, erroneous or ambiguous results. The
cost and limited availability of suitable NMR instruments is
still an obstacle; one that hopefully will diminish with time.

    Chloroaniline Residues. Covalent bond formation between
pesticide residues and humic compounds was not covered in
Stevenson's review (7) and we have some ongoing work to report
in this area although, to date, spectrometric techniques played
only a minor role in this investigation. The biodegradation
of phenylamide herbicides results in release of aniline moieties
(45,46,47). Studies with 14-C labeled 3,4-dichloroaniline (DCA)
and 4-chloroaniline, compounds representative of the aniline
moieties of several phenylamide herbicides, showed that these
chloroanilines are subject to absorption as well as to covalent
binding in soil (48). The mineral part of soil plays only a
minor role in absorption, the greater portion of the anilines
becomes attached to the soil organic matter. The nature of the
non-covalent binding was not investigated in detail, but the
basic character of the anilines suggests ion exchange and
hydrogen bonding mechanisms. This reversible absorption, by
bringing the aniline molecules in intimate contact with the
humic acid molecules is probably of importance also for the
subsequent covalent binding. At 5 ppm application rate, 85-90%
of the applied aniline is covalently bound within 5 days. About
50% of the bound DCA is released in unchanged form after acid
or alkaline hydrolysis; 50% of the radioactivity remains attached
to the humic compounds. We have no direct evidence to prove
that this attached radioactivity still represents intact chloroaniline, but we believe this to be a reasonable assumption.
    To elucidate the nature of both the hydrolyzable and the
non-hydrolyzable aniline binding, we compared IR-spectra of DCA,
of humic acid and of the DCA-humic acid complex (Figure 4).
The complex contained less than 0.25% DCA by weight, and no
obvious changes in spectrum, as compared to the untreated humic
acid, could be discerned. Being unsuccessful in this preliminary
spectrometric approach, we turned to model reactions in order
to gain insight into the possible mechanisms of the binding.

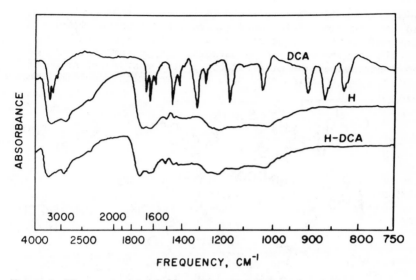

Figure 4. IR spectra of 3,4-dichloroaniline (DCA), humic acid (H), and their complex (H-DCA) (recorded in KBr)

Under ambient conditions we obtained hydrolyzable binding of chloroanilines to aldehydes and to quinones in form of anils and anilinoquinones, respectively (Figure 5). For the non-hydrolyzable binding of anilines, on theoretical basis, a number of reactions (Figure 6) can be suggested from the

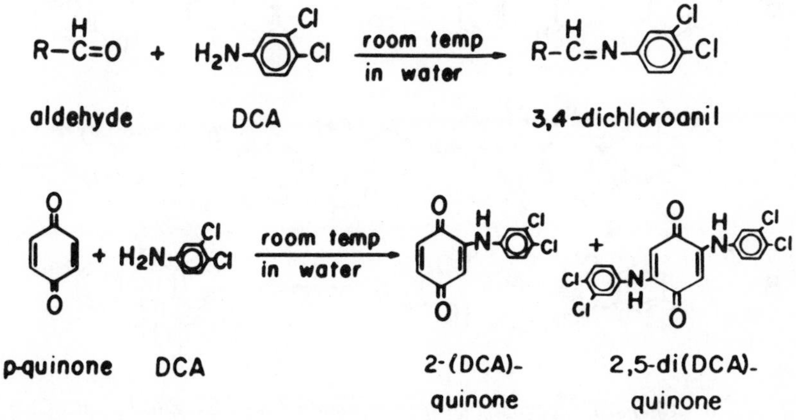

Figure 5. Reactions of 3,4-dichloroaniline (DCA) with aldehydes and p-quinone

available chemical literature (49). We are now in the process
of testing some of the more relevant reactions as models for
the nonhydrolyzable attachment of DCA to humic acid. Our
preliminary results are encouraging. Spectrometric methods
will undoubtedly be of great use in the characterization of
these model reaction products, and armed with specific infor-

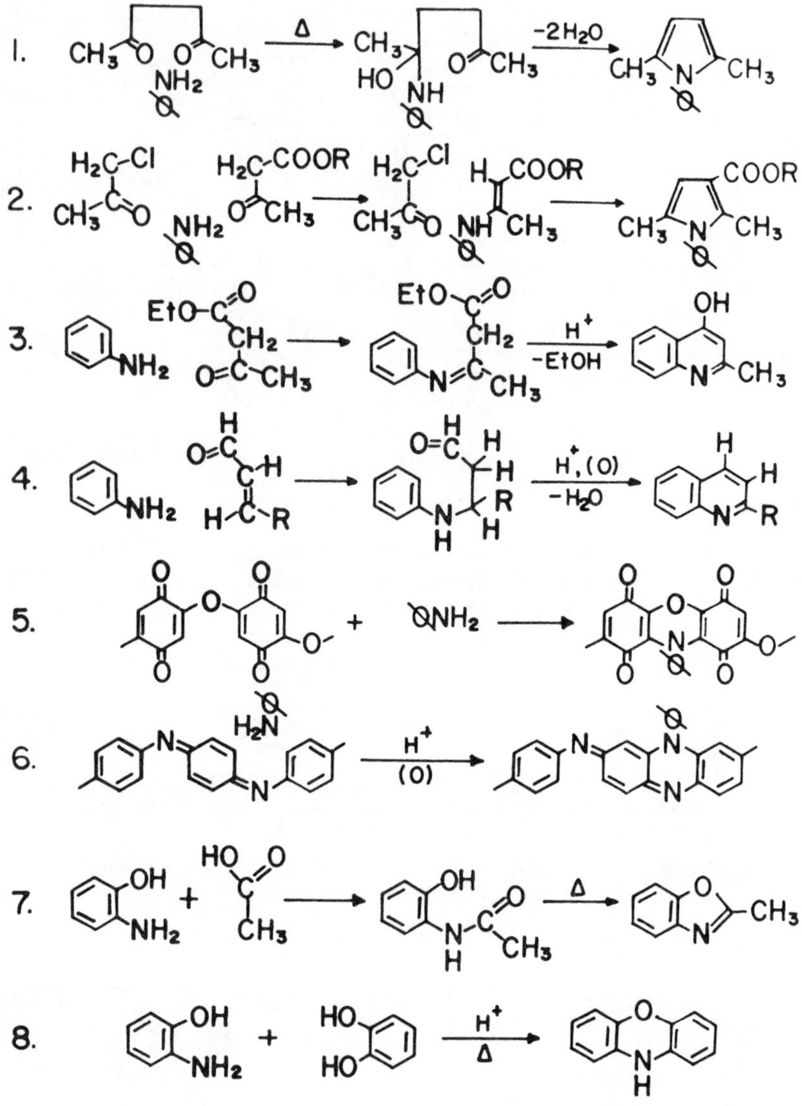

Heterocyclic Chemistry

*Figure 6. Chemical reactions that lead to non-hydrolyzable attachment of ani-
lines (49)*

mation we may be more successful in the spectrometric investigation of DCA-humic acid complexes in the future.

While much work remains to be done, we believe that the
phenomenon of the covalently bound aniline residues may resemble
a very familiar but in its details still obscure natural process
of ammonia and amino acid attachment to humus that occurs both
in hydrolyzable and in non-hydrolyzable forms (50). The
mechanisms by which phenoxazines are formed from 4-methylcatechol
and ammonia (Figure 7,A), and phenazines are formed from quinones and ammonia (Figure 7,B) resulting in heterocyclic non-
hydrolyzable nitrogen (51) have very obvious analogies to the

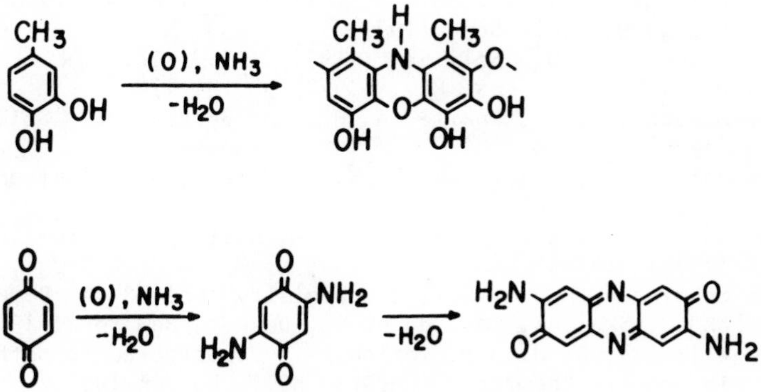

Figure 7. Reactions of 4-methylcatechol (A) and of p-quinone *(B) with ammonia, leading to non-hydrolyzable incorporation of nitrogen into phenoxazine (A) and phenazine (B) type compounds. After Ref. 51*

type of reactions we propose for aniline binding. If we look
at DCA as ammonia tagged by a stable and easily recognizable
chlorophenyl ring, we can imagine that our applied research on
DCA binding to humus may eventually contribute to the understanding of much more fundamental aspects of soil chemistry.

LITERATURE CITED.

1. Swain, F. M. "Geochemistry of Humus". In: Organic Geo-
    chemistry. (I.A. Berger, ed.) pp. 81-147, Pergamon Press,
    New York, 1963.
2. Kononova, M. M. "Soil Organic Matter." Pergamon Press,
    New York, 1966.

3. Haider, K. and Martin, J. P. Soil. Biol. Biochem. (1970) 2:145-156.
4. Martin, J. P. and Haider, K. Soil Sci. (1971) 111:54-63.
5. Stevenson, F. J. In: C. A. Black (ed.) "Methods of Soil Analysis" pp. 1409-1421, American Society of Agronomy, Madison, 1965.
6. Stevenson, F. J. and Butler, J. H. In: Organic Geochemistry - Methods and Results. (G. Eglington and M. T. J. Murphy, eds.) pp. 534-557, Springer, New York 1969.
7. Stevenson, F. J. Bioscience (1972) 22:643-650.
8. Tan, K. H. Soil Sci. Soc. Amer. Proc. (1975) 39:70-73.
9. Mathur, S. P. Soil Sci. (1972) 113:136-139.
10. Felbeck, G. T. Jr. Adv. Agron. (1965) 17:327-368.
11. Felbeck, G. T. Jr. Soil Sci. (1971) 111:42-48.
12. Flaig, W. Soil Sci. (1971) 111:19-33.
13. Haworth, R. D. Soil Sci. (1971) 111:71-78.
14. Hayes, M.H.B. Res. Rev. (1970) 32:131-174.
15. McLaren, A. D., Pukite, A. H., and Barshad, I. Soil Sci. (1975) 119:178-180.
16. Haider, K., Frederick, L. R. and Flaig, W. Plant and Soil (1965) 22:49-64.
17. Ladd, J. N. and Butler, J. H. A. Austral. J. Soil Res. (1966) 4:41-54.
18. Schnitzer, M. In:"Soil Biochemistry" Vol. 2 (A. D. McLaren and J. Skujins, eds. pp. 60-95, Dekker, New York, 1971.
19. Williams, D. H. and Fleming, I. "Spectroscopic Methods in Organic Chemistry" McGraw Hill, London 1973.
20. Nakanishi, K. "Infarared Absorption Spectroscopy - Practical." Holden-Day, San Francisco, 1962.
21. Stevenson, F. J. and Goh, K. M. Soil Sci.(1972) 113:334-345.
22. Schnitzer, M. and Skinner, I. M. Soil. Sci. (1965) 99:278-284.
23. Stevenson, F. J. and Goh, K. M. Soil Sci. (1974) 117: 34-41.
24. Schnitzer, M. Soil Sci. (1974) 117:94-102.
25. Schnitzer, M. and Skinner, S. I. M. Soil Sci. (1963) 96:86-93.
26. Schnitzer, M. and Hoffman, I. Soil Sci. Soc. Amer. Proc. (1964) 28:520-525.
27. Steelink, C. and Tollin, G. In: Soil Biochemistry (A. D. McLaren and G. Peterson, eds.) pp. 147-169, Marcel Dekker, New York, 1967.
28. Schnitzer, M. and Skinner, S. I. M. Soil Sci. (1969) 108:383-390.
29. Atherton, N. M., Cranwell, P. A., Floyd, A. J., and Haworth, R. D. Tetrahedron (1967) 23:1653-1667.
30. Cheshire, M.V. and Cranwell, P.A. J. Soil Sci.(1972)23:424-430.
31. Haworth, R. D. Soil Sci. (1971) 111:71-79.
32. Bible, R. H. Jr. "Interpretation of NMR Spectra" Plenum Press. New York, 1965.

33. Barton, D. H. R. and Schnitzer, M. Nature (1963) 198:217-218.
34. Felbeck, G. T. Jr. Soil Sci. Soc. Amer. Proc. (1965) 29:48-55.
35. Bailey, G. W. and White, J. L. J. Agr. Food Chem. (1964) 12:324-333.
36. Upchurch, R. P. Res. Rev. (1966) 16:46-85.
37. Matsuda, K. and Schnitzer, M. Bull. Env. Contam. Toxicol. (1971) 6:200-204.
38. Ogner, G. and Schnitzer, M. Science (1970) 170:317-318.
39. Ballard, T. M. Soil Sci. Soc. Amer. Proc. (1971) 35:145-147.
40. Khan, S. U. Can. J. Soil Sci. (1973) 53:199-204.
41. Khan, S. U. J. Env. Qual. (1974) 3:202-206.
42. Sullivan, J. D. Jr. and Felbeck, G. T. Jr. Soil Sci. (1968) 106:42-52.
43. Li, G.-C. and Felbeck, G. T. Jr. Soil Sci. (1972) 113:140-148.
44. Levy, G. C. and Nelson, G. L. "Carbon 13 Nuclear Magnetic Resonance for Organic Chemists" Wiley-Interscience, New York, 1972.
45. Geissbühler, H. In: Degradation of Herbicides. (P. C. Kearney and D. D. Kaufman, eds.) pp. 79-111, Dekker, New York, 1969.
46. Herrett, R. A. In: Degradation of Herbicides. (P. C. Kearney and D. D. Kaufman, eds.) pp. 113-145, Dekker, New York, 1969.
47. Bartha, R. and Pramer, D. Adv. Appl. Microbiol. (1970) 13:317-341.
48. Hsu, T.-S. and R. Bartha. Soil Sci. (1974) 116:444-452.
49. Joule, J. A. and Smith, G. F. "Heterocyclic Chemistry" van Nostrand Reinhold, London, 1972.
50. Nömmik, H. Plant and Soil (1970) 33:581-595.
51. Lindbeck, M. R. and Young, J. L. Anal. Chim. Acta (1965) 32:73-80.

ACKNOWLEDGEMENT: This paper of the Journal Series, New Jersey Agricultural Experiment Station, New Brunswick, N.J., was supported by RR,NE-63 and Hatch funds.

20

Classification of Bound Residues Soil Organic Matter: Polymeric Nature of Residues in Humic Substance

R. W. MEIKLE, A. J. REGOLI, and N. H. KURIHARA
Dow Chemical U.S.A., Ag-Organics Research, 2800 Mitchell Drive, Walnut Creek, Calif. 94598

D. A. LASKOWSKI

Dow Chemical U.S.A., Ag-Organics Department, Midland, Mich. 48640

In the course of studying the decomposition of radioactive, foreign organic compounds in soil, we invariably find radioactivity associated with the soil organic matter. This associated radio-activity has generally not been identified because of the difficulty of working with the material.

As the incubation time for the organic compound in soil increases, the amount of radioactivity in the soil organic matter also generally increases. Consequently, it has become increasingly important to have some notion of how this radioactivity is combined structurally with the soil organic matter.

The organic matter of soils consists of a mixture of plant and animal products in various stages of decomposition, of substances synthesized biologically and/or chemically from the breakdown products, and of microorganisms and small animals and their decomposing remains. To simplify this very complex system, organic matter is usually divided into two groups: (a) nonhumic substances and (b) humic substances.

Nonhumic substances include compounds of known chemical characteristics. To this class of compounds belong carbohydrates, proteins, peptides, amino acids, fats, waxes, resins, pigments and other low-molecular-weight organic substances. In general, these compounds are relatively easily attacked by microorganisms in the soil and have a relatively short survival rate.

The bulk of the organic matter in most soils consists of humic substances. These are amorphous, brown or black, hydrophilic, acidic, polydisperse substances of molecular weights ranging from several hundreds to tens of thousands. Based on their solubility in alkali and acid, humic substances are usually divided into three main fractions: (a) humic acid (HA), which is soluble in dilute alkaline solution but is precipitated by acidification of the alkaline extract; (b) fulvic acid (FA), which is that humic fraction which remains in the aqueous acidified solution, i.e., it is soluble in both acid and base; and (c) the humic fraction that cannot be extracted by dilute base and acid, which is referred to as humin.

There is increasing evidence that the chemical structure and properties of the humin fraction are similar to those of HA, and that its insolubility arises from the firmness with which it combines with inorganic soil and water constituents. Data available at this time suggest that structurally the three humic fractions are similar to each other, but that they differ in molecular weight, ultimate analysis, and functional group content. The FA fraction has a lower molecular weight but a higher content of oxygen-containing functional groups per unit weight than do HA and the humin fraction. While the fractionation scheme is arbitrary -- the fractions are still molecularly heterogeneous -- it has nonetheless been widely accepted.

The ability of synthetic cross-linked polydextran gels to separate molecules by their molecular size has become increasingly important in the study of polymeric substances (1). This report describes the use of Sephadex® gels to fractionate radioactive humic substances extracted from soil after the soil has been incubated with radioactive ditalimfos fungicide, 0,0-diethyl-phthalimido-1-^{14}C-phosphonothioate.

Experimental

Three soil samples were used in this study and their physical properties are described in Table I. Mechanical analyses were carried out using the hydrometer method (2). Soil pH was measured in water at a 1:1 soil:solution ratio with a glass electrode assembly (3). Organic matter content of soil was determined using a wet combustion method (4). The moisture content of the soils at 1/3 bar tension was also determined (5).

TABLE I

Some properties of the soils used in the study of ditalimfos decomposition.

Soil textural classification and source	Sand, %	Silt, %	Clay, %	Organic carbon, %	Soil moisture content at 1/3 bar tension	pH
Loam, Davis, California	46	35	18	0.86	21.75	6.4
Sandy Loam, No. Dakota	66	22	12	2.2	22.52	7.3
Silty Clay Loam, Geneseo, Illinois	14	54	32	4.2	26.31	5.8

These soils were incubated with radioactive ditalimfos and then analyzed for this compound and its decomposition products at appropriate times. The results of this study will be reported elsewhere.

The soils were first extracted with acidified ether to remove extractable radioactive compounds, rinsed with water, and dried at ambient temperature. Humic substances were obtained from the extracted soils by shaking 2-g. soil samples for 18 hours at ambient temperature with 3g. of DOWEX® A-1 chelating resin (sodium form) of 50 to 100 mesh and 25 ml. of water. The total nominal capacity of the resin was 2.6 meq./3g. The soil suspensions were centrifuged at 12,000 x G. Aliquots of these radioactive solutions were examined by gel chromatography, separation into humin-humic acid-fulvic acid fractions, and dialysis.

Gel chromatography. The polydextran gels (Sephadex G-50 and G-100) were prepared as recommended by the manufacturer (6).

The column used was 2.6 x 70 cm. The void volume (V_o) was determined empirically by using Blue Dextran 2000 (Pharmacia). V_o as shown on the figures indicates the first excluded Blue Dextran fraction. The total bed volume (V_t) was obtained by water calibration of the column before packing the bed. V_t for the G-50 columns was 248 ml and for the G-100 column, 266 ml. The gel did not compact during elution. Five-ml. fractions of column effluent were collected. The flow rate was maintained at 0.5 ml/min.

The buffer systems used as eluant were 0.025M sodium borate (pH 9.1) for the G-50 gel and 0.1M sodium hydroxide for the G-100 gel.

The fractionation ranges of the gels are reported (6) to be as follows: G-50, solutes with molecular weights from 500 to 10,000; correspondingly for G-100, solutes with molecular weights from 1,000 to 100,000. These values are based on calibrated dextrans (Pharmacia).

Over a considerable range, the elution volume (V_e) of a polymer from a dextran gel column is approximately a linear function of the logarithm of the molecular weight (7,8,9,10). The gel columns were calibrated for molecular weight using samples of calibrated dextrans (Dextran T®, Pharmacia). Dextran in the eluted fractions was determined by the method of Dubois, et al (11). A linear regression of ln molecular weight on the "gel affinity constant" (K_{av}) allowed the calculation of apparent molecular weights of eluted radioactivity from the elution volume (V_e). The constant, K_{av}, is defined by Laurent and Killander (9) and it is related to V_e as follows: $K_{av} = (V_e-V_o)/(V_t-V_o)$. K_{av} is independent of column geometry and packing density. This constant we define as the "gel affinity constant" where its magnitude bears a direct relationship to the affinity of the eluted molecule for the gel.

Fractionation of soil extracts. A traditional fractionation
of the soil resin extracts with alkali into humic acid (precipitated
from alkaline solution by acid), fulvic acid (that part of the
alkaline solution not precipitated by acid), and humin (organic
material not soluble in alkali) was performed as described by
Schnitzer and Kahn (12).

Dialysis of soil extracts. The soil resin extracts were also
submitted to dialysis against running tap water in cellulose
acetate at ambient temperature.

Results and Discussion

Of the large number of extractants that have been tested,
dilute aqueous sodium hydroxide has been the most commonly used
and quantitatively the most effective reagent for extracting
humic substances from soils. When the incubated soils listed
in Table I were extracted with hot 1N sodium hydroxide solution
or with DOWEX A-1 chelating resin and water, after first being
extracted with acidified ether to remove extractable radioactive
compounds, we found that the two extraction methods were equally
efficient at removing radioactive humic substances. These results
are shown in Table II.

TABLE II

Extraction of radioactivity from soils using DOWEX A-1 Chelating
Resin and 1N sodium hydroxide[a]

Soil textural classification and source	Incubation conditions		Radioactivity,% in..[b]	
	time,days	temp,°C	resin extract	NaOH extract
Loam, Davis, California	56	15	95	82
Loam, Davis, California	40	35	76	80
Sandy Loam, No. Dakota	33	25	82	88
Silty Clay Loam, Geneseo, Illinois	175	15	91	85
		Ave.	86	84
		St'd. error	4	2

a/ Soils had been incubated with ditalimfos-^{14}C (5 ppm) and subsequently extracted with ether/0.1N HCl (1.5/1.0 v/v).
b/ These values are % of that present in soil after acidified
ether extraction. The initial values were 37%, 31%, 35% and
30% of the applied radioactivity in the soils as listed in the
table.

Resin extraction of soil results in efficient removal of poly-
valent cations that bind organic substances in soil. This increases the dispersity of humic substances and also increases their
solubility by disrupting the hydrogen bonds of the fixed metallic
cations. Sodium hydroxide accomplishes much the same thing but
is a more severe reagent. Thus, extraction of soils with a
chelating resin will usually result in less degradation to soil
organic matter (13).

When aliquots of the resin soil extracts were submitted to
gel chromatography the results shown in Figures 1 to 5 were obtained. In each case, a portion of the radioactive material placed on
the column was eluted in two main fractions. The apparent molecular weights and percent recovery based on applied radioactivity
are indicated on the figures.

It is recognized that the molecular weights shown in these
figures are only approximate. Manufacturers use dextrans to
calibrate their polydextran cross-linked gels. If the humic
substance molecules are more asymmetric than the dextrans used
for calibration, as seems likely, then any particular grade of
gel will exclude lower molecular weight humic substances than
the nominal value would indicate. Put another way: For equal
molecular weight substances, a higher degree of molecular asym-
metry is equivalent to a larger size. Thus, the apparent molecular
weight values in these figures are probably high.

The evidence clearly indicates there are at least two radioactive polymer fractions in each of the soil samples. The North
Dakota soil (Figure 3) appears to have five additional radioactive
fractions but this degree of separation would need confirming.
The range of apparent molecular weights for these polymer fractions
is 2100 to >10,000. However, when the North Dakota soil sample
extract was submitted to gel chromatography using a gel with less
cross-linking, thus extending the exclusion limit of the gel, it
was found that the high molecular weight fraction in the extract
could be assigned an apparent molecular weight >100,000 (Figure 5).
The apparent molecular weight range of humus is reported to be
600 to 300,000 (14).

To further characterize the radioactive polymeric substances
in the DOWEX A-1 resin extracts of soil, a sample of the North
Dakota soil extract was separated into fulvic acid, humic acid
and humin using the tradition fractionation scheme described by
Schnitzer and Kahn (12). The proportion of radioactivity in
humic acid to that in fulvic acid was 1.8:1.0. A hot, 1N
sodium hydroxide extraction of this same soil, followed by
separation into humic acid and fulvic acid, resulted in a radioactive humic acid-fulvic acid proportion of 0.6:1.0. In
the one case, where soil was extracted with a chelating resin,
the radioactive humic acid fraction was high relative to the
radioactive fulvic acid; in the other -- extraction with hot
sodium hydroxide -- the radioactive humic acid fraction was
low. The reason for this reversal is that hot sodium hydroxide
causes greater degradation of the humic acid polymers (high
molecular weight) than does the chelating resin. The resulting

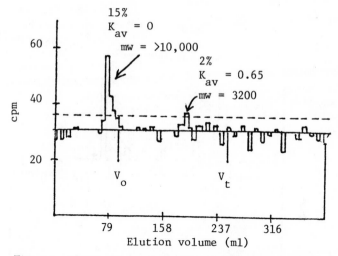

Figure 1.   *Elution diagram for radioactive polymers in chelating resin extract of Davis, Calif. soil (15°C): Sephadex G-50, 0.025M sodium borate, pH 9.1*

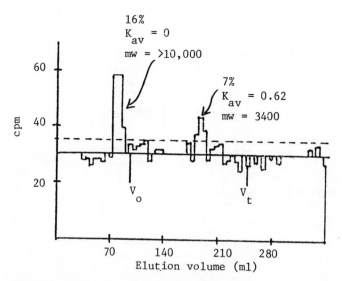

Figure 2.   *Elution diagram for radioactive polymers in chelating resin extract of Davis, Calif. soil (35°C): Sephadex G-50, 0.025M sodium borate, pH 9.1*

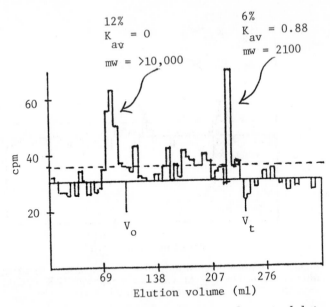

Figure 3. *Elution diagram for radioactive polymers in chelating resin extract of North Dakota soil (25°C): Sephadex G-50, 0.25M sodium borate, pH 9.1*

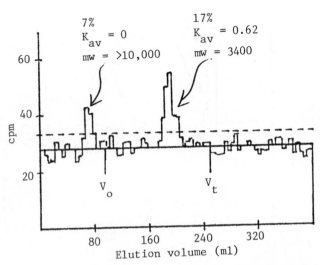

Figure 4. *Elution diagram for radioactive polymers in chelating resin extract of Illinois soil (15°C): Sephadex G-50, 0.025M sodium borate, pH 9.1*

decomposition products have a lower molecular weight and tend
to fractionate as fulvic acids. Dormar (15) has shown
that extraction of organic matter with chelating resin provides
humic substances with minimum alteration.

The humic and fulvic acid fractions separated from the DOWEX
A-1 resin extract of the North Dakota soil were each submitted to
gel chromatography and the results appear in Figures 6 and 7. We
see high molecular weight radioactive material in the humic acid
fraction and it comprises the major part of the moveable radioactivity in this fraction. The lower molecular weight radioactive material appears in the moveable portion of the fulvic
acid fraction with some overlap of 2300 Dalton polymers into
the humic acid fraction. Thus the molecular weight distribution
of radioactive fractions in the soil extracts follows the pattern
expected for fractionation of humic substances.

When aliquots of the DOWEX A-1 resin soil extracts of each
soil were dialyzed through cellophane (cellulose acetate) an
average 53% of the radioactive material was retained by the
membrane. That portion of the north Dakota soil extract retained
by the cellophane membrane was submitted to gel chromatography
using gel G-50. The results are shown in Figure 8. We see that
the radioactive polymers with $K_{av} > 0$, apparent molecular weight,
<10,000, diffused through the membrane, while those with an appar-
ent molecular weight range >10,000 were retained and appear in
Figure 8 (see Figure 3 for comparison). This is another demons-
tration that a portion of the radioactivity in the resin extracts
of soil is associated with non-dialyzable, high molecular weight
humic substances.

The recovery of radioactive material from the Sephadex gel
columns varied from 17% to 31% of that put on the column. In the
case of the dialysis experiment, only 17% of the radioactivity
applied to the column appear in the eluate as a single peak
in Figure 8. Apparently a large part (83%) of the high molecular
weight (>10,000) radioactive material is in some way strongly
adsorbed by the gel. When this gel was removed from the column
and segments were assayed for radioactivity, 93% of the retained
activity was found in the first inch and 100% in the first 5 inches.
Sephadex gels are known to adsorb some proteins (16), aromatic and
heterocyclic compounds (17), and humum molecules (14). This
phenomenon probably accounts for the low recovery of radioactive
material from the gel columns used in our work.

It has been shown in our work with ditalimfos-^{14}C/soil that
the specific radioactivity of the humus fractions, as dpm/mg. of
carbon, bears an inverse relationship to molecular weight. The
data showing the relationship are reproduced as Table III.

These changes in the specific activity of the soil organic
carbon fractions are consistent with the concept that the more
soluble fractions have a more rapid turnover. Thus, if humin
represents organic carbon that is formed and broken down more
slowly than fulvic acid, for example, then a smaller proportion
of the total carbon of the humin will be "new" carbon containing
[14]C.

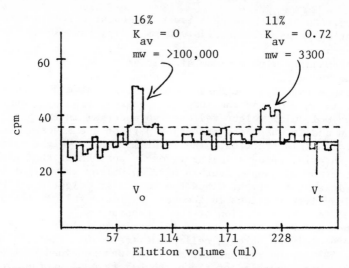

Figure 5.   Elution diagram for radioactive polymers in chelating resin extract of North Dakota soil: Sephadex G-100, 0.1M sodium hydroxide

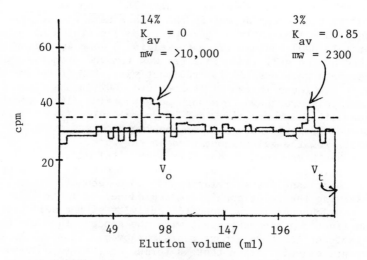

Figure 6.   Elution diagram for radioactive humic acid from chelating resin extract of North Dakota soil: Sephadex G-50, 0.025M sodium borate, pH 9.1

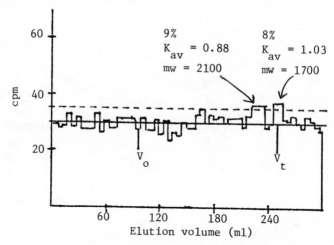

Figure 7. *Elution diagram for radioactive fulvic acid from chelating resin extract of North Dakota soil: Sephadex G-50, 0.025M sodium borate, pH 9.1*

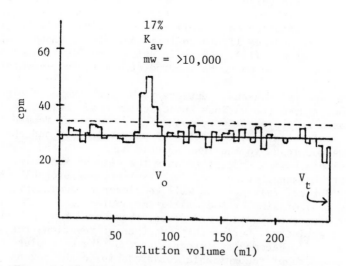

Figure 8. *Elution diagram for radioactive polymers retained by membrane after dialysis of chelating resin extract of North Dakota soil: Sephadex G-50, 0.025M sodium borate, pH 9.1*

TABLE III

Specific activities, dpm/mgC, for fractionated soil organic matter
after incubation of soils with ditalimfos-[14]C.

| Soil Sample | FRACTION Increasing molecular weight————————> | | |
	(fulvic acid)	humic acid)	(humin)
Davis soil, 15°	2080	2074	467
Davis soil, 35°	2162	1904	481
No. Dakota soil, 25°	641	582	281
Illinois soil, 15°	638	281	46

Formation of humic substances in soil is a dynamic process
occurring through the action of microbes on plant material (18).
Macromolecules are formed at the expense of carbohydrates of
plant origin. These macromolecules include bacterial gums, alginic
acid, pectic acid, and other less well-defined polymeric carboxylic
acids. Aromatic polyphenols formed by way of oxidation of quinones
can condense with amino acids to ultimately give humic-like substances. Basidiomicetes as well as other microscopic fungi have
been found to degrade lignin to form appreciable amounts of humic
acid-like polymers (19). Phenolic units from ^{14}C-labeled phenolase
lignin have been shown to be incorporated into fungi-synthesized
polymers (20).

The general consensus appears to be that there is a genetic
relation between the various humic substances. Fulvic acid is
considered to represent poly-condensation material formed from
simpler molecules. Continuation of polymerization and chemical
modification leads to the less soluble humic acid and eventually
to insoluble humin, thought to have the highest molecular weight
and most resistant structure. The earlier, and probably more
rapidly formed, fulvic acids will be closer to equilibrium with
the ^{14}C pool of simpler and smaller molecules than will materials
farther down the sequence and would, therefore, have a higher
specific activity. During this sequence of reactions the incorporated ^{14}C becomes an integral part of the molecular
structure without recognizable relationship to the parent
molecule from which it is derived.

The rate of humin degradation is very slow (21). Sorenson
(22) studied the degradation of labeled glucose and cellulose in

three soils. After a rapid initial breakdown, half-lives of 5 to
9 years were reported for the remaining ^{14}C to be degraded. These
data imply that, even with readily metabolized compounds, incorporation into humic substances occurs and limits the extent to
which complete degradation to CO_2 proceeds. Likewise, pesticide
molecules degrade and the products ultimately become incorporated
in humic materials. These macromolecules so formed are indistin-
guishable from those derived from carbon compounds natural to soil.
Other authors have demonstrated the formation of humin from readily
decomposable organic compounds (23,24).

In summary, we have shown that when an organic compound
incorporated in soil is decomposed, a part of the decomposition
products ultimately become associated with the soil organic
material. These products are sometimes referred to as "bound
material". In reality a large part of the soil organic matter
can be solubilized with reagents such as hot aqueous sodium
hydroxide or DOWEX A-1 chelating resin and water. The later
is preferred because it is much less destructive to the organic
matter.

Further, we have shown that a part of the decomposition
products are combined with the extracted organic material in
such a way that the products are an integral part of the poly-
molecular structure of the organic material.

Finally, we have shown that the fractions of soil organic
material, commonly known as fulvic acid, humic acid and humin,
contain incorporated decomposition products. These macromolecules can be separated into radioactive fractions having
apparent molecular weights ranging from 2100 to >100,000.

LITERATURE CITED

1. Altgelt, K. H., and Segal, L., "Gel Permeation Chromatography",
    Dekker, New York (1971).
2. Day, P. R., Particle fractionation and particle-size analysis,
    pages 545-566 in C. A. Black (ed), "Methods of Soil Analysis",
    Amer. Soc. of Agron., Inc., Madison, Wisc. (1965).
3. Peech, M., Hydrogen-ion activity, pages 914-925 in C. A. Black
    (ed), "Methods of Soil Analysis", Amer. Soc. of Agron., Inc.,
    Madison, Wisc. (1965).
4. Allison, L. E., Organic Carbon, pages 1367-1378 in C. A.
    Black (ed), "Methods of Soil Analysis", Amer. Soc. of Agron.
    Inc., Madison, Wisc. (1965).
5. Richards, L. A., Physical condition of water in soil, pages
    131-137 in C. A. Black (ed), "Methods of Soil Analysis",
    Amer. Soc. of Agron., Inc., Madison, Wisc. (1965).
6. Anon., "Sephadex-gel filtration in theory and practice",
    Pharmacia Fine Chemicals, Inc., 800 Centennial Ave.,
    Piscataway, N. J. 08854. (1966).
7. Carnegie, P. R., Nature (1965) 206, 1128.
8. Andrews, P., Biochem. J. (1965), 96, 595.

9.   Laurent, T. C., and Killander, J., J. Chromatog. (1964),
     14, 317.
10. Carnegie, P. R., Biochem. J. (1965) 95, 9P.
11. Dubios, M., Gilles, K. A., Hamilton, J. K., Rebers, P. A.,
     and Smith, F., Anal. Chem. (1956), 28, 350.
12. Schnitzer, M., and Kahn, S. U., "Humic Substances in the
     Environment", p. 17, Dekker, New York, (1972).
13. Bremner, J. M., Organic nitrogen in soils, pages 93-149 in
     W. V. Bartholomew and F. E. Clark (eds), "Soil Nitrogen",
     Amer. Soc. Agron., Madison, Wisc. (1965).
14. Gjessing, E. T., Nature (1965), 197, 1091.
15. Dormaar, J. F., Bull. Ass. Fr. Etude Sol (1973), (2), 71-9.
     Chem. Abstr. (1974), 80, 107113t.
16   Glazer, A. N., and Wellner, D., Nature (1962), 194, 862.
17. Gelotte, B., J. Chromatography (1960), 3, 330.
18. McLaren, A. D., Science (1961), 141, 3586.
19. Hurst, H. M. and Burger, W. A., Lignin and Humic Acids,
     pages 260-286 in "Soil Biochemistry", McLaren, A. D., and
     Peterson, G. H. (eds), Dekker, New York (1967).
20. Martin, J. P., and Haider, K., Soil Sci. (1971), 111, 54.
21   Stevenson, I. L., Biochemistry of Soil, Page 242, in
     "Chemistry of the Soil No. 160", Bear, F. E. (ed),
     Reinhold Publ. Corp., New York (1964).
22. Sorenson, L. H., Soil Biol. Biochem. (1972), 4, 245.
23. Chekalar, K. I., and Illyuvieva, V. P., Pochvovedenie
     No. 5, pp. 40-5 (1962).
24. Sinha, M. K., Plant and Soil (1972), 36, 283.

Chemical Extraction of Certain Trifluoromethanesulfonanilide Pesticides and Related Compounds from the Soil

SURESH K. BANDAL, HENRY B. CLARK, and JAY T. HEWITT

3M Co., 3M Center, St. Paul, Minn. 55101

The 3M Company, Saint Paul, Minnesota, has discovered three promising new agrichemicals of the N-aryl 1,1,1-trifluoromethanesulfonamide class of compounds:

Perfluidone (DESTUN® Herbicide)

Fluoridamid (SUSTAR® Plant Growth Regulator)

MBR 12325 (Experimental Herbicide/Plant Growth Regulator)

Perfluidone shows a dramatic control of nutsedge (Cyperus sp.) and is also herbicidal to a variety of important grassy and broadleaf weeds. Perfluidone has been granted a temporary permit for use on cotton by the EPA, and petitions for full registration for use on cotton and for establishment of negligible residue tolerance in cottonseed have been submitted. Fluoridamid is fully registered by the EPA for certain applications as a turf growth retardant. MBR 12325 is undergoing development as a grass and ornamental plant growth retardant and as an agent for enhancing sugar content in sugarcane.

Because these compounds are either applied directly to soil or are applied foliarly, in which case an appreciable amount of compound would be expected to eventually lodge on the soil, it was necessary to study the degradation of each compound in soil. In addition, it was of interest to study the behavior in soil of several known or potential soil metabolites of these three compounds in order to elucidate the reactivity of the various functional groups present. The ten compounds which were studied are shown in Figures 1 to 4; for convenience, they are divided into four groups based on similarities of chemical structure. The known soil metabolites are Compound II (Figure 1), which is the major soil metabolite of perfluidone (1) and Compound VII (Figure 3) which has been shown to be a major soil metabolite of MBR 12325 (2).

Materials

All chemicals were of greater than 99% purity. Appropriate amounts of nonradiolabeled chemicals were mixed with the corresponding carbon-14 labeled compounds to yield the desired specific activities. Perfluidone and metabolite α were uniformly labeled on the trisubstituted ring; all others were uniformly labeled on the benzene ring. All radiolabeled compounds were of greater than 99% radiochemical purity. The specific activities of the various compounds are listed in Table I.

The soil used was a sandy loam obtained from Brainerd, Minnesota, and contained 57% sand, 32% silt, 11% clay, and 2.0% organic matter. The pH of the soil was 6.5.

Methods

Thin-layer Chromatography (tlc). Silica gel F_{254} precoated chromatoplates (20 x 20 cm, MN brand,

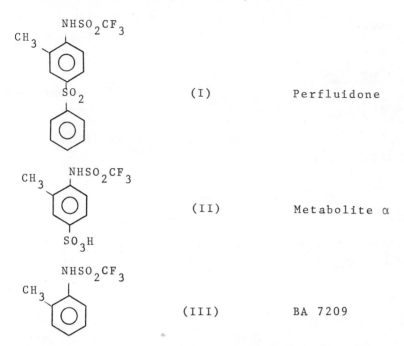

(I) Perfluidone

(II) Metabolite α

(III) BA 7209

Figure 1. *Group 1 compounds: 1,1,1-trifluoromethanesulfonanilides containing no other —NHR groups*

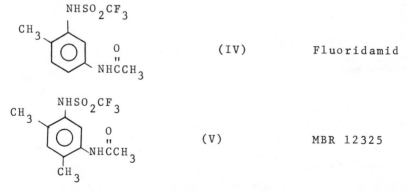

(IV) Fluoridamid

(V) MBR 12325

Figure 2. *Group 2 compounds: 1,1,1-trifluoromethanesulfonanilides containing a 3-acetamido group*

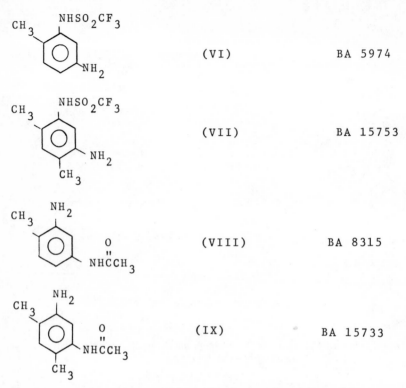

(VI)   BA 5974

(VII)   BA 15753

(VIII)   BA 8315

(IX)   BA 15733

Figure 3. Group 3 compounds: 1,1,1-trifluoromethanesulfonanilides or acetanilides containing one free —NH₂ group

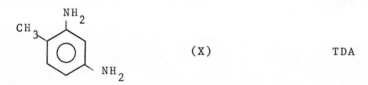

(X)   TDA

Figure 4. Group 4 compound: compound containing two free —NH₂ groups

Table I. Specific Activities of the Compounds Used and TLC Solvent Systems Used for Cleanup and Two-Dimensional TLC Analysis.

Compound	Designating Number	Specific Activity, dpm/μg	TLC Solvent System	
			Cleanup and First Development in 2-Dimensional Analysis	Second Development in 2-Dimensional Analysis
Perfluidone	I	1993	A	C
Metabolite α	II	972	D	E
BA 7209	III	958	A	C
Fluoridamid	IV	199	B	C
MBR 12325	V	830	B	C
BA 5974	VI	1355	B	C
BA 15753	VII	1975	B	C
BA 8315	VIII	1957	B	C
BA 15733	IX	2006	B	C
TDA	X	2248	B	C

obtained from Brinkmann Instruments, Incorporated,
Westbury, New York) were used at 0.25 and 0.50 mm
gel thickness for analysis and at 1.0 and 2.0 mm gel
thickness for preliminary cleanup of soil extracts.
The solvent systems used and their alphabetical
designations are as follows:
   A: Ethyl acetate-toluene-chloroform-formic acid
        (1-1-1-0.06),
   B: Ethyl acetate-acetic acid (49-1),
   C: Chloroform-methanol-acetic acid (45-5-1),
   D: Chloroform-methanol-acetic acid-water
        (25-15-4-2), and
   E: n-Butanol-water-acetic acid (3-1-1).
   Two-dimensional cochromatography of a radiolabeled component, detected by radioautography
(Kodak No-Screen X-ray Film) with the authentic
unlabeled compound in two different solvent systems,
was considered sufficient to constitute tentative
identification of the component. The tlc solvent
systems used for each compound are also listed in
Table I. Quantitative data were obtained by scraping
radioactive gel regions, as detected by radioautography, into scintillation vials for radiocarbon
content determination by direct liquid scintillation
counting (lsc). Each vial was recounted after addition
of a known amount of toluene-^{14}C to determine counting
efficiency.

   Preparation of Soil Samples. For each compound,
for each time period studied, 100 gram portions of
soil (dry weight basis) were gently packed into glass
jars (7 cm high X 7 cm in diameter). Each compound
was applied to the soil surface as a solution of 1 mg
of the diethanolamine or potassium salt of the compound
in 5 ml of distilled water (pH ∿8) so that the
concentration of the compound in soil was 10 ppm. The
treated soil samples were watered periodically but
allowed to dry out between waterings so that simulated
field conditions of alternate wetting and drying were
achieved. The soil samples were held in the greenhouse
and GROLUX^R (Model FR 96T12-GRO-VHO-WS, Sylvania
corporation, Salem, Massachusetts) reflectorized wide
spectrum lamps were suspended at a distance of approximately 30 inches from the top of the soil with a
day-night cycle of 18:6 hours.

   Soil Analysis. For each compound, at desired time
intervals ranging up to two months after soil treatment,
duplicate soil samples were separately soaked with 15%
(V/W) of distilled water for 12 hours and then soxhlet-

extracted for 16 hours with an acetonitrile-water
azeotrope (5-1); a t_o sample was similarly extracted
but without soaking. The radiocarbon content of
each extract was determined by submitting an aliquot
to lsc. The extracts were evaporated to dryness in
vacuuo using a rotary evaporator with a water bath
maintained at 45°C or less to minimize thermal decomposition. The concentrated extracts were then analyzed
by tlc and radioautography.

The solvent extracted soil samples were air-dried
and put into one-pint jars, mechanically rotated for
at least 24 hours to insure thorough mixing, and a
sample (approximately 1-1.5 grams) analyzed by combustion-lsc to determine residual unextracted radiocarbon content. The difference between the amount of
radiocarbon applied and the sum of extractable plus
unextractable radiocarbon was calculated, which
represented the loss of radiocarbon by volatilization
of parent compound or derived products, such as $^{14}CO_2$.

The soil was then fractionated into humic acid,
fulvic acid, and humin fractions according to the
method of Stevenson (3). The radiocarbon content of
each of these fractions was determined by lsc or
combustion-lsc. The humic acid fraction was dried
over P_2O_5 and weighed; the fulvic acid fraction was
lyophilized, dialyzed against distilled water to
remove sodium chloride, lyophilized again, and weighed.
The percentage of radiocarbon in each fraction was
then calculated.

Results

Acetonitrile-Water Extraction Analysis. Figures
5 to 14 show, for the various compounds at various
time periods, the radiocarbon accounted for in the
soxhlet extracts, the amounts of extracted radiocarbon
accounted for as parent compound and as polar components (i.e., those that did not move from the
origin in tlc analysis using non-polar solvent systems)
the amount of unextractable radiocarbon, and the amount
of radiocarbon lost as volatile components.

The two Group 1 compounds containing an -SO$_2$R
group, Compounds I and II, showed the least reactivity
of the compounds studied, both compounds exhibiting
nearly identical behavior in soil. Both showed high
extractabilities of radiocarbon from soil (about 90%
after two months) with nearly all of the radiocarbon
represented by parent compound. Low levels of polar
extractable components and unextractable radiocarbon
were found, and radiocarbon loss as volatile components

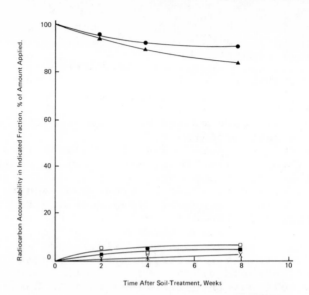

Figure 5. Radiocarbon accountability at various time periods after soil-treatment with radiolabeled perfluidone. ●, *radiocarbon extracted;* ▲, *parent-14C extracted;* ■, *loss as volatiles;* □, *unextractable ("bound") radiocarbon;* χ, *extractable 14C-polar substances.*

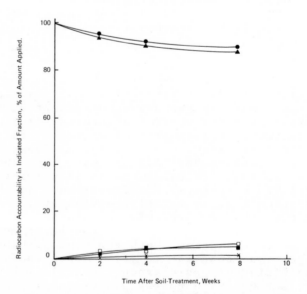

Figure 6. Radiocarbon accountability at various time periods after soil-treatment with radiolabeled metabolite α. ●, *radiocarbon extracted;* ▲, *parent-14C extracted;* ■, *loss as volatiles;* □, *unextractable ("bound") radiocarbon;* χ, *extractable 14C-polar substances.*

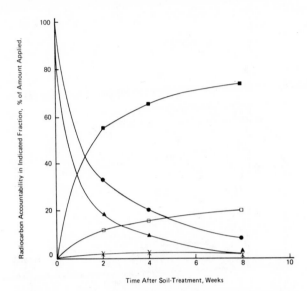

Figure 7.   Radiocarbon accountability at various time periods after soil-treatment with radiolabeled BA 7209. ●, *radiocarbon extracted;* ▲, *parent-¹⁴C extracted;* ■, *loss as volatiles;* □, *unextractable ("bound") radiocarbon;* χ, *extractable ¹⁴C-polar substances.*

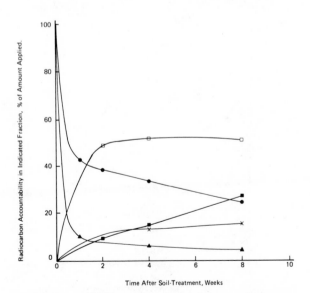

Figure 8.   Radiocarbon accountability at various time periods after soil-treatment with radiolabeled fluoridamid. ●, *radiocarbon extracted;* ▲, *parent-¹⁴C extracted;* ■, *loss as volatiles;* □, *unextractable ("bound") radiocarbon;* χ, *extractable ¹⁴C-polar substances.*

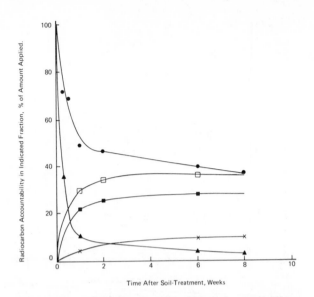

Figure 9. Radiocarbon accountability at various time periods after soil-treatment with radiolabeled MBR 12325. ●, *radiocarbon extracted;* ▲, *parent-^{14}C extracted;* ■, *loss as volatiles;* □, *unextractable ("bound") radiocarbon;* χ, *extractable ^{14}C-polar substances.*

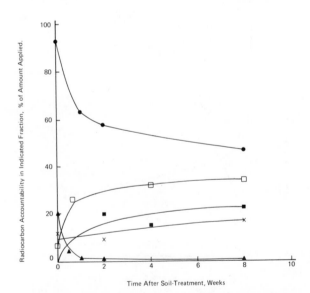

Figure 10. Radiocarbon accountability at various time periods after soil-treatment with radiolabeled BA 5974. ●, *radiocarbon extracted;* ▲, *parent-^{14}C extracted;* ■, *loss as volatiles;* □, *unextractable ("bound") radiocarbon;* χ, *extractable ^{14}C-polar substances.*

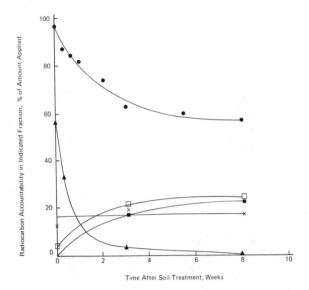

Figure 11. Radiocarbon accountability at various time periods after soil-treatment with radiolabeled BA 15753. ●, radiocarbon extracted; ▲, parent-^{14}C extracted; ■, loss as volatiles; □, unextractable ("bound") radiocarbon; χ, extractable ^{14}C-polar substances.

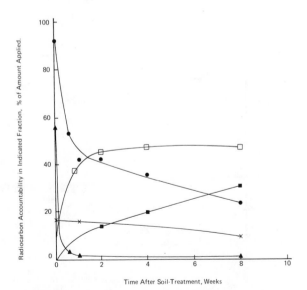

Figure 12. Radiocarbon accountability at various time periods after soil-treatment with radiolabeled BA 8315. ●, radiocarbon extracted; ▲, parent-^{14}C extracted; ■, loss as volatiles; □, unextractable ("bound") radiocarbon; χ, extractable ^{14}C-polar substances.

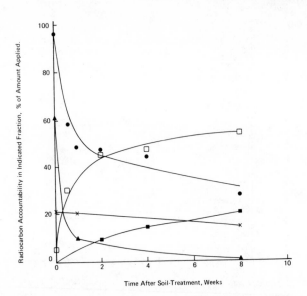

Figure 13. Radiocarbon accountability at various time periods after soil-treatment with radiolabeled BA 15733. ●, *radiocarbon extracted;* ▲, *parent-*[14]*C extracted;* ■, *loss as volatiles;* □, *unextractable ("bound") radiocarbon;* χ, *extractable* [14]*C-polar substances.*

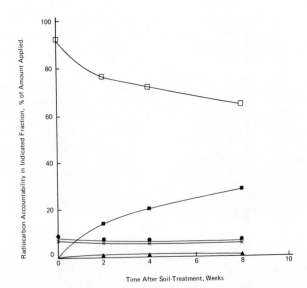

Figure 14. Radiocarbon accountability at various time periods after soil-treatment with radiolabeled TDA. ●, *radiocarbon extracted;* ▲, *parent-*[14]*C extracted;* ■, *loss as volatiles;* □, *unextractable ("bound") radiocarbon;* χ, *extractable* [14]*C-polar substances.*

was quite low (5% or less after two months).

The third Group 1 compound, Compound III, exhibited far different behavior, with the major route of dissimilation being loss by volatilization ($\sim$70% after two months). It should be pointed out that in a separate experiment (2) using biometric flasks according to a modification of the method of Bartha and Pramer (4), only a relatively minor amount of $^{14}CO_2$ was collected through two weeks; therefore, it appears that under field conditions a very large amount of Compound III may volatilize. Because of this, radiocarbon extractability was far less than with Compounds I and II, with about one-half of the extracted radiocarbon being accounted for as parent compound. Radiocarbon "binding" was also higher, with about 20% of radiocarbon being present as unextractable fraction after two months.

The two Group 2 compounds showed parallel behavior. Radiocarbon extractability dropped sharply for about one week, then leveled off; parent compound accounted for about one-fourth to one-third of extractable radiocarbon. The amount of "binding" concomitantly rose sharply, then leveled off. Radiocarbon loss was about 25% after two months. Compound V, which contains a 4-methyl group, showed less binding than Compound IV.

The four Group 3 compounds, each of which has one free amino group, all behaved similarly in soil. Extractabilities of radiocarbon and radiocarbon "binding" were similar to the two Group 2 compounds, the major difference being in the very rapid disappearance of parent compound with the Group 3 compounds. Even at t_o (actually about 15 minutes) the recoveries of parent compounds were considerably less than the expected 100%, ranging from 20 to 60%, and recoveries rapidly dropped thereafter. (In contrast, with all of the Group 1 and 2 compounds all of extracted radiocarbon at t_o was due to parent compound). With the two Group 3 compounds not having a 4-methyl group, disappearance of parent compound was complete in less than a week. With a 4-methyl group present, disappearance of parent compound was delayed somewhat but was still complete after eight weeks.

The Group 4 compound was the most reactive; only about 10% of the radiocarbon was extractable, and this value remained about constant through two months. No parent compound was detected at any time period, including t_o. Radiocarbon loss, however, was comparable to that of the Group 2 and 3 compounds.

Fractionation of Soil Organic Matter. Table II
shows the recovery of soil radiocarbon as humic acid,
fulvic acid, and humin fractions two weeks and two
months after soil application of various radiolabeled
compounds. The values represent percent of the total
radioactivity present in the soxhlet-extracted soil,
i.e. the unextractable fraction. It is apparent that
with time the amount of radiocarbon in the humic
acid decreased, and that, concomitantly, there was
an increase in the amount of radiocarbon associated
with the fulvic acid fraction. The only exception
to this observation was MBR 12325.

Discussion

The results of the Group 1 compounds suggest
that the $-NHSO_2CF_3$ group, under the present experimental conditions, does not bind appreciably to soil;
Compounds I and II are readily extracted from soil
even two months after soil-treatment, and Compound
III readily volatilizes from the soil. Compound III
undergoes a higher degree of conversion to other soil
products than Compounds I and II, suggesting that the
$-SO_2R$ group, lacking in III, confers stability towards
soil degradation under these conditions.
In the two m-phenylenediamine compounds with both
amino groups blocked, fluoridamid and MBR 12325,
radiocarbon extraction is lower and soil-binding
of radiocarbon is higher than with the Group 1 compounds, suggesting that the acetamido group may be
important for "binding" to soil particles. In the
four m-phenylenediamine compounds with one free and
one blocked amino group, the similar binding behavior
to the Group 2 compounds plus the rapid disappearance
of parent compound suggest that binding for all six
compounds occurs via a free $-NH_2$ group; in the case
of the Group 2 compounds this is obtained by cleavage
of the N-C bond of the acetamido group. The fact
that the Group 1 compounds do not bind indicates that
the N-S bond of the $-NHSO_2CF_3$ group is not readily
cleaved.
The behavior of TDA reinforces the theory that
free $-NH_2$ groups are responsible for the soil "binding"
behavior of the compounds studied. With two free $-NH_2$
groups, binding was almost instantaneous and no
parent compound was detected at any time.
The relationship between the amounts of radiocarbon associated with humic and fulvic acids at the
two time periods studied supports the findings of
Schnitzer (5) in that the amounts of extractable

Table II. Fractionation of the Unextractable Soil
Radiocarbon into Humic Acid, Fulvic Acid
and Humin Two Weeks and Two Months After
Soil-Application of Indicated Radiolabeled
Compound.

Compound	Radiocarbon Recovery, % of Total Present in Solvent-Extracted Soil					
	Humic Acid		Fulvic Acid		Humin	
	2 Wk	2 Mo	2 Wk	2 Mo	2 Wk	2 Mo
No Free $-NH_2$ Group						
Perfluidone	--[a]	22.7	--[a]	47.0	--[a]	30.3
Metabolite α	--[a]	17.5	--[a]	38.4	--[a]	44.1
BA 7209	38.9	23.6	29.9	46.2	31.2	30.2
Fluoridamid	42.8	41.6	23.4	37.0	33.8	21.4
MBR 12325	33.7	42.8	44.7	34.9	21.7	22.3
Free $-NH_2$ Group(s) Present						
BA 5974	57.1	38.2	36.9	53.2	6.0	8.6
BA 15753	56.2	33.3	40.0	54.2	3.8	12.5
BA 8315	48.2	29.2	28.1	67.8	23.7	3.0
BA 15733	46.6	25.9	34.1	58.2	19.3	15.9
TDA	38.7	40.4	36.4	41.0	24.9	18.6

[a] Fractionation not done.

humic acids were negatively correlated with the process of humification, whereas the amounts of extractable fulvic acids were positively correlated. It is
possible that with time the degradation products of
metabolism of these compounds parallel the process
of humification and are incorporated mainly in the
fulvic acid fraction. Except for MBR 12325, the findings on the radiocarbon distribution between humic
and fulvic acids observed in the present study are
in accordance with Alexandrova (6) who states that
humic acid breaks down into fulvic acid, although
fulvic acid does not polymerize into humic acid.
The generally held belief that the generic relationship between various fractions of the soil organic
matter, where the process of humification proceeds
from fulvic acids to humic acids to insoluble humic
substances (humin), was not clearly evident in the
present investigation.

Literature Cited

(1) Bandal, S. K., H. B. Clark, and S. C. Anderson.
1973. Paper presented at the 167th National Meeting
of the American Chemical Society, Los Angeles,
California
(2) Bandal, S. K., H. B. Clark and J. T. Hewitt.   1975.
Unpublished results.
(3) Stevenson, F. J. 1965. Methods of Soil Analysis,
edited by B. A. Black, Am. Soc. Agron., Volume 2,
pages 1409-1421.
(4) Bartha, R. And D. Pramer. 1965. Soil Sci., 100,
68.
(5) Schnitzer, M. 1967. Can. J. Soil Sci., 47, 245.
(6) Alexandrova, L. N. 1966. Intern. Soc. Soil Sci.
Trans. II, IV, Comm., Aberdeen, p. 73.

Biological Unavailability of Bound Paraquat Residues in Soil

D. RILEY and W. WILKINSON
ICI Plant Protection Division, Jealott's Hill Research Station, Bracknell Berkshire, U. K.

B. V. TUCKER
Chevron Chemical Co., Ortho Division, Richmond, Calif. 94804

1. INTRODUCTION

The herbicidal properties of paraquat [1, 1'-dimethyl-4, 4'-bipyridylium ion] were discovered at ICI's Jealott's Hill Research Station, England, in 1955. Paraquat is now used commercially worldwide. It is normally manufactured as the dichloride salt.

$$\left[CH_3 - N \bigcirc\!\!\!-\!\!\!\bigcirc N - CH_3 \right]^{2+} 2Cl^-$$

Paraquat is a broad spectrum, rapidly acting contact herbicide which is highly effective against grasses and most broad leaved species. A unique property of paraquat is its rapid and complete adsorption onto soil. Paraquat that reaches the soil is rendered unavailable to plant roots. It is used to kill emerged weeds anytime before planting a crop or before the crop emerges. Consequently, paraquat is widely used in agriculture for preplant and preemergence weed control and it has an important use in minimum tillage farming systems. It is also used for weed control between trees and as a directed spray in row crops. Due to its desiccating properties, it finds wide application as a harvest aid.

The properties of paraquat have been reviewed by Calderbank (1) and Akhavein and Linscott (2). This paper summarizes studies on the biological unavailability of 'bound' paraquat residues in soil.

2. NATURE AND AMOUNTS OF PARAQUAT SOIL RESIDUES

Paraquat is normally applied as a spray at rates of 0.1 to 2 kg/ha.* Some of the paraquat reaches the soil directly and is adsorbed by clay minerals or organic matter. The remainder is intercepted by the target weeds. Paraquat adsorbed on plant surfaces is subject to photochemical decomposition by sunlight. The main photochemical degradation products are 4-carboxy-1-methylpyridinium chloride and methylamine hydrochloride (3). The former has a low toxicity and is rapidly degraded in soil and culture solutions (1, 4, 5). Methylamine occurs naturally and is readily degraded (5).

Paraquat adsorbed on glass slides or thin-layers of soil is photochemically degraded in sunlight or under UV lamps (1). However, photochemical degradation of significant amounts on soil surfaces in the field has not been clearly demonstrated.

The amount of paraquat eventually reaching the soil obviously depends on factors such as density of weed cover and sunlight intensity to cause paraquat photodegradation on plant surfaces. Analysis of soils from over 50 sites has shown that the amount of paraquat which reaches the soil can range from 10 to almost 100% of that applied. If 50% of a 1 kg/ha application reached the soil, this would result in 0.5 μg paraquat/g soil if the residues were uniformly incorporated into the top 15 cm.

In field experiments, ^{14}C labelled paraquat was sprayed onto a grass sward, onto bare soil, or incorporated into the soil. After 1 year there was no significant degradation of paraquat in the soil; at least 90% of the ^{14}C labelled residues were paraquat (B.C. Baldwin - unpublished data). Because paraquat is firmly bound to soil (see below) it is immobile unless the adsorbent itself moves.

Paraquat initially adsorbed onto plant debris incorporated into soil or onto soil organic matter transfers to clay minerals which adsorb paraquat much more strongly (1). In dilute suspensions the transfer from soil organic matter to clay is rapid (6). Also when montmorillonite clay was mixed with a moist peat soil containing available paraquat residues the paraquat was rapidly deactivated (7). This shows that in moist soils, as well as slurries, paraquat rapidly transfers from weak adsorption sites on the clay minerals. This is not surprising since much of the soil organic matter is closely associated with the clay sur-

---

* Note: Throughout this paper rates and concentrations refer to paraquat cation.

faces. On peat soils containing only small amounts of clay
several weeks may be required for the transfer of paraquat from
organic matter to the clay (8).

The adsorption of paraquat on soils, clays and organic matter
has been extensively investigated (6, 7, 9 to 27). Calderbank (1)
reviewed the nature of paraquat adsorption in soil. Hayes et al
(26) and Khan (27) reviewed the mechanism of paraquat adsorption on clays and organic matter, respectively. The quantity of
paraquat adsorbed by soils is always less than the base exchange
capacity for inorganic cations, such as ammonium ions.
Furthermore, although some paraquat can be displaced by high
concentrations of ammonium ions when the adsorption sites of the
soil are saturated with paraquat, displacement is never complete
and a portion of the paraquat is firmly fixed to the soil. As the
quantity of adsorbed paraquat decreases, displacement becomes
progressively more difficult and more concentrated salt solutions
are required. Malquori and Radaelli (17) compared the relative
effectiveness of NH_4^+, K^+, Ca^{2+}, Mg^{2+} and Na^+ for releasing
paraquat previously adsorbed on five different clay minerals at
varying concentrations. K^+ and NH_4^+ were, in general, more
effective than the other cations studied, but no paraquat was released when its concentration on the clay was below a certain
limit, which varied considerably with the type of mineral.

The only effective means of displacing paraquat from soil
when it is present in relatively low concentrations, even from
very sandy soils, is to reflux the sample with strong acids, e.g.
18 N, sulfuric acid. Boiling with strong sulfuric acid represents
more than elution. The structure of the clay is partially des-
troyed and the binding sites are thus eliminated. Even very
small amounts, below 1.0 μg/g paraquat, can be quantitatively
recovered from soils when treated in this way.

On a molecular scale, the phenomenon of firm adsorption is
associated with the shape and charge distribution of the paraquat
ion. The two pyridine rings of paraquat can rotate about the in-
terring bond and readily assume a coplanar structure - a pre-
requisite of herbicidal activity. This flat configuration undoubtedly facilitates their interaction with the clay mineral surfaces.
Paraquat is highly polarizable and its normal charge distribution
is distorted in the vicinity of the negatively charged clay surfaces
thus charge transfer complexes are formed (12), reinforcing the
normal coulombic attraction forces. Paraquat is readily displaced quantitatively from cation exchange resins (28) as distinct
from clay minerals which further suggests that other adsorption
forces, in addition to coulombic forces, are involved in the para-

quat-clay system.

Published data and a large number of unpublished experiments conducted by ICI Plant Protection Division in the United Kingdom and the Chevron Chemical Company in the U.S.A. have shown that the paraquat soil residues resulting from normal applications of paraquat are firmly bound. Clays can firmly bind up to about 50,000 μg paraquat/g clay (50 meq/100 g). The amount and strength of paraquat adsorption by soil depends on the amount and type of clay minerals present. However, almost all agricultural soils contain sufficient clay to firmly bind 50 to 5000 μg paraquat/g soil. The only exceptions are very sandy and peat soils containing less than 5% clay; however, even many of these soils will firmly bind at least 50 μg/g. Peats adsorb large amounts of paraquat but the adsorption is not as strong as with clays and most of the paraquat is desorbed with 5M NH_4Cl (24).

3. UNAVAILABILITY OF BOUND PARAQUAT SOIL RESIDUES TO PLANTS

Paraquat is an extremely active herbicide when applied to plant roots (grown in nutrient solution) as well as their leaves. This is illustrated in Figure 1. Pure solutions of paraquat dichloride, with and without Hoagland and Snyder nutrient solutions (29), were bioassayed with pregerminated wheat seedlings (Triticum aestivium cultivar Kolibri). The results given in Figure 1 are the mean of results from three bioassay experiments conducted at different times; at each time each treatment was replicated three times. The seedlings failed to grow in solutions containing more than 1 μg paraquat/ml. In solutions below 1 μg/ml increasing the paraquat concentration gave corresponding reduction in the elongation of both roots and shoots. The effect was especially marked on the roots, the activity - concentration curve having a much sharper cut-off at a slightly lower concentration than with the shoots. The smaller effect on shoot elongation was probably due to poor translocation of paraquat (B.C. Baldwin, unpublished data). There was no chlorosis or necrosis of the leaves of treated plants except that when plants were extremely stunted (above 0.1 μg/ml) they then became chlorotic and died. The effect as measured by root and shoot elongation was greater than the effect as measured by the dry weight of the seedlings. Successive bioassays on the same paraquat solutions showed that the paraquat concentration in the solutions decreased 2-3 fold during each bioassay. The lowest concentration of paraquat which had a statistically significant effect (p = 0.05) on root length ranged from 0.005 to 0.02 μg/ml and for shoots it ranged from 0.01 to 0.1 μg/ml. The concentration of paraquat which resulted in a 50% reduction in the lengths of roots ranged from 0.005 to 0.04 μg/ml and for shoots ranged from 0.03 to 0.10 μg/ml. The values depended on growth conditions and whether or not nutrients were added during the bioassay.

Plant roots are in intimate contact with soil particles as well as with the soil solution. There is evidence for a bridge of "mucigel" between the roots and soil particles (30), which might provide a route by which chemicals adsorbed on soil surfaces can move to roots without entering the soil solution. It has been suggested that nutrients such as potassium can be adsorbed by roots directly from adsorption sites on soil particles (30, 31) and it is conceivable that adsorbed pesticides can also be taken up by direct contact. However, it has been shown that some herbicides, such as atrazine, are supplied to roots mainly via the soil solution (32).

During the past 15 years paraquat has been applied to millions of hectares throughout the world. There have been an insignificant number of reports of paraquat having any residual herbicidal activity in the soil. Therefore, it must be concluded that the paraquat soil residues are unavailable to plants. It has also been shown in glasshouse experiments that 'bound' paraquat is unavailable to plants (25). In most soils, paraquat has to be applied at several hundred to several thousand times the normal rate of application before the residues have any effect on plants. After such extremely high rates of application paraquat can be detected in the equilibrium soil solution (16). Also some of the adsorbed paraquat residues can be desorbed with high concentrations of inorganic salt solutions, such as 5M ammonium chloride (24).

We have used two different approaches to characterize the availability of paraquat residues to plants:

A Determination of the capacity of soils to reduce the concentration of paraquat in the equilibrium solution below phytotoxic levels.

B Determination of the capacity of soils to bind paraquat so tightly that it is not desorbed with saturated (5M) ammonium chloride.

A. Capacity of Soils to Reduce Paraquat in the Equilibrium Solution Below Phytotoxic Levels

These experiments showed that any residual activity of

paraquat, resulting from extremely high rates of application
to soil, are due to uptake of paraquat from the soil solution;
paraquat adsorbed on soil particles is not taken up by plants.
Details of 4 English soils studied are given in Table I.

Paraquat adsorption isotherms were determined by shaking
10 g of soil with 200 ml paraquat dichloride solutions in 0.01M
$CaCl_2$. The 0.01M $CaCl_2$ solution was used to simulate the salt
concentration often found in the soil solution (33). After shaking
for 16 hours, the suspensions were centrifuged and the supernatant solutions analyzed colorimetrically after reduction with alkaline dithionite solution (1). Concentrations in the range 0.01 to
0.1 μg/ml were determined after the paraquat had been concentrated 10 fold using Zerolit 225 cation exchange resin. Equilibrium soil solutions were also bioassayed with wheat seedlings
as described above, without any addition of nutrients, other than
those extracted from the soil.

Glasshouse Studies

Soil bioassays were also performed on the 4 soils given in
Table I. Paraquat dichloride solutions (200 ml) were thoroughly
mixed with 2 kg samples of each of the 4 soils. Each treatment
was replicated 3 times. The soils were placed in plastic pots
and arranged in a randomized block design in the glasshouse.
After leaching out excess chloride salts (there was negligible
leaching of paraquat) and the addition of nitrogen, phosphorous
and potassium nutrients, the pots were bioassayed with a series
of crops. Between each crop the soil was air dried and mixed.
Each crop was grown for about 6 weeks before harvest. Dry
weights of shoots and in some cases length and dry weight of
roots (after washing) were measured. The crops grown were
wheat (Triticum aestivum, cultivar Kolibri), radish (Raphanus
sativas, cultivar Sutton scarlet Globe), peas (Pisum sativum -
cultivar Meteor) and lettuce (Lactuca sativa, cultivar Suttons
Unrivalled).

After these bioassays had been completed samples of each
treated soil were bioassayed with the aquatic weed Lemna
polyrhiza which has been reported to be very sensitive to paraquat (34). Soil samples (10 g) were mixed with 100 ml distilled
water plus 1 ml Hoagland and Snyder nutrient solution (29).
Samples of Lemna were floated on the surface of the mixture.
The increase in the dry or fresh weight of the Lemna was determined after 7 days growth. Samples of the treated Sandy Hills

TABLE I

PROPERTIES OF SOILS USED IN PARAQUAT ADSORPTION AND DEACTIVATION STUDIES

Name	Type	pH	Organic Matter (%)	Clay (%)	Silt (%)	Sand (%)	Cation Exchange Capacity (meq/100g at pH 7.0)
Sandy Hills	Sandy Loam	6.7	1.7	11	10	79	4.5
Broadricks	Sandy Loam	5.8	1.9	17	18	65	10.0
Tarlton	Calcareous loam	7.8	4.8	21	48	31	27.8
Methwold	Fen peat (Muck)	7.3	42.3	-	-	-	65.0

soil were also bioassayed with <u>Lemna</u> in the presence of 0.05 and 0.1 M $CaCl_2$; the <u>Lemna</u> did not grow in the 0.1M $CaCl_2$. Pure solutions of paraquat dichloride plus nutrient solution were similarly bioassayed.

Samples (400 g) of the treated Sandy Hills and Methwold soils were bioassayed with S23 Perennial Ryegrass (<u>Lolium perenne</u>) in the presence of 0, 0.5 and 0.25 M $CaCl_2$. The $CaCl_2$ concentration in the soil solution was monitored and kept constant by periodic additions of $CaCl_2$ solution, to replace $CaCl_2$ leached from the pots during watering.

Figure 2 shows the amount of paraquat adsorbed by the 4 soils when the concentration of paraquat in the equilibrium soil solution is in the range 1-20 μg/ml. Figure 3 shows the amount of paraquat adsorbed with 0.005 to 0.2 μg/ml in the equilibrium solution. Clearly the soils have very different adsorption isotherms. For example, the peat (Methwold) adsorbs larger amounts of paraquat than the Broadricks and Tarlton soils when the concentration of paraquat in the equilibrium solution is 5 μg/ml, but the peat (Methwold) adsorbs lower amounts of paraquat than the Broadricks and Tarlton soils when the equilibrium solution contains less than 0.1 μg/ml.

Results of the bioassay of the equilibrium solutions are given in Figure 4. Only results for the root growth are given because they were affected more than shoots; as they were in pure paraquat dichloride solutions. Typical results for the soil bioassays are given in Figure 5 (first wheat crop on all soils) and Figure 6 (wheat, pea, and lettuce crops on Broadricks soil); all the soil bioassay results are summarized in Table II. The residual activity of the paraquat decreased slightly during the first few bioassays. This was probably due to a slow equilibration of the paraquat with the adsorption sites. Analysis of the soils showed it was not due to degradation of the paraquat. The residual activity of paraquat was very similar on all the plant species tested.

The difference between the sensitivities of roots and shoots to paraquat was not as great in the soil bioassay as in the solution bioassay. This was probably because the emerging shoots as well as roots were in contact with the residual paraquat in the soil bioassay, but not in the solution bioassay. Also, in solution cultures severely stunted roots can maintain normal shoot growth, while in the more austere soil environment the stunted roots may not be able to supply the shoots with sufficient water and nutrients.

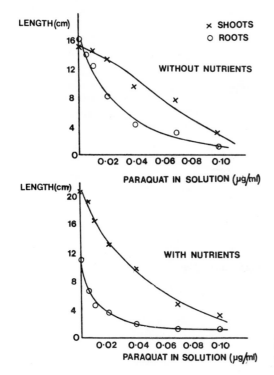

Figure 1. *Effect of paraquat and nutrients on the length of shoots and roots of 14-day-old wheat seedlings*

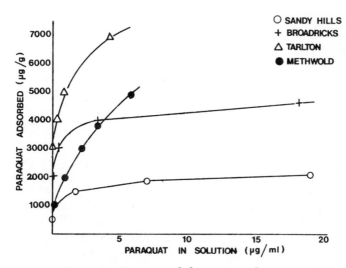

Figure 2.   *Paraquat soil absorption isotherms*

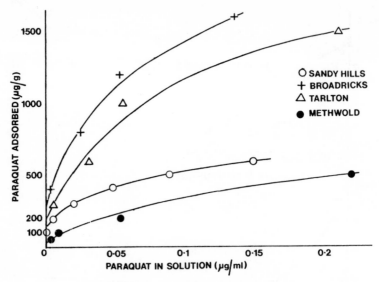

Figure 3.   Paraquat soil adsorption isotherms

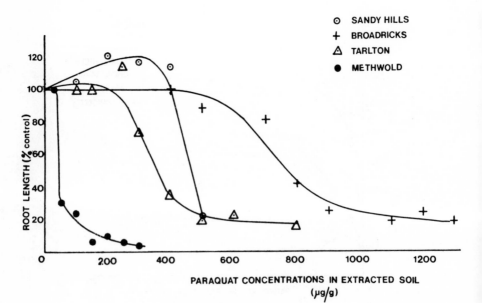

Figure 4.   Wheat bioassay of equilibrium solutions from soils treated with paraquat

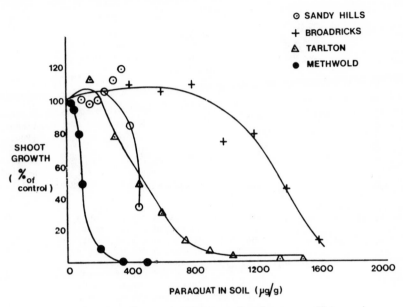

Figure 5. *Wheat bioassay of soils treated with paraquat (first crop)*

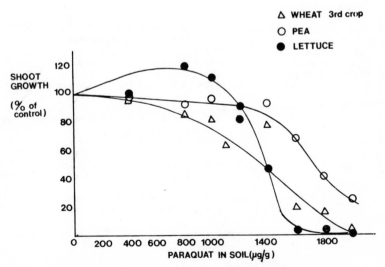

Figure 6. *Bioassay of Broadricks soil treated with paraquat*

TABLE II

CAPACITIES OF SOILS TO DEACTIVATE PARAQUAT

Soil	Bioassay	Date of Bioassay	Observed effect of paraquat	Lowest concentration of paraquat which significantly ($P=0.05$) affected crop (μg paraquat/g soil)
Sandy Hills	Soil treated	March 1971		
	Wheat 1st crop	April 1971	Increased dry wt.shoots	300
			Decreased dry wt.shoots	400
	Wheat 2nd crop	June 1971	Increased dry wt.shoots	300
			Decreased dry wt.shoots	> 450
			Increased dry wt.roots	300
			Decreased dry wt.roots	> 450
			Decreased length roots	300
	Wheat 3rd crop	July 1971	Decreased dry wt.shoots	500
			Decreased length roots	450
	Wheat 4th crop	Sept.1971	Decreased dry wt.shoots	500
	Wheat 5th crop	Jan. 1972	Decreased dry wt.shoots	550
			Decreased dry wt.roots	550
			Decreased length roots	500
	Radish 1st crop	Mar.1972	Decreased dry wt.whole plant	> 700
	Radish 2nd crop	May 1972	Decreased fresh wt.whole plant	600
			Decreased dry wt.whole plant	600
	Lemna	July 1972	Decreased dry wt.	550
	Lemna (+0.05MCaCl$_2$)	July 1972	Decreased dry wt.	500
	Ryegrass	June 1973	Decreased dry wt.shoots	600
			Decreased length roots	550
	Ryegrass(+0.05MCaCl$_2$)	June 1973	Decreased dry wt.shoots	600
			Decreased length roots	550
	Ryegrass (+0.25MCaCl$_2$)	June 1973	Decreased dry wt.shoots	600
			Decreased length roots	450
Broadricks	Soil treated	Dec. 1971		
	Wheat 1st crop	Feb. 1972	Decreased dry wt.shoots	1400
	Wheat 2nd crop	April 1972	Decreased dry wt.shoots	1200
	Radish	June 1972	Decreased fresh wt.whole plant	1600
			Decreased dry wt.whole plant	1600
	Lemna	Aug.1972	Decreased fresh wt.	400
	Pea	Oct. 1972	Decreased dry wt.shoots	1800
			Decreased length roots	1400
	Wheat 3rd crop	Feb. 1973	Decreased dry wt.shoots	1400
			Decreased length roots	1000
	Lettuce	June 1973	Decreased dry wt.shoots	1400
			Decreased length roots	1200
Methwold	Soil treated	May 1972		
	Wheat 1st crop	June 1972	Decreased dry wt.shoots	75
			Decreased length roots	50

TABLE II. CAPACITIES OF SOILS TO DEACTIVATE PARAQUAT (CONT.)

Soil	Bioassay	Date of Bioassay	Observed effect of paraquat	Lowest concentration of paraquat which significantly (P=0.05) affected crop (μg paraquat/g soil)
Meth-wold (Cont.)	Wheat 2nd crop	Aug. 1972	Decreased dry wt.shoots	75
			Decreased length roots	50
	Lemna	Sept.1972	Decreased fresh wt.	200
	Soil treated	May 1972		
	Wheat 3rd crop	Sept. 1972	Decreased dry wt.shoots	75
			Decreased length roots	50
	Wheat 4th crop	Oct. 1972	Decreased dry wt.shoots	200
			Decreased length roots	50
	Wheat 5th crop	Jan. 1973	Decreased dry wt.shoots	100
			Decreased length roots	75
	Wheat 6th crop	Mar. 1973	Decreased dry wt.shoots	200
			Decreased length roots	75
	Ryegrass	June 1973	Decreased dry wt.shoots	200
			Decreased length roots	200
	Ryegrass(+0.05MCaCl$_2$)	June 1973	Decreased dry wt.shoots	200
			Decreased length roots	100
	Ryegrass(+0.25MCaCl$_2$)	June 1973	Decreased dry wt.shoots	100
			Decreased length roots	25
Tarl-ton	Soil treated	Aug. 1971		
	Wheat 1st crop	Mar. 1972	Decreased dry wt.shoots	300
	Wheat 2nd crop	April 1972	Decreased dry wt.shoots	450
	Radish	Aug. 1972	Decreased fresh wt.whole plant	1350
			Decreased dry wt.whole plant	> 1500
	Wheat 3rd crop	Oct. 1972	Decreased dry wt.shoots	1050
			Decreased length roots	450
	Wheat 4th crop	Dec. 1972	Decreased dry wt.shoots	1350
			Decreased length roots	450
	Wheat 5th crop	Jan. 1973	Decreased dry wt.shoots	600
			Decreased length roots	600
	Lemna	Sept.1973	Decreased fresh wt.	> 1500

Comparison of Figures 4 and 5 shows that the effects of paraquat on growth were similar in separated equilibrium soil solutions and in the total system of soil plus equilibrium solution. Therefore, the paraquat bound to the soil was not available to the plants and the phytotoxicity observed was due to free paraquat in the soil solution. The concentration of bound paraquat in the soil was approximately 10,000 times greater than the paraquat concentration in the soil solution when phytotoxicity was first observed.

Under our conditions, the Lemna bioassay had a similar sensitivity to that of the wheat bioassay; 0.02 µg/ml paraquat severely reduced growth. In the Lemna bioassay of the soils, roots were suspended in the equilibrium solution above the soil but came into contact with the soil toward the end of the 7-day growth period. Results for the Lemna bioassay were similar to those of the wheat bioassays of both soils and equilibrium solutions (Table II).

When the concentration of paraquat in soil is very high it is possible to displace some of the adsorbed paraquat with high concentrations of inorganic cations (1). Nevertheless, the effect of soil residues on the growth of Lemna was not increased in the presence of 0.05 M $CaCl_2$. Lemna would not grow in the presence of higher concentrations of $CaCl_2$. 0.05 and 0.25 M $CaCl_2$ only slightly reduced the capacities of Sandy Hills and Methwold (peat) soils to deactivate paraquat with the exception of 0.25 M $CaCl_2$ on the Methwold peat; 0.25M $CaCl_2$ itself reduced the growth of ryegrass by about half. The 0.25M concentration of $CaCl_2$ is higher than the salt concentration found in most soil solutions (33) and surrounding fertilizer bands (35). Therefore, there is no danger of inorganic cations in soils displacing bound paraquat residues.

The determination of the capacity of soils to deactivate paraquat by bioassaying soils treated with high rates of paraquat in low volumes of treatment solution is not practical because it takes at least several months for the paraquat to equilibrate in the soil. However, in dilute slurries equilibration is complete after overnight shaking and the equilibrium solution can then be analyzed photometrically or bioassayed as described above. The paraquat 'Strong Adsorption' capacity of a soil determined by bioassaying equilibrium solutions with wheat is defined as the concentration of soil-adsorbed paraquat when the concentration of paraquat in the equilibrium solution is sufficient to reduce the length of 14-day old wheat roots by 50%; this solution concentration is about 0.01 µg paraquat/ml (Table III). The concentration

TABLE III

'STRONG ADSORPTION' CAPACITIES OF SOILS

Soil	'Strong Adsorption' Capacity (1) (µg paraquat/g soil)	Paraquat concentration in soil solution which reduced length of wheat roots by 50% (µg paraquat/ml)	kg paraquat/ha required to saturate 'Strong Adsorption' Capacity of top 15 cm soil
Sandy Hills	400	0.05	720 (2)
Tarlton	300	0.01	540 (2)
Broadricks	800	0.02	1440 (2)
Methwold	40	0.004	30 (3)
Pure paraquat dichloride solution		0.005 - 0.04	

1. Mean of 2 determinations

2. Assuming a bulk density of 1.2 g/cc

3. Assuming a bulk density of 0.5 g/cc

of paraquat which reduced root length by 50% is only slightly higher than that which first significantly affects root length (see above). 50% reduction of root growth is used to measure the 'Strong Adsorption' capacities of soils because it can be easily determined by visual examination of the data, whereas the first effect concentration depends very much on the variability of the experiment and the data have to be statistically analyzed.

Bioassay of the equilibrium solutions and thus 'Strong Adsorption' capacity values, gives a good estimate of the capacity of soils to deactivate paraquat (compare Tables II and III). If anything, the 'Strong Adsorption' capacity values tend to underestimate the capacity of soils to deactivate paraquat. This solution bioassay technique has been widely used by ICI Plant Protection Division to determine the capacity of soils to deactivate paraquat.

Field Trials

The relationship between 'Strong Adsorption' capacity values and the capacity of soils to deactivate paraquat in the field has been studied. Paraquat was applied at rates of 1/2, 1 and 4 times the 'Strong Adsorption' capacity found by the equilibrium soil solution bioassay technique. The effect of paraquat soil residues on crop growth and crop residues were accurately predicted; residues below the 'Strong Adsorption' capacity being unavailable to plants.

One experiment was conducted at Frensham in England. The loamy sand soil contains 9% clay, 8% silt and 83% sand and 2.0% organic matter; it has a pH of 6.6 and cation exchange capacity of 5 meq/100 g at pH 7.0. The 'Strong Adsorption' capacity of the soil is 120 µg paraquat/g soil. Paraquat adsorption isotherms, determined in dilute slurries as described above, are given in Figure 7.

The trial was laid out in 3 blocks. Each block was split into two halves and each half was split into 4 plots, 40 m x 5 m. Four plots of one-half of each block were treated with 0, 15, 33 and 120 kg/ha of paraquat; this was then lightly incorporated into the top 2 cm soil using a hand rake. The 4 plots on the other half of each block were treated with 0, 90, 198 and 720 kg/ha of paraquat; this was thoroughly rotovated into the top 15 cm. The paraquat was applied in November, 1971, and the plots were left undisturbed through the winter. In the spring, weeds were controlled by an overall spray of 0.5 kg/ha paraquat. In March, 1972, the trial area was seeded with barley (Hordeum vulgare -

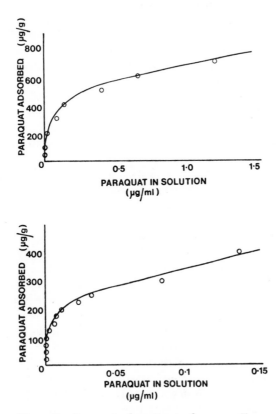

Figure 7. *Paraquat adsorption isotherms on Frensham soil*

cultivar Julia) by direct drilling with a triple disc drill, i.e., by
zero tillage. In August, when the crop had reached maturity,
samples of grain and straw were analyzed for paraquat residues.
A second crop of barley was similarly grown in 1973. Also in
1973, large soil samples were taken from the top 15 cm of the
deep incorporated treatments. These soil samples were used to
grow carrots (Daucus carota - cultivar New Model Red Core)
outdoors in 20 cm diameter, 20 cm deep pots. The carrots were
harvested 119 days after sowing, thoroughly washed and analyzed
for paraquat residues.

    In 1972, the 90 kg/ha incorporated and 15 and 33 kg/ha sur-
face treatments only slightly affected early growth of barley and
the crop quickly and completely recovered (Table IV). The 20
kg/ha surface treatment and 198 kg/ha incorporated treatment
had a significant effect on barley growth and reduced the number
of plants. However, the remaining plants grew normally and
although grain yields were slightly less than on the control plots
there was no statistical difference in the yield at the 95% confidence level. There was an almost complete crop failure on the
720 kg/ha incorporated treatment. However, a few plants survived and they grew normally; they were harvested for residue
analysis. The residual activity of the paraquat was less in 1973
than 1972. This was due to a slow equilibrium of the large
amounts of paraquat with adsorption sites; there was no significant degradation of paraquat during the experiment. In 1973, the
15 and 33 kg/ha surface treatments and 90 and 198 kg/ha incorporated treatments had no significant effects on barley growth.
The 120 kg/ha surface treatment slightly affected barley growth
but did not significantly affect grain yield (Table IV). The 720 kg/
ha incorporated treatment was again severely phytotoxic but
many more plants survived and grew normally than in 1972.

    Residues in barley grain and straw were negligible, except
for the 720 kg/ha incorporated treatment (Table IV). This could
have been due to contamination with traces of soil containing
large paraquat residues as well as adsorption by the crop. Residues in the carrots were very low, i.e., 0.04 µg/g or less
(Table IV)

    A further trial was conducted on Broadrick's field at Jealott's
Hill Research Station. Details of the soil are given in Table I.
It has a 'Strong Adsorption' capacity of 800 µg paraquat/g soil.
Two 6 x 6 m plots were treated with 5 annual applications of 112
kg/ha paraquat; this resulted in residues of 1000 µg/g in the top
2.5 cm soil. S23 perennial ryegrass sown in the top 2.5 cm of
soil failed to germinate or was severely stunted. However, after

TABLE IV. YIELD AND PARAQUAT RESIDUES OF CROPS GROWN IN LOAMY SAND TREATED WITH HIGH RATES OF PARAQUAT FRENSHAM, ENGLAND. BARLEY RESULTS ARE MEANS OF 3 REPLICATE PLOTS, AND CARROT RESULTS THE MEAN OF 6 REPLICATE PLOTS.

Depth Paraquat incorporated in soil	2 cm				15 cm			
Paraquat applied (kg/ha)	0	15	33	120	0	90	198	720
1972 Barley grain yield (kg dryweight/ha)	2596	2826	2807	2296	2604	2474	2281	(a)
1972 Paraquat residues in grain (μg/g air dry weight)	<0.01	<0.01	0.02	<0.01	<0.01	<0.01	<0.01	0.06
1972 Paraquat residues in straw (μg/g air dry weight)	<0.1	NA	NA	<0.1	<0.1	<0.1	0.1	2.2
1973 Barley grain yield (kg dry weight/ha)	3118	3512	3512	4300	2929	3260	3055	632
1973 Paraquat residues in carrot roots (μg/g fresh weight)	-	-	-	-	<0.01	0.01	0.04	NG

Grain yield on the treated plots were not significantly different to the controls (P=0.05) except the 720 kg/ha treatment.

(a) = insufficient crop to combine harvest

NA = not analyzed

NG = no growth

thorough hand digging to a depth of about 10 cm, ryegrass grew
normally. Another two 6 x 6 m plots were treated with 560 kg/
ha and another two plots with 1700 kg/ha. The paraquat was
thoroughly rotovated into the top 10 cm of soil; control plots were
also rotovated. Average paraquat residues in the top 10 cm of
soil were 700 µg/g and 2,000 µg/g for the 560 and 1700 kg/ha
treatments, respectively. Growth of S23 ryegrass sown 4 month
months later was slightly retarded on the 560 kg/ha treatments,
compared to the controls; however, it completely recovered and
grew normally. Ryegrass failed to grow on the 1700 kg/ha treatments. The plots remained bare for about 1 year but after 2
years they were partially colonized by weeds, particularly
clover (Trifolium spp) and knotgrass (Polygonum ariculare).
    These trials show that bioassay of the equilibrium soil solu-
tion and thus 'Strong Adsorption' capacity values accurately pre-
dicts the capacity of soils to deactivate paraquat under field conditions.

B. Capacity of Soils to Bind Paraquat so Tightly that it is not
    Desorbed with Saturated (5M) Ammonium Chloride

    In the preceding section the capacity of soil to adsorb para-
quat from an aqueous equilibrium solution was discussed. In
this section, characteristics of the adsorbed paraquat will be
discussed.
    Two general types of adsorbed paraquat in soil have been
defined. 'Loosely Bound' paraquat is classified as adsorbed
paraquat that can be desorbed with saturated (5M) ammonium
chloride. 'Tightly Bound' paraquat is classified as adsorbed
paraquat that cannot be desorbed with saturated ammonium chloride, but can be released from soil by refluxing with 18 N sulfuric acid (24).
    Glasshouse and field trials have shown that 'Tightly Bound'
paraquat is not available to plants. 'Loosely Bound' paraquat can
potentially become available to plants. However, the 'Tightly
Bound' capacity of most soils is very high, compared to normal
rates of application.

Glasshouse Studies

    Corn (Zea Mays L. cultivar. Golden Cross Bantam) and
beans (Phaseolus vulgaris L. cultivar Idaho 1-11) were grown in
four different soils each treated at four different levels of paraquat. The treatment levels for each soil were based on their experimentally determined 'Tightly Bound' and 'Loosely Bound'

paraquat capacities.

The four treatment levels were designed to be:
1. Below the 'Tightly Bound' capacity.
2. Slightly above the 'Tightly Bound' capacity.
3. Well above the 'Tightly Bound' capacity.
4. Near the 'Loosely Bound' capacity.

The soil characteristics and their 'Tightly Bound' and 'Loosely Bound' paraquat capacities are given in Table V. The 'Tightly Bound' capacity for the sand was 20 µg/g, for the loamy sand was 90 µg/g, for the loam 190 µg/g and for the muck was 80 µg/g. The 'Loosely Bound' capacity for the sand was 600 µg/g, for the loamy sand was 1400 µg/g, for the loam was 1800 µg/g and for the muck was greater than 8,000 µg/g. These capacities were determined by a procedure described earlier (24) using leaching with saturated ammonium chloride and refluxing with 18 N sulfuric acid.

For the glasshouse test, the soil treatment levels varied from 5 µg/g to 4,000 µg/g (Table VI). Soil from each treatment was analyzed for "unbound" paraquat by leaching with water, then leached with saturated ammonium chloride to determine 'Loosely Bound' paraquat, and finally refluxed with 18 N sulfuric acid to determine 'Tightly Bound' paraquat.

At the lowest treatment level, essentially all of the paraquat was 'Tightly Bound' in the loamy sand and loam and approximately 50% was 'Tightly Bound' in the sand and muck. As the paraquat treatment was increased, the quantity of 'Tightly Bound' paraquat reached a relatively constant value, but the quantity of 'Loosely Bound' paraquat continued to increase. At the higher levels of treatment, unbound paraquat was also present in the soils.

At the high treatment levels where unbound paraquat was present, the germination of the corn and beans was inhibited. The soil concentration of paraquat that inhibited germination varied greatly with soil type. Paraquat at 200 µg/g in the sandy soil gave complete inhibition of germination, while 1200 µg/g paraquat was required in the loamy sand, 2500 µg/g paraquat in the loam, and 4000 µg/g in the muck. Even at 4000 µg/g paraquat in the muck, some bean seedlings did emerge.

When the majority of the paraquat in the soil was 'Loosely Bound', the plants became stunted and chlorotic. These symptoms were observed in the sand at 20 µg/g paraquat, in the loamy sand at 500 µg/g paraquat, in the loam at 1000 µg/g paraquat and in the muck at 1000 µg/g paraquat. Thus, the amount of 'Loosely Bound' paraquat in soil available for uptake varies with soil

TABLE V

CHEMICAL AND MECHANICAL ANALYSIS OF SOILS

Soil Type	% Clay	% Silt	% Sand	% Organic Matter	pH	Cation Exchange Capacity meq./100 g	Soil Origin	Paraquat 'Tightly Bound' Capacity (μg/g)	Paraquat 'Loosely Bound' Capacity (μg/g)
Muck	-	-	-	approx. 100	3.3	112.7	Ocoee, Florida	80	> 8000
Sand	1	2	97	0.5	7.0	1.4	Ocoee, Florida	20	600
Loamy Sand	2	13	85	1.5	6.6	3.0	Clovis, California	90	1400
Loam	13	40	47	2.8	4.8	18.2	Moorestown, New Jersey	190	1800

TABLE VI. DIFFERENT FORMS OF PARAQUAT IN SOIL ACCORDING TO STRENGTH OF ADSORPTION

Soil Type	Soil fortification		μg Paraquat/g dry soil			Gross Plant Symptoms
	μg Paraquat/g dry soil	kg/ha in top 15 cm	'Unbound' Leached with water	'Loosely Bound' Leached with saturated ammonium chloride	'Tightly Bound' residue extracted by refluxing 18 N sulfuric acid	
Sand	5	9	0	3	2	Minor Stunting
	20	36	0	15	2	Eventual Kill
	200	360	16	159	4	No germination
	500	900	200	181	5	No germination
Muck	20	9	0	12	10	None
	100	45	0	49	14	None
	1000	450	0	930	15	Stunted
	4000	1800	5	3603	101	Partial Germination
Loamy Sand	20	36	0	1	18	None
	100	180	0	34	50	Minor Stunting
	500	900	0	328	86	Eventual Kill
	1200	2160	1	901	85	No germination
Loam	20	36	0	0	17	None
	500	900	0	222	174	Minor Stunting
	1000	1800	0	648	170	Eventual Kill
	2500	4500	360	1528	203	No germination

*Calculations based on bulk densities of 1.2 g/cc for mineral soils and 0.3 g/cc for muck.

types. Similar symptoms were observed with 15 μg/g 'Loosely Bound' paraquat in the sand and 600 μg/g 'Loosely Bound' paraquat in the loam. By comparing these bioassay results with those given in Section A above, it can be concluded that when a soil contains 'Loosely Bound' paraquat there are traces of paraquat (0.005 - 0.05 μg/ml) in the equilibrium soil solution.

The sand, loamy sand and muck were treated with ^{14}C-methyl paraquat at their lowest treatment level in the unlabelled paraquat experiment. Three weeks after planting, the plants grown in the sand contained 0.02 to 0.05 μg/g ^{14}C calculated as paraquat. The corn and beans grown in the loamy sand and muck contained 0.01 or less than 0.01 μg/g paraquat (Table VII). Thus, when most of the paraquat is 'Tightly Bound' there is little or no availability to plant uptake (25).

Field Trials

Field trials were performed using excessively high paraquat soil treatments to study the availability of bound paraquat residues. Trials were in Moorestown, New Jersey, and Fresno, California. The characteristics of the soils and their adsorption capacities for paraquat are given in Table VIII. Both soils were sandy loams and the paraquat 'Tightly Bound' capacity was approximately 300 μg/g for the California soil and 1000 μg/g for the New Jersey soil. This would be equivalent to approximately 670 kg/ha in the top 15 cm for the California soil and 2240 kg/ha for a 15 cm depth for the New Jersey soil.

Plots were treated at three different levels of paraquat at each test site. The treatment rates were 1/2, 1-1/2 to 2 and 2-1/2 to 5 times the 'Tightly Bound' capacity. The treatment rates in California were 336, 1400 and 3360 kg/ha for a 15 cm depth and in New Jersey were 1120, 3360 and 5600 kg/ha for a 15 cm depth. The plots were rototilled right after treatment and several times a year thereafter. Crops were grown 1 to 3 years after the soil was treated.

Crops grown in soil treated with paraquat below the 'Tightly Bound' capacity developed similar to those grown in untreated soil. In soil treated at 1-1/2 to 2 times the 'Tightly Bound' capacity, some crop injury was observed. At soil treatments of 2-1/2 to 5 times the 'Tightly Bound' capacity, gross plant symptoms were observed. Visual evaluations of crops planted in the California trial one year after soil treatment are given in Table IX. At 336 kg/ha in the top 15 cm, which was 1/2 of the 'Tightly Bound' capacity, no detectable phytotoxicity was observed on rye, barley, alfalfa, white clover, turnip, dichondra, radish or

TABLE VII

UPTAKE OF ^{14}C PARAQUAT FROM TREATED SOILS*

| Soil Type | Soil Fortification µg/g Paraquat | Three Weeks After Planting | | | | Six Weeks After Planting | | | | Eight Weeks After Planting | |
| | | Corn | | Bean | | Corn | | Bean | | Bean Pod | |
		Net counts/min	µg/g Paraquat	Net counts/min	µg/g Paraquat	Net counts/min	µg/g Paraquat	Net counts/min	µg/g Paraquat	Net counts/min	µg/g Paraquat
Sand	5	160	0.02	387	0.05	139	0.02	**	**	**	**
Muck	20	40	0.01	23	0.00	46	0.01	**	**	**	**
Loamy Sand	20	17	0.00	86	0.01	7	0.00	53	0.01	31	0.00

* Limit of detection is 0.01 µg/g.
 Background was 45 counts/min.
 Corn and bean plant values are averages of triplicate analyses.
 Bean pod values are averages of duplicate analyses.
** Not analyzed.

TABLE VIII

CHARACTERISTICS OF SOILS FROM HIGH RATE FIELD TRIALS

Test Location	% Clay	% Silt	% Sand	Soil Type	% Organic Matter	Cation Exchange Capacity (meq/100 g)	pH	Paraquat 'Tightly Bound' Capacity (µg/g)	Paraquat 'Loosely Bound' Capacity (µg/g)
Moorestown, New Jersey	14	12	74	Sandy Loam	1.1	16	5.7	1125	7600
Fresno, California	11	36	53	Sandy Loam	0.3	14	5.8	338	3266

TABLE IX

CROP GROWTH IN PARAQUAT HIGH RATE SOIL TRIALS AT FRESNO, CALIFORNIA*

C R O P   I N J U R Y (1)

Paraquat Soil Treatment (kg/ha)	Rye 11/30	Rye 2/18	Barley 11/30	Barley 2/18	Alfalfa 11/30	Alfalfa 2/18	Wht Clover 11/30	Wht Clover 2/18	Turnip 11/30	Turnip 2/18	Dichondra 11/30	Dichondra 2/18	Radish 11/30	Radish 2/18	Cabbage 11/30	Cabbage 2/18
336	0.0	0.0	0.5	0.0	0.0	0.0	0.0	0.0	0.0	0.0	0.0	--	0.0	0.0	0.0	1.0
1400	0.8	2.8	2.3	4.5	0.5	2.0	0.3	1.8	1.0	1.5	0.0	--	1.0	1.0	0.0	1.5
3360	5.3	8.0	7.8	9.3	1.8	4.8	2.0	4.5	4.5	7.5	3.0	--	5.8	9.0	3.0	8.8
Control	0.0	0.0	0.0	0.0	0.0	0.0	0.0	0.0	0.0	0.0	0.0	--	0.0	0.0	0.0	0.0

(1) Rating 0-10; 0 = no crop injury, 10 = complete kill.
* Soil treated 9/17/70. After treatment soil was rototilled to 15 cm depth.
  Crops planted 10/29/71.

cabbage. At 1400 kg/ha in the top 15 cm, which was 2 times the
'Tightly Bound' capacity, phytotoxicity to alfalfa, clover, turnip,
dichondra, radish and cabbage were commercially acceptable,
but injury to rye and barley was not acceptable. At 3360 kg/ha
in the top 15 cm, which was 5 times the 'Tightly Bound' capacity,
the crop injury was severe.

Three years after the soil was treated, the crop growth at the
different paraquat soil treatments was similar to that observed
at the 1 year interval.

Paraquat residues found in the sample crops are given in
Table X. Crops grown in soil treated at 1/2 the 'Tightly Bound'
paraquat soil capacity contained less than 0.01 to 3.8 µg/g paraquat. Crops grown in soil treated at 1-1/2 to 2 times the 'Tightly
Bound' paraquat soil capacity contained less than 0.01 to 17 µg/g
paraquat. Crops grown in soil treated at 2-1/2 to 5 times the
'Tightly Bound' paraquat soil capacity contained 0.07 to 42 µg/g
paraquat. The lowest residues were found in shelled soybeans
and cotton fuzzy seed while the highest residues were found in
mature potato vines. A portion of the residue from the potato
vines could be due to soil contamination and not uptake. Interpretation of crop residues from field trials of excessively high
soil treatments is difficult, since trace soil contamination can
cause significant crop residues. Nevertheless, the residue data
show that crops grown in soil treated at excessively high rates of
paraquat contain small amounts, if any, of paraquat. That is,
paraquat is bound to the soil and not available to plants.

C. Comparison of 'Strong Adsorption' Capacity and 'Tightly Bound' Capacity of Soils

The 'Strong Adsorption' capacity procedure determines the
amount of paraquat adsorbed by a soil before phytotoxic trace
amounts are present in the equilibrium solution (about 0.01 µg
paraquat/ml). The 'Tightly Bound' capacity procedure determines the capacity of a soil to adsorb paraquat so firmly that it
cannot be desorbed by saturated (5M) ammonium chloride; it can
only be released by refluxing the soil with 18 $\underline{N}$ sulfuric acid.

Both the 'Strong Adsorption' capacity and the 'Tightly Bound'
capacity of 4 soils have been determined. Details of the soils
(Frensham, England; Moorestown, New Jersey; Fresno, California) are given above. The Ocoee soil was a sand (approximately 98% sand) from Florida. The adsorption capacities
determined by the two methods are given in Table XI. The 'Strong
Adsorption' and Tightly Bound' capacities for a given soil did not

TABLE X

PARAQUAT RESIDUES IN CROPS FROM HIGH RATE SOIL TRIALS

Test Location	Crop	Crop Part Analyzed	Months After Soil Treatment	Soil Treatment (kg/ha) in 15 cm Depth	Paraquat Found in Crop (µg/g)*
	Soybeans	Shelled Beans	13	1120	0.08
				3360	0.04
				5600	0.07
		Immature Vines	33	1120	0.18
				3360	0.41
				5600	1.5
Moores-	Potatoes	Mature Vines	34	1120	3.8
town,				3360	17
New Jersey				5600	42
		Tubers	34	1120	0.13
				3360	0.85
				5600	0.48
	Volunteer Weeds	Foliage	33	1120	0.40
				3360	0.26
				5560	0.12
	Carrots	Tops	34	1120	0.29
				3360	1.9
				5600	. -
		Roots	34	1120	0.74
				3360	1.1
				5600	-
Fresno,	Volunteer Weeds	Foliage	33	336	0.08
California				1400	1.0
				3360	7.0
	Cotton	Fuzzy Seed	37	336	0.00
				1400	0.00
				3360	-
		Leaves & Twigs	37	336	0.07
				1400	0.07
				3360	-

* Average of 2 determinations.

differ by a factor of more than 2. This was expected because both values accurately predict the capacity of a soil to deactivate paraquat.

Throughout the remainder of this report the term - bound paraquat soil residues - will be used to describe soil residues below the 'Strong Adsorption'/'Tightly Bound' capacities of soil.

4. UNAVAILABILITY OF BOUND PARAQUAT SOIL RESIDUES TO EARTHWORMS

Extensive field and laboratory studies have shown that bound paraquat has no effect on worms and it is not absorbed by worm bodies. Very large paraquat soil residues, which result in 'free' paraquat in the equilibrium soil solution (greater than about 0.01 µg/ml) can reduce the number of worms. The reduced plant cover is as likely to be the cause of a reduction as direct paraquat toxicity.

Earthworm samples are obtained by expelling them from areas 60 cm square with 4.5 L of 0.2% formaldehyde solution or occasionally by hand sorting of soil. They have been sampled on 12 sites where paraquat has been used applied at normal or excessive rates, but only two are recorded in this paper. These are sites on which much paraquat soil adsorption work has been done, where the soils have a moderate and low clay content, and on which paraquat has been applied at grossly excessive rates.

One trial was done on Broadricks field at Jealott's Hill Research Station. Details of the soil have already been given. It has a 'Strong Adsorption' capacity of 800 µg paraquat/g soil. The field was covered with a thin grass sward. The trial was laid out in October 1964 as two replicate blocks, containing 6 m square plots separated by 1.5 m guard rows. Three random plots on each block were rotovated and sprayed 2 weeks later with 0, 2.24 and 112 kg/ha paraquat. After a further week, they were seeded with ryegrass. Three undisturbed plots on each block were sprayed with the same rates of paraquat, and ryegrass sown one week later by direct drilling (zero tillage). In May 1965, the 112 kg/ha plots were reseeded as the grass was thin and patchy.

Earthworms were assessed in October 1965, 12 months after application, and the results are shown in Table XII. Only two samples per treatment were taken, so that considerable variation might be expected. Nevertheless, it is clear that the population has not been adversely affected even by the excessive rate of 112 kg/ha.

TABLE XI

COMPARISON OF 'STRONG ADSORPTION' AND
TIGHTLY BOUND' CAPACITIES OF SOILS

Soil	'Strong Adsorption' capacity (µg paraquat/g soil)	'Tightly Bound' capacity (µg paraquat/g soil)
Moorestown	1500	1100
Frensham	120	200
Fresno	500	300
Ocoee	6	3

TABLE XII

EARTHWORMS FROM THE BROADRICKS SITE 12 MONTHS AFTER PARAQUAT APPLICATION
EXPRESSED AS NUMBER AND WEIGHT/M^2, EACH FIGURE DERIVED
FROM THE TOTAL EXPELLED FROM TWO 0.37 m^2 AREAS

Cultivation Before Paraquat Application	None			Rotovated		
Paraquat applied (kg/ha)	0	2.24	1.12	0	2.24	112
Total number/m^2	44	83	40	81	43	100
Total biomass (g/m^2)	20	40	24	35	29	73

At the Frensham, England site, another trial was done; details of which have previously been described. The earthworm population was assessed in November 1972, a year after application of the paraquat and three months after the barley harvest. The results were statistically analyzed after logarithmic transformation. The detransformed means are given in Table XIII.

Examination of the effects of cultivation, averaged over the rates, showed the numbers of Allolobophora caliginosa and A. chlorotica to be highly significantly lower (P = 0.01) on the plots which had been deeply rotovated. This is taken to be a residual effect of physical damage to near-surface species caused by triple rotovated. The combined species total showed a reduction of similar significance, being weighted largely by A. caliginosa. The significance of the biomass reduction under rotovation was lower (P = 0.05). There was no such difference in the species having deep burrows, Lumbricus terrestris and A. longa/nocturna, and it could be argued that they were able to take refuge below rotovation depth. Other workers have also found that the number of worms on uncultivated soil, where paraquat has been used to control weeds, tend to be greater than when the soil has been ploughed and cultivated (36). Caution must be excercised in interpreting results obtained by expulsion methods, as the soil structure can affect efficiency.

It is less easy to ascribe causes to differences observed on plots of the three paraquat rates within the same cultivation type. The lower numbers occurred principally on plots where the excessively high rates had brought about lighter vegetation cover. It is hardly surprising that the 720 kg/ha rate, bare for much of the year with a consequent reduction in organic material, should have an impoverished earthworm population. The 198 and 120 kg rates had a reduced cover of both weeds and barley plants in the early part of the year. It is known that ground cover and soil moisture are important factors affecting earthworm distribution, but direct toxic effects of 'free' paraquat in the soil solution cannot be ruled out.

The numerically dominant immature A. caliginosa present the clearest picture, correlating with vegetation sparseness. The greatest reduction, on the 720 kg rate, and that on the 120 kg rate, are highly signifcant (P = 0.01) and that on the 198 kg rate significant (P = 0.05) compared with their controls. Numbers of A. caliginosa adults and A. chlorotica on the same plots are low, but sample distribution is such that there is no statistical significance.

TABLE XIII

EARTHWORMS FROM THE FRENSHAM SITE 12 MONTHS AFTER APPLICATION OF PARAQUAT AND CULTIVATION, AND 7 MONTHS AFTER DIRECT DRILLING OF BARLEY SEED. EXPRESSED AS NO/M², CALCULATED FROM THE LOGARITHMICALLY TRANSFORMED MEANS OF SAMPLES FROM SIX 0.36m² AREAS

Depth Paraquat Incorporated into Soil	2 cm				15 cm			
Paraquat Applied (kg/ha)	0	15	33	120	0	90	198	720
Lumbricus terrestris (adult)	1	1	3	2	1	1	1	1
Lumbricus terrestris (immature)	5	6	5	1*	5	4	6	1*
Allolobophora longa/nocturna (immature)+	3	1*	1*	0**	5	1**	1**	0**
A. caliginosa (Savigny) (adult)	5	6	3	3	2	1	4	1
A. caliginosa (1mm.)X	79	93	77	23**	51	43	16*	8**
A. chlorotica (Savigny) (adult and immature)	6	9	15	2	3	5	1	0
Total number	103	119	111	33**	71	56	31*	11**
Total biomass (g/m²)	49	45	48	21*	29	32	21	10*

* Significantly different from untreated of same cultivation type (p = 0.05).
** Highly significantly different (P=0.01)
+ Adults too few for statistical analysis and not included.
X May include a few speciments of A. rosea (Savigny).

Numbers of <u>Lumbricus terrestris</u> were small, but the low
catch on the 720 and 120 kg plots was found to be statistically significant.

<u>Allolobophora</u> <u>longa</u> and <u>A</u>. <u>nocturna</u> are often difficult to
separate, and have been treated as one. It can be seen from the
talbe that the numbers of immatures from all treated plots are
all significantly lower than their controls. On the face of it here
would seem to be an effect of paraquat, but experience on other
sites, where such an effect has not been recorded, leads us to
consider this very unlikely. The numbers at Frensham are
small; therefore less reliable, and the few adult <u>longa/nocturna</u>
found were equally distributed between the 33 kg rate and its
control.

The combined totals follow the same pattern of signficance as
the immature <u>A</u>. <u>caliginosa</u>. The biomass reductions are less
strongly marked in both amount and statistical significance, being
being influenced by the heavier <u>L</u>. <u>terrestris</u> and <u>A</u>. <u>longa/</u>
<u>nocturna</u>.

As earthworms are an important means whereby a chemical
can be transferred from the soil to vertebrate life, tests were
made to discover whether paraquat is carried by worms and if
so where it is situated and if it is accumulated.

Paraquat residues in worms were determined at 8 trial sites.
Paraquat was applied at the normal rate (1 or 2 kg/ha) and at 112
kg/ha. Crops of grass or cereals were grown by direct drilling
(zero tillage) or after the soil had been rotovated or ploughed.
Worms were collected by the formaldehyde method, at intervals
between 2-12 months after paraquat application, and analyzed for
paraquat residues. A total of 14 different application rate/crop/
method of cultivation/trial site treatments were studied. Typical
worm residue data, for the Broadricks trial, are given in Table
XIV. Typical soil residue data are given in Table XV. Details of
the Broadricks trial have already been stated. In all trials,
paraquat residues in worms were the same, or more generally
much less, than in the soil. Thus, there is no tendency for
paraquat to accumulate in worms.

A further investigation was made to determine whether the
paraquat residues actually enter the body of the worm or whether
they remain bound to soil in the gut. Large numbers of worms
were collected from the two 112 kg/ha treatments in the Broadricks trial and bulked together. The worms, mostly <u>Lumbricus</u>
<u>terrestris</u>, were opened and the gut contents removed by a
strong jet of water. The gut contents and cleaned worm bodies
were analyzed for paraquat residues. The gut contents contained

TABLE XIV

PARAQUAT RESIDUES IN WORMS FROM THE BROADRICKS TRIAL

Paraquat Applied (kg/ha)	Cultivation Treatment Before Paraquat Application	Time After Paraquat Application (Months)	Paraquat Residues in Whole Worms (µg/g Fresh Weight)
2.24	None	12	0.60
2.24	Rotovated	12	0.83
112	None	12	62.9
112	Rotovated	5	90.8
112	Rotovated	12	28.7

TABLE XV

PARAQUAT RESIDUES IN BROADRICKS SOIL (µG PARAQUAT/G SOIL);
PARAQUAT WAS APPLIED OCTOBER 26, 1964

Paraquat Application Rate (kg/ha)	Cultivation Treatment Before Paraquat Application	Date Soil Sampled (Day/Month/Year)	Soil Depth (cm)			
			0-2.5	2.5-5	5-7.5	7.5-10
2.24	Rotovated	12.3.65	5.7	1.1	0.1	0.1
2.24	Rotovated	10.2.66	4.2	1.5	0.7	0.2
2.24	None	12.3.65	6.6	0.4	0.3	0.3
2.24	None	1.3.66	3.5	1.2	0.8	0.3
112	Rotovated	12.3.65	268	42	6	2
112	Rotovated	10.2.66	192	90	34	9
112	None	12.3.65	356	14	7	2
112	None	1.3.66	206	45	20	7

111 µg paraquat/g while the cleaned bodies contained only 0.28 µg
paraquat/g fresh weight, and even this could have been due to
imperfect washing out of the gut. Bound paraquat is clearly not
significantly absorbed by the worm bodies. This conclusion is
supported by experiments on the rapid elimination of paraquat
from living worms. Specimens were collected from a straw-
berry field at Fernhurst, England which had received 5 annual
2 kg/ha applications, the last dose having been about 3 months
before. Some whole worms were analyzed and found to contain
1.71 µg/g fresh weight of paraquat. Other living worms were
placed in untreated soil for 36 hours, then recovered and analyzed. There were no detectable residues ($<$ 0.1 µg/g) in the
worms.

It is concluded that bound paraquat soil residues are not absorbed by worm tissue. Also, it is extremely unlikely that any
such paraquat ingested by a worm predator could affect it in any
way.

5. UNAVAILABILITY OF BOUND PARAQUAT SOIL RESIDUES TO MICROARTHROPODS

Eleven trials have shown that bound paraquat residues in
soils have no effect on microarthropods. Very large paraquat
residues, which result in 'free' paraquat in the equilibrium soil
solution (greater than about 0.01 µg/ml), can reduce the number
of microarthropods, but the effect may be due to changes in
vegetation cover rather than a direct toxic effect of the paraquat.

A soil microarthropod sample is normally obtained by taking
4 soil cores 25 mm diameter x 150 mm deep and bulking them.
This is done in preference to taking single large cores as it reduced sampling error caused by faunal aggregations. The soil is
stored at -15°C before the microarthropods are extracted by a
washing, flotation and differential wetting technique. Mites and
springtails form the bulk of the samples, and of these the Mesostigmata, mainly predators, and the Cryptostigmata and Collembola, principally detritus and fungus feeders, are currently
chosen for identification to species level. In some of the early
studies the groups were not so detailed.

Of the 11 sites where soil microarthropod observations on
paraquat treated plots have been made, only two are quoted here.
As with the earthworm studies, results for only a few sampling
occasions are given. Groups selected for the tables include
some which are exclusively euedaphic, principally hemiedaphic,
and those of mixed habits or no known soil depth preference.

One of the studies was done on Broadricks field at Jealott's Hill Research Station. Brief details of the trial have been given in the earthworm section. The condensed results of assessment on three occasions during the first year are shown in Table XVI. The two-month sampling may be considered less reliable than later work, as single large cores were used for sampling, and the extraction technique was still being developed. The figures were not subjected to statistical analysis.

The results show no more variation between treatments than might be expected from the sampling regime and plot histories. The number of Astigmata (Tyrophagus spp. and Schwiebia sp.) on the non-cultivated 112 kg plots after 12 months is exceptionally high, but this group is notoriously erratic in occurrence.

Another trial, of which details have already been given, was done at Frensham. Microarthropods were sampled at the same time as the earthworms, and the same general remarks about the vegetation and physical effects apply. All Mesostigmata, Cryptostigmata and Collembola were identified to species, and other animals to higher categories. For the purpose of this paper 9 species of different life-forms and habits have been selected, and 6 main groups. Analysis of variance was done on logarithmically transformed data. Population estimates based on the corrected means are shown in Table XVII.

The effect of rotovation averaged over rates is much less marked than for the earthworms. In only three groups (Rhodacarus roseus, total Mesostigmata and Lepidocyrtus cyaneus) was the small reduction in numbers found to be statistically significant. It has been discovered in some other trials that disturbance of the habitat by ploughing and cultivation has an appreciable effect on many soil microarthropods. Disturbance effects would be expected to be less on the relatively unstratified sandy soil at Frensham.

The combined species of Mesostigmata, a predatory group, show a considerable and significant reduction only on the 720 kg rate. Several individual species are absent or present only in small numbers on these plots, but uneven distribution precludes statistical significance. Rhodacarellus silesiacus presents some somewhat anomalous results. The low numbers on the 120 kg rate are statistically significant compared with the untreated plots of the same cultivation type. On the 90 and 198 kg rates the numbers are significantly higher than their control. A link might exist with the nematodes (not sampled) on which this euedaphic species is believed to feed.

TABLE XVI

MICROARTHROPODS FROM THE BROADRICKS SITE ASSESSED AT INTERVALS UP TO A YEAR AFTER PARAQUAT APPLICATION. EXPRESSED AS HUNDREDS/m^2, CONVERTED FROM THE UNTRANSFORMED MEANS OF 6 SAMPLES PER TREATMENT, EACH FROM 4 SOIL CORES OR THEIR EQUIVALENT

Cultivation Before Application	NONE									ROTOVATED								
Paraquat Applied (kg/ha)	0			2.24			112			0			2.24			112		
Months After Application	2	6	12	2	6	12	2	6	12	2	6	12	2	6	12	2	6	12
E Rhodacaridae	3	53	22	5	77	43	0	56	25	8	40	31	5	33	26	2	51	25
Total Mesostigmata	18	61	30	6	90	52	2	63	42	12	47	50	10	53	33	3	70	43
Total Cryptostigmata	28	58	51	26	110	49	39	66	90	49	79	63	28	65	34	21	73	43
Total Astigmata	98	59	3	49	42	6	35	21	171	40	6	3	42	15	17	46	35	22
Total Prostigmata	8	3	29	0	15	8	0	5	13	5	14	13	0	6	3	1	20	7
Hypogastruridae	2	13	24	2	15	22	0	8	48	4	19	22	0	7	14	1	11	31
E Onychiuridae	5	361	235	35	239	468	11	195	357	12	225	469	132	148	418	124	128	334
Isotomidae	0	20	25	3	4	14	1	8	53	0	21	8	0	5	31	2	8	40
H Entomobryidae	0	3	3	1	5	3	1	1	5	0	4	3	3	3	3	1	1	8
H Sminthuridae	0	2	1	0	2	1	0	3	1	0	4	1	0	5	0	0	5	0
Total Collembola	7	398	288	41	265	507	13	214	464	16	273	503	136	168	465	127	154	413
Total Microarthropods+	165	580	409	122	529	635	91	369	793	127	425	641	221	310	561	91	353	553

E = euedaphic groups.   H = Hemiedaphic.   + includes remaining insects and the myriapods.

TABLE XVII

MICROARTHROPODS FROM THE FRENSHAM SITE 12 MONTHS AFTER APPLICATION OF PARAQUAT AND CULTIVATION, AND 7 MONTHS AFTER DIRECT DRILLING OF BARLEY SEED. (EXPRESSED AS HUNDREDS/m^2, CALCULATED FROM THE LOGARITHMICALLY TRANSFORMED MEANS OF 6 SAMPLES PER TREATMENT, EACH SAMPLE FROM 4 SOIL CORES.)

Depth Paraquat Incorporated into Soil	2 cm				15 cm			
Paraquat Applied (kg/ha)	0	15	33	120	0	90	198	720
Veigaia planicola (Berlese)	2	6	1	1	4	3	2	0
Hypoaspis aculeifer (Canestrini)	3	3	2	4	3	6	6	0
E Rhodacarus roseus Oudemans	13	16	19	24	4	12	24	10
E Rhodacerellus silesiacus (Willmann	15	9	15	4*	2	11*	18**	0
Total Mesostigmata	77	70	85	68	46	56	76	15*
Total Cryptostigmata	1	1	0	1	2	0	0	0
Total Astigmata	12	30	44	208**	50	16	85	58
Total Prostigmata	31	25	25	61	19	23	23	5
E Tullbergia spp	342	199	290	225	177	314	281	82*
H Cryptopygus thermophilus (Axelson)	147	74	109	38*	183	90	60	8**
E Isotomodes productus (Axelson	9	7	4	13	15	14	19	4
E Folsomia candida Willem	24	30	29	8*	46	21	28	5**
H Lepidocyrtus cyaneus Tullberg	8	8	7	11	7	7	3	0**
Total Collembola	649	458	596	386*	676	569	461	116**
Total microarthropods+	812	693	825	913	810	710	673	240**

* Significantly different from untreated of same cultivation type (P = 0.05).
** Highly significantly different from untreated of same cultivation type (P = 0.01).
E = euedaphic species.
H = hemiedaphic.
+ = includes remaining insects and the myriapods.

There is a paucity of Cryptostigmata on the site, and numbers are too low and irregular to be meaningful.

The Astigmata, often associated with decaying vegetable matter, are represented at Frensham chiefly by Tyrophagus spp. The large population on the 120 kg/ha plots is highly significantly greater than that on the control plots. Comparison may be made with the large numbers on the 12 month direct drilled high rate plots at Broadricks, but no definite explanation can be offered.

The Prostigmata, of which Pyemotidae is the dominant family here, has low numbers on the 720 kg plots, but not significantly lower than on the control plots.

Among the euedaphic Collembola Tullbergia spp. have significantly lower numbers on the 720 kg rate. At this paraquat rate Isotomodes productus was low but not significantly lower than on the control plots. The effect was stronger with Folsomia candida, where the reduction on the 120 kg rate was significant $(P = 0.05)$ and on the 720 kg rate highly significant $(P = 0.01)$.

Cryptopygus thermophilus, abundant at Frensham, is a surface-dwelling isotomid highly sensitive to vegetation cover. The very great and highly significant reductions on the almost bare 720 kg plots was therefore predictable. The numbers were also low on the 120 and the 198 kg rates, though the latter was not statistically significant. Lepidocyrtus cyaneus, another surface species, but not common in November, was absent altogether on the 720 kg plots.

Taken as a whole, the Collembola on the 720 kg rate were about a fifth of the control plots population, on the 120 kg more than half, and on the 198 kg rate marginally but not significantly less.

In conclusion, it may be said that no consistent effects on soil microarthropods have been detected at normal application rates of paraquat in any of the 11 trials. Such changes as have been recorded are associated with regimes involving mechanical disturbance or excessive rates causing alterations in vegetation cover.

6. UNAVAILABILITY OF BOUND PARAQUAT SOIL RESIDUES TO MICROORGANISMS

Bound paraquat residues in soil are unavailable to microorganisms. Microorganisms differ greatly in their sensitivity to paraquat; the lowest concentration which inhibits growth or activity of microorganisms in liquid cultures ranges from 0.1 to 50,000 µg/ml paraquat. In soils, residues up to several hundred µg/g have no significant effects on microorganisms, even paraquat sensitive organisms such as <u>Azotobacter</u> spp. Soils contain microorganisms, particularly <u>Lipomyces</u> spp., which rapidly degrade paraquat in liquid cultures. However, in soil bound paraquat is unavailable to microorganisms; consequently, these residues are degraded extremely slowly or not at all.

Effect of Paraquat on Microorganisms

The effect of paraquat on a wide range of microorganims has been extensively studied in both liquid cultures and soil (<u>37</u> to <u>66</u>). In liquid cultures the lowest concentration which affects microorganisms ranges from about 0.1 to 50,000 µg/ml paraquat. For example, some strains of <u>Azotobacter</u> are completely inhibited by 0.5 µg/ml (<u>50</u>), some fungi will grow in 1,000 µg/ml (<u>64</u>) and <u>Corynebacterium fascians</u> can be "trained" to tolerate 50,000 µg/ml (<u>62</u>).

The effect of paraquat on the number and activity of microorganisms in soil is very much less than in liquid cultures. For example, Szegi (<u>53</u>) showed that the activity of 20 cellulolytic microscopic fungi were severely reduced by 2-5 µg/ml paraquat. However, in sand and Chernozem soils the lowest concentration of paraquat which significantly reduced decomposition of cellulose was 500 and 1000 µg/g paraquat, respectively (<u>53</u>). The published data also shows that several hundred µg/g paraquat has to be applied to most soils before it significantly affects the number and activity of soil microorganisms. Any effects at lower concentrations are small and transitory. Thus, it can be concluded that bound paraquat has no significant effects on soil microorganisms. Even when traces of "free" paraquat residues are present in the soil solution, effects on microorganisms are often small. For example, at a field trial site at Ocoee, Florida the soil has exceptionally low 'Tightly Bound' and 'Strong Adsorption' capacity of 3 and 6 µg/g, respectively (Table XI). However, applications of paraquat at 5.6, 22.4 and 112 kg/ha had only small effects on the numbers of actinomycete, bacterial and fungal propagules (<u>37</u>); these small effects could have been caused by reduced plant growth rather than direct toxic effects of paraquat.

The lack of effects of paraquat soil residues on microorganisms is well illustrated by studies on nitrogen transformations in soil.

Tu and Bollen (<u>55</u>, <u>56</u>) showed that ammonification of native soil organic matter and added peptone was not appreciably affected by 0.25, 0.5, 2.5 or 25 µg/g paraquat in two silt loams, a silty clay loam and a sandy loam.

In liquid culture solutions paraquat is highly inhibitory of the oxidation of NH_4^+ to NO_2^- and NO_3^- by Nitrosomonas and Nitrobacter (39). However, many workers have shown that the conversion of added NH_4^+ to NO_3^- and nitrification of soil organic matter is not significantly affected by paraquat residues. Tu and Bollen (55, 56) showed that 0.25 to 25 µg/g paraquat in the 4 soils listed in the previous paragraph did not significantly affect the conversion of NH_4^+ to NO_3^-. Debona and Audus (39) also showed that 50 and 100 µg/g paraquat in an unclassified soil did not significantly affect the conversion of NH_4^+ to NO_2^- by Nitrosomonas and NO_2^- to NO_3^- by Nitrobacter. 500 µg/g slightly reduced nitrification and 1000 µg/g reduced nitrification by about two-thirds. Nitrite did not accumulate in the soil suggesting the reduced nitrification was due to paraquat affecting Nitrosomonas rather than Nitrobacter. Anderson and Drew (65) showed that 4 and 40 µg/g paraquat in a calcareous loam had no significant effects on the conversion of NH_4^+ to NO_2^- and NO_3^-. Thorneburg and Tweedy (54) also showed that 6 and 12 µg/g paraquat in a silt loam soil did not affect the conversions of NH_4^+ to NO_3^-.

Giardina et. al (42) studied the effect of paraquat on the urease activity of soil; urease is responsible for the conversion of urea to NH_4^+. They showed that 100, 200 and 400 µg/g paraquat in a clay loam soil did not affect its urease activity.

Nitrogen fixing organisms vary widely in their sensitivity to paraquat in liquid cultures. The growth and N-fixing capabilities of some strains of Azotobacter spp are severely inhibited by as low as 0.1 µg/ml while many strains of Rhizobium spp are not affected by 50 µg/ml and some Rhizobium spp will tolerate more than 100 µg/ml (44, 50, 52). In soil the effect of paraquat on nitrogen fixing organisms is much less than in liquid cultures. Szegi et. al. (52) found that 1000 µg/g paraquat did not affect the numbers of Azotobacter in calcareous sand and chernozem soils. Anderson and Drew (65) showed that the application of 1.1 and 11 kg/ha paraquat to a grass-clover sward on a calcareous loam soil had only small effects on the number of Azobacter spp in the soil. They attributed the small effects to changes in the vegetation cover and consequent changes in soil moisture rather than a direct effect of paraquat.

It is concluded that bound paraquat does not have any significant effects on nitrogen transformations in soil.

Effect of Microorganism on Paraquat

Soils contain organisms capable of degrading paraquat in nutrient cultures; for example, Lipomyces spp (62), Coryne-bacterium fascians (62), Clostridium pasteurianum (62) and an unidentified bacterial isolate (64). Also, it has been reported that actinomycetal isolates degrade paraquat (67); however, we have been unable to confirm this. Lipomyces spp (probably Lipomyces starkeyi) is particularly effective and readily de-grades paraquat under a wide range of temperature and pH values (68). It will grow and degrade paraquat in a wide range of nutrient broths and can use paraquat as its sole nitrogen source; however, it requires an alternative carbon source for synthesis of the paraquat-catabolizing system (68). Lipomyces spp is widely distributed in soils throughout the world (J.R. Anderson, ICI Plant Protection Division unpublished data).

The persistence of paraquat in soils has been extensively studied (25, and 70 to 75) plus very extensive unpublished ICI and Chevron trials in both the laboratory and field with both ^{14}C labelled and unlabelled paraquat). These trials show that bound paraquat is not degraded, even when the soils contain organisms, such as Lipomyces spp, capable of degrading paraquat in culture solutions. It is concluded that bound residues are not available to microorganisms.

The following pot experiment illustrates typical results of studies on the persistence of bound paraquat residues in soil. Four soils (Table XVIII) were treated with paraquat and placed in 5L plastic buckets with drainage holes in the bottom. The samples were kept outside unprotected from the weather and re-ceived a minimum of 7 cm of water per month. At intervals, samples were extracted with 18 $\underline{N}$ H_2SO_4 and the paraquat measured colorimetrically after a resin cleanup and reduction with alkaline dithionite (24). Results are summarized in Table XIX. Three of the 4 soils treated with 4 to 50 µg/g paraquat showed no loss of paraquat after 4 1/2 years.

One sand sample treated at 18 µg/g paraquat showed a 25% loss of paraquat after 4 1/2 years. However, leaching data indi-cates the loss in the sand was probably due to leaching. Two other sand samples treated at 4 and 11 µg/g paraquat did not show any loss in 4 1/2 years.

TABLE XVIII

CHARACTERISTICS OF SOILS USED IN OUTDOOR POT EXPERIMENT

Soil Origin	Soil Type	% Clay	% Sand	% Silt	% Organic Matter	pH	Cation Exchange Capacity (Meq/100 g Soil)
Florida	Sand	1	97	2	0.5	7.0	1.4
Florida	Muck	-	-	-	Approx. 100	3.3	112.7
Calif.	Loam A	16	47	37	1.3	6.5	33.2
Calif.	Loam B	12	46	42	3.3	5.6	22.4
Hawaii	Silt Loam	26	4	70	3.8	7.7	16.2

TABLE XIX

PARAQUAT SOIL RESIDUES IN OUTDOOR POT EXPERIMENT

(μg paraquat/g soil)[a]

Sample			MONTHS AFTER TREATMENT							
			0	1-2	3-4	7-9	14-16	22-24	38-40	53-55
	(c) 1		11.2	12.9	11.7	11.3	11.7	10.2	13.0	-
Sand	(b) 2		4.41	4.10	-	3.83	4.01	4.77	4.08	4.15
	(b) 3		18.2	19.5	-	14.5	12.7	10.1	12.4	14.1
Muck	(c) 4		47.0	53.7	55.1	53.0	46.7	-	-	-
Loam A	(c) 5		14.5	16.0	16.4	18.5	13.7	13.2	17.4	18.8
	(c) 6		14.0	15.6	12.9	17.5	15.0	17.1	13.9	16.6
Loam B	(b) 7		6.29	4.48	-	3.53	4.57	5.08	3.21	4.52
	(b) 8		24.3	21.6	-	16.5	14.6	15.8	16.8	18.7
Silt Loam	(b) 9		4.34	5.54	-	5.98	4.93	3.79	5.18	3.91
	(b) 10		19.4	18.2	-	16.8	15.4	14.3	18.3	17.0

(a) Average of duplicate analyses.

(b) Paraquat bis-methyl sulfate.

(c) Paraquat dichloride

7. LONG TERM CONSEQUENCES OF
REPEATED PARAQUAT APPLICATIONS

The long term safety of the repeated use of paraquat is well illustrated by a trial conducted in South Africa on an extremely sandy soil which contains only 1% clay. In an eight year period over 20 applications (total 15.6 kg/ha paraquat) have been made to control weeds in a commercial vineyard. The soil residues have had no effect on the vines or barley and Lemna test crops.

The experiment was started in 1965, to study methods of controlling weeds in vines. The trellised Barlinka table grapes, 27 years old in 1973, are irrigated. A natural weed cover is allowed to develop in the winter. Paraquat is used to control weeds in the summer; occasionally the herbicide diquat is also applied. Between 2 and 4 applications of paraquat are used each season; rates of application ranged from 0.3 to 1.0 kg/ha.

The soil contains 98% sand, 1% clay and $<$ 1% organic matter and has a pH of 6.0. It has a 'Strong Adsorption' capacity of 70 ug paraquat/g soil (0.08 meq/100 g soil) as determined by the bioassay of the soil equilibrium solution method (see Section 3). Paraquat residues in the soil have been measured at least once per year. Twenty samples were taken at random, with a 3.8 cm diameter corer from each plot and the cores divided into 0.25, 2.5-5, 5-10 and 10-15 cm deep horizons. The samples from each depth were bulked together.

Annual application rates and total amount paraquat residues in the top 15 cm soil are given in Table XX. The concentration of paraquat residues in different soil horizons are given in Table XXI.

The paraquat residues had no significant effects on the vines or grape yields. Samples of leaves, grapes and twigs were taken in March, 1973. No paraquat residues were detected ($<$0.05, $<$0.03 and $<$0.03 µg/g in leaves, grapes and twigs, respectively). Soil samples collected from the top 2.5 cm soil in May, 1970 were bioassayed in the glasshouse. Samples (10 g) containing 0, 7.3 and 8.9 µg paraquat/g were bioassayed with Lemna as described above and samples (150 g) containing 0 and 12.4 µg/g were bioassayed with barley (cultivar Zephyr). The paraquat residues had no effect on the growth of the plants. Thus, even on this extremely sandy soil the continued use of paraquat does not leave any harmful residues in the soil. On most other soils the safety margin is even higher (see Section 3 above).

TABLE XX

PARAQUAT APPLICATION RATES AND SOIL RESIDUES
WEED CONTROL IN IRRIGATED VINES - RHODES FRUIT FARM, SOUTH AFRICA

Date	Paraquat applied to Date (kg/ha)	Paraquat residues* in top 15 cm soil (kg/ha)	Paraquat residues in top 15 cm soil expressed as percentage of that applied
End of 1st season April 1966	2.2	0.5	23
End of 2nd season April 1967	4.9	1.0	20
End of 3rd season April 1968	6.6	2.7	41
End of 4th season April 1969	8.5	4.4	52
End of 5th season April 1970	9.9	4.5	45
End of 6th season April 1971	11.6	6.7	58
End of 7th season April 1972	13.0	6.4	49
End of 8th season April 1973	15.6	5.6	36

*Calculated by assuming the soil had a bulk density of 1.2 g/cc. Values given are the means of 4 replicate plots.

TABLE XXI

PARAQUAT SOIL RESIDUES (μg PARAQUAT/g SOIL). WEED CONTROL IN IRRIGATED VINES - RHODES FRUIT FARM, SOUTH AFRICA. VALUES ARE THE MEANS OF 4 REPLICATE PLOTS.

Depth Sampled (cm)	Date Sampled									
	16/9/65	1/4/66	18/4/67	21/9/67	15/3/68	8/5/69	6/5/70	26/5/71	17/5/72	7/5/73
0-2.5	0.5	1.5	2.9	7.4	7.8	9.5	9.5	13.8	12.1	10.2
2.5-5	0.3	0.2	0.4	0.9	0.8	3.4	3.4	5.0	5.5	3.6
5-10	0.1	< 0.1	< 0.1	0.3	0.2	0.8	0.8	1.4	1.5	2.0
10-15	< 0.1	< 0.1	< 0.1	0.2	< 0.1	0.1	0.3	0.4	0.4	0.5

8. CONCLUSIONS

1. Bound paraquat residues in soil are not available to living organisms, such as plants, earthworms, microarthropods and microorganisms.
2. There is no risk of bound paraquat being "freed" and becoming available to living organisms.
3. Soils can bind and deactivate many times the residues resulting from normal applications of 0.1 to 2 kg/ha. Most soils deactivate hundreds of kg paraquat/ha in the top 15 cm alone.
4. Laboratory determined 'Strong Adsorption' and 'Tightly Bound' capacities accurately predict the capacities of soils to deactivate paraquat under field conditions. If anything, these laboratory determinations tend to underestimate the capacity of soils to deactivate paraquat.
5. Plants are among the most sensitive living organisms to "free" paraquat; concentrations as low as 0.01 µg/ml in the soil solution being phytotoxic. Thus, plants act as sensitive indicators of "free" paraquat.
6. If soils contain "free" paraquat, for example, because of accidental spillage of the herbicide concentrate, it can be deactivated by ensuring that it contacts further adsorption sites. This may be achieved either by cultivating and mixing more soil with the paraquat treated layer, or by the addition of montmorillonite clays, such as bentonite. Montmorillonite can deactivate up to 5% of own weight of paraquat so that the rates of application of clay mineral which are required are not excessive. At worst they are of the same order of magnitude as rates of application of lime for acidity correction.
7. Although the work discussed in this paper was concerned with residues in soil the same conclusions apply to muds in aquatic environments (69).

It is concluded from these facts that the continued and repeated use of paraquat will not leave harmful residues in soil.

ACKNOWLEDGEMENTS

We thank our large number of colleagues who carried out many of the experiments described in this paper and for their comments on drafts of the paper.

LITERATURE CITED

1. Calderbank, A., Advances in Pest Control Research (1968) 8, 127-235.
2. Akhavein, A. A. and Linscott, D. L., Residue Reviews (1968) 23, 97-145.
3. Slade, P., Weed Res. (1966) 6, 158-167.
4. Wright, K. A. and Cain, R. B., Soil Biol. Biochem. (1969) 1, 5-14.
5. Orpin, C. G., Knight, M. and W. C. Evans, Biochem. J. (1972) 127, 833-844.
6. Burns, R. G. and Audus, L. J., Weed Res. (1970) 10, 49-58.
7. Damanakis, M., Drennan, D.S.H., Fryer, J.D. and Holly K., Weed. Res. (1970) 10, 264-277.
8. O'Toole, M.A., Irish, J. Agr. Res. (1965) 4, 231-233.
9. Best, J.A. Weber, J.B. and Weed, S.B., Soil Sci., (1972), 114, 444-450.
10. Burns, I. G., Hayes, M. H. B. and Stacey, M., Weed Res. (1973), 13, 79-90.
11. Coats, G.E., Funderburk, H.H., Lawrence, J.M. and Davis, D.'E., Weed Res. (1966), 6, 58-66.
12. Haque, R., Lilley, S. and Coshow, W.R., J. Colloid Interface Sci., (1970), 33, 185-188.
13. Khan, S.U., Can. J. Soil Sci. (1973) 53, 199-204.
14. Khan, S.U., J. Soil Sci. (1973) 24, 244-248.
15. Knight, B.A.G. and Denny, P.J. Weed Res. (1970) 10, 40-48.
16. Knight, B.A.G. and Tomlinson, T.E., J. Soil Sci. (1967) 18, 233-243.
17. Malquori, A. and Radaelli, L., Ric. Sci. (1966) 36, 1094-1095.
18. Radaelli, L. and Fusi, P., Agrochimica (1968) 12, 558-566.
19. Tomlinson, T.E., Knight, B.A.G., Bastow, A.W. and Heaver, A.A., Soc. Chem. Industry (London, 1968) Monograph No. 29, 317-333.
20. Weber, J.B., Perry, P.W. and Upchurch, R.P., Soil Sci. Soc. Am. Proc. (1965) 29, 678-688.
21. Weber, J.B. and Weed, S.B., Soil Sci. Soc. Am. Proc. (1968) 32, 485-487.
22. Weber, J.B., Meek, R.C. and Weed, S.B., Soil Sci. Soc. Am. Proc. (1969) 33, 382-385.
23. Weed, S.B. and Weber, J.B., Soil Sci. Soc. Am. Proc. (1969) 33, 379-382.

24. Tucker, B.V., Pack, D.E. and Ospenson, J.N., J. Agr. Food Chem. (1967) 15, 1005-1008.
25. Tucker, B.V., Pack, D.E., Ospenson, J.N., Omid, A. and Thomas, W.D., Weed Sci. (1969) 17, 448-451.
26. Hayes, M.H.B., Pick, M.E. and Toms, B.A., Residue Reviews (1975) 57, In press.
27. Khan, S.U., Residue Review (1974) 52, 1-26.
28. Weber, J.B., Ward, T.M. and Weed, S.B., Soil Sci. Soc. Am. Proc. (1968) 32, 197-200.
29. Steward, B.C. (Editor). "Plant Physiology. Vol III. Inorganic Nutrition of Plants." Academic Press (1963).
30. Jenny, H. and Grossenbacher, K., Soil Sci. Soc. Am. Proc. Proc. (1963) 27, 273-277.
31. Jenny, H., Plant and Soil (1966) 25, 265-289.
32. Walker, A., Pesticide Sci. (1972) 3, 139-148.
33. Barber, S.A., Soil Sci. (1962) 93, 39-49.
34. Damanakis, M., Weed Res. (1970) 10, 77-80.
35. Isensee, A.R. and Walsh, L.M., J.Sci. Food Agric. (1972) 23, 509-516.
36. Schwerdtle, F.Z., Pflanzenkrankheit und Pflanzenschutz (1969) 76, 635-641.
37. Camper, N.D., Moherek, E.A. and Huffman, J. Weed Res. (1973) 13, 231-233.
38. Central Coffee Research Institute India Annual Report (1969/70) p. 160-161.
39. Debona, A.C. and Audus, L.J., Weed Res. (1970) 10, 250-263.
40. Dey, S.K., Sarkar, S.K. and Bhattacharya, N.G. Two and a Bud (1969) 16, 73-75.
41. Edwards, C.A., Proc. 10th British Weed Control Conf. (1970), p. 1052-1062.
42. Giardina, M.C., Tomati, U. and Pietrosanti, W., Meded. Fac. Landbouwwetensch. (1970) 35, 615-626.
43. Giardina, M.C., Tomati, U. and Pietrosanti, W., Nuovi Ann. Ig. Microbiol. (1973) 24, 191-196.
44. Grossbard, E. "White Clover Research", edited by J. Lowe. Occasional Symposium No. 6 - British Grassland Society (1970).
45. Jones, G.D. and Williams, J.R., Trans. Br. Mycol. Soc. (1971) 57, 351-357.
46. Kokke, R., Antonie van Leeuwenhoek (1970) 36, 580-581.
47. Rodrigues-Kabana, R., Curl, E.A. and Funderburk, H.H., Phytopathology (1966) 56, 1332-1333.

48. Rodrigues-Kabana, R. , Curl, E.A. and Funderburk, H.H. ,
     Phytopathology (1967) 57, 911-915.
49. Langkramer, O.,VII Congres International de la Protection
     des Plantes. Paris 1970 Resumes No. 157-C3.10, p. 798-
     799.
50. Manninger, E. , Bakondi, E. and Takats, T. , Symposia
     Biologica Hungarica (1972) 2, 401-404.
51. Satyanarayana, G. and Sarkar, S.K. , Two and a Bud (1968)
     15, 163.
52. Szegi, J. , Gulyas, F. , Manninger, E. and Zamory, E. ,
     10th Intern. Cong. Soil Science (1974) 3, 179-184.
53. Szegi, J. , Symposia Biologica Hungarica (1972) 2, 349-354.
54. Thorneburg, R.P. and Tweedy, J.A. , Weed Sci. (1973) 21,
     397-399.
55. Tu, C.M. and Bollen, W.B. , Weed Res. (1968) 8, 28-37.
56. Tu, C.M. and Bollen, W.B. , Weed Res. (1969) 8, 38-45.
57. Wallnofer, P.,Z. PflKrankh. PflPath. PflSchutz. (1968) 75
     218-224.
58. Walter, B., Z. PflKrankh. PflPath. PflSchutz. (1970) 75
     Sonderheft V, 29-31.
59. Wilkinson, V. and Lucas, R.L. , Trans. Br. Mycol.Soc.
     (1969) 53, 297-299.
60. Wilkinson, V. and Lucas, R.L. , New Phytol. (1969) 68,
     709-719.
61. Wilkinson, V. and Lucas, R.L. , Weed Res. (1969) 9, 288-
     295.
62. Baldwin, B.C. , Bray, M.F. and Geoghegan, M.J. Bio-
     chem. J. (1966) 101, 15P.
63. Funderburk, H.H. and Bozarth, G.A. , J. Agri. Food
     Chem. (1967) 15, 563-567.
64. Bozarth, G.A. , Funderburk, H.H. , Curl, E.A. and Davis,
     D.E., Proc. 5th Weed Conf. 18, 615.
65. Anderson, J.R. and Drew, E.A. , Zentralblatt fur Bakteri-
     ologie, Parasitenkunde, Infektionskrankheiten und Hygiene
     Abt. II (1975) In Press.
66. Anderson, J.R. and Drew, E.A. , Zentralblatt fur Bakteri-
     ologie, Parasitenkunde, Infektionskrankheiten und Hygiene
     Abt. II (1975) In Press.
67. Namdeo, K.N. , Indian J. Experimental Biol. (1972) 10, 133-
     135.
68. Anderson, J.R. and Drew, E.A. , J. Gen. Microbiol. (1972)
     70, 43-58.
69. Calderbank, A. , Outlook in Agriculture (1972) 7, 51-54.

70. Calderbank, A., Environmental Effects of the Herbicide Paraquat. 3rd International Symposium; Chemical and Toxicological Aspects of Environmental Quality. Tokyo; 1973.
71. Calderbank, A. and Tomlinson, T.E., Outlook on Agri. (1968) 5, 252-257.
72. Soerenson, O., Wasser (1971) 38, 17-26.
73. Fryer, J.D., Hance, R.J. and Ludwig, J.W., Weed Res. (1975) 15, 189-194.
74. Way, M.J., Newman, J.F., Moore, H.W. and Knaggs, F.W., J. App.Ecol. (1971) 8, 509,532.
75. Khan, Marriage, and Saidak, Can. J. Soil Sci., (1975), 55, 73-75.

23

Fixed and Biologically Available Soil Bound Pesticides

J. B. WEBER

North Carolina State University, Raleigh, N. C.

Many types of pesticides and their metabolites become complexed with soil colloids over a period of time and cannot be separated without first destroying the architecture of the soil colloids. These are considered to be "irreversibly bound residues" and the ultimate fate of these materials is unknown. Certain pesticides or metabolites are less strongly bound to soil colloids or are bound in a more accessible fashion. These substances may be extracted from soil by the use of salt solutions such as $\underline{N}$ NH_4Ac, or organic solvents such as methanol. They are biologically available, however, their availability is greatly reduced.

Biologically-available bound residues have been characterized as follows: a) extractable from soil with $\underline{N}$ acid, alkaline, or salt solutions, or with polar, or nonpolar, organic solvents; b) the extractable fraction is related to that which is biologically active; and c) the chemical may be readily chromatographed and separated from soil fractions. Irreversibly-bound residues have been characterized as follows: a) nonextractable from soil with $\underline{N}$ acid, alkaline, or salt solutions or with organic solvents; b) extractable from soil upon destruction of soil colloids by use of HF acid, or oxidation or combustion methods; and c) inseparable from soil colloids by chromatographic techniques.

Biological availability of soil-bound pesticides is dependent upon the adsorption mechanisms involved, the type of organisms and soil colloid, and edaphic factors such as soil pH, and kinds and amounts of nutrients or other substances present. Cationic pesticides, such as diquat and paraquat, are ionically bound to negatively charged organic and inorganic soil colloids. They are readily desorbed from organic colloids and certain types of clay minerals with salt solutions, and substantial amounts are biologically available. Only very small amounts can be desorbed from montmorillonic clay minerals with salt solution and only those molecules adsorbed on external surfaces are biologically available. Adsorptivity and biological availability of basic pesticides, such as the $\underline{s}$-triazine herbicides, are dependent upon the

354

pH of the soil systems. Greater adsorption and lower availability occurs under acid conditions than under neutral or alkaline conditions. Most acidic pesticides, with the exception of those possessing the elements phosphorus and arsenic, are only weakly adsorbed on soil colloids and are generally biologically available. Highly hydrophobic pesticides, such as the dinitroanilines, exist in the undissolved form in the soil and gradually become associated with the lipophillic portion of organic soil colloids. In the bound form, only small amounts are biologically active.

24

An Experimental Approach to the Study of the Plant Availability of Soil Bound Pesticide Residues

F. FUHR

Arbeitsgruppe Radioagronomie, Kernforschungsanlage Julich GmbH, D 517 Julich, Postfach 1913, West Germany

It is the general object of a tracer experiment with pesticides to collect extensive information on their fate in the soil/ plant system. We therefore conduct standardized lysimeter experiments under outdoor conditions using either topsoil (1 m) or undisturbed soil columns (0.25 m) to simulate the practical application to field crops. After one growing season, precise results can be obtained on uptake and translocation by plants as well as metabolism in plants and soil, translocation in the soil profile, and loss due to mineralization and/or volatilization.

The same experimental plot is then used to study the uptake by the succeeding crops of remaining pesticide residues, as well as their further fate in the soil. In parallel experiments, the portion of soil bound residues and their nature can be determined by chemical methods. Using specially designed experimental pots, their turnover rates in soil and their uptake by roots can be studied under standard climatic conditions.

Methabenzthiazuron, a broad spectrum herbicide (N,N'-dimethyl-N'-(2-benzthiazolyl)-urea) for weed control in cereal crops, proved to be a very stable compound in such a lysimeter experiment with spring wheat. At harvest time 83% of the applied [14]C (benzthiazolyl-2-^{14}C) was still in the soil, preferentially in the upper 5 cm layer, and almost 90% of the acetone-extractable portion was still representing methabenzthiazuron. Increasing with time, a relatively high percentage of the labeled residues could not be extracted from the soil with organic solvents. However, where polar solvents were used in soil organic matter analysis (e.g. 0.1 N NaOH) most of the remaining ^{14}C could be extracted and the labeled compounds were found predominantly in the fulvic acid fraction. There are indications that these compounds are adsorbed rather than bound by soil organic matter.

The following crops, rye and carrots, took up about 1% of the ^{14}C residues from soil and, compared to the methabenzthiazuron treated wheat crop, the rye contained only 1/10 of residual compounds derived from methabenzthiazuron.

Methabenzthiazuron was not detected in the plant material.

Using the experimental approach described above, and in combination with analytical results, it was possible to collect extensive information on the biological availability of methabenzthiazuron residues in soil.

25

Degradation of the Insecticide Pirimicarb in Soil—Characterization of "Bound" Residues

I. R. HILL

ICI Plant Protection Division, Jealott's Hill Research Station, Bracknell, Berkshire, England

The degradation of pirimicarb (2-dimethylamino-5,6-dimethylpyrimidin-4-yl dimethylcarbamate), a fast acting selective aphicide, has been studied in several sterile and non-sterile soils maintained in controlled laboratory conditions. Each soil type was treated with 1 Kg ai/ha of the insecticide radiolabeled at the pyrimidine ring-2-carbon at the carbamate-carbonyl-carbon, and incubated 'aerobically' (at 40% of moisture holding capacity) and after flooding the soil with water to a depth of 2.5 cm. Pirimicarb and any degradation products were extracted directly from the moist or wet soils (air drying can result in 'loss' of radioactivity) by refluxing in acetone:water (5:1) (selected after studies of the efficiency of a range of solvents). Chromatographic analyses of the soil extracts indicated that pirimicarb degradation proceeded via cleavage of the carbamate moiety from the ring, N-dealkylation of the 2-dimethyl= amino group and pyrimidine ring cleavage. Degradation was most rapid in alkaline soils.

After extraction the soils were freeze-dried, ground to a powder in a hammer mill and thoroughly mixed by rotation. As this procedure 'homogenized' the soil sample, making subsequent analyses more accurate, and did not result in any loss of radio-activity, all measurements of unextractable radioactivity were made by dry combustion. Radioactivity remaining after exhaustive extraction was considered to be 'bound'.

In 'aerobically' incubated soils considerable amounts of the ring-2-^{14}C, but relatively little (less than 10% of that applied) of the carbamate-^{14}C, radiolabel were present as 'bound' residues. Within an incubation period of 1 year, approximately 20-60% of the applied ring-^{14}C became 'bound' in a range of 9 soil types of pH 5.5-8.1 and organic matter content 1.7-10.4%. In one alkaline soil (Gore, pH 7.8), in which the insecticide was very rapidly degraded (50% in 1-2 weeks, 90% in 3-5 weeks), over half of the applied ring-^{14}C was 'bound' within 5 weeks incubation. In sterile soils (autoclaved and γ-irradiated) both the rate of pirimicarb degradation and the accumulation of 'bound' residues were considerably less than in their non-sterile equivalents.

An acid soil (Broadricks, pH 5.5) and an alkaline soil (Gore) were also incubated under flooded conditions. In Broadricks, the amount of 'binding' was very similar to that observed under aerobic conditions. However, in Gore 20-30% of both the ring- and carbamate- radiocarbons became 'bound' during an incubation period of 20 weeks.

In most soils the amount of 'bound' radioactivity increased progressively throughout the incubation period. In Broadricks treated with ring-2-^{14}C pirimicarb and incubated 'aerobically' for 2 years, 70% of the applied radioactivity was 'bound'. In Gore, whether incubated 'aerobically' or in flooded soil, the amount of 'bound' radioactivity reached a maximum within 5-10 weeks, but thereafter declined steadily for the remainder of the incubation period. There was no simple relationship between the amount or rate of formation of 'bound' residues and either the rate of pirimicarb degradation or the soil pH and organic matter content.

Attempts were made to extract and characterize the 'bound' pesticide residues in the following extracted, freeze-dried soils (composite samples from soils incubated for a minimum of 5 weeks):

a. Gore 'aerobically' incubated with ring-2-^{14}C pirimicarb (GAR);
b. Broadricks 'aerobically' incubated with ring-2-^{14}C pirimicarb (BAR);
c. Gore incubated in flooded state with ring-2-^{14}C pirimicarb (GFR); and
d. Gore incubated in flooded state with carbamate-^{14}C pirimicarb (GFC).

Extractions were carried out with acetone:water (5:1), acid, alkali and salt solutions by shaking, refluxing and ultrasonication. For all but the ultrasonicated treatments extractions were carried out in 'enclosed' vessels through which a stream of air was passed. The effluent air was bubbled through tubes containing, in succession, methoxyethanol and ethanolamine or 0.5 M sodium hydroxide, to trap products evolved into the air stream during extraction.

Soils GAR and BAR contained approximately 40% and 35%, respectively, of the radiolabeled carbon originally applied to the soil as pirimicarb. Virtually all of this 'bound' radioactivity was extracted with 0.1 M NaOH and with 0.1 M Na$_4$P$_2$O$_7$, 60-80% with 0.5 M H$_2$SO$_4$ and only 10-33% with 1.0 M HCl. No more than 1% of the radioactivity was evolved during acid extraction and no more than 3% during alkaline or salt extraction. After acidification (to pH 1.0) of the NaOH extracts of GAR and BAR, almost all of the radioactivity was present in the acid-alkali soluble fulvic acid component. Any 'bound' pyrimidin-4-yl carbamate may well be hydrolyzed to a hydroxypyrimidine during NaOH extraction which, together with any 'bound' hydroxypyrimidines, would fractionate with the fulvic acids. Refluxing of GAR and BAR with acetone:water extracted 63% and 78%,

respectively, of the 'bound' radioactivity. Clearly the freeze
drying and/or grinding had altered the state of these residues,
for previously these soils had been exhaustively extracted
in the same solvent. From chromatographic analyses of these
acetone:water extracts, 8% and 18% of the 'bound' radioactivity
in GAR and BAR respectively was identified as pirimicarb, 5% and
less than 1% as its 2-methylformamido and 2-methylamino derivatives, and 8% and 38% as hydroxypyrimidines formed by decar-
bamoylation of the parent and its 2-methylamino degradation
product. Thus in both GAR and BAR approximately 5% of the originally applied radioactivity was present in the 'bound' residues
as identified pyrimidin-4-yl carbamates. This compares approximately with the total 'bound' radioactivity in soils incubated
'aerobically' with carbamate-^{14}C-pirimicarb.

Subsequently, Gore soil treated with ring-^{14}C-pirimicarb has
been used to investigate the effect of sequential acetone:water
extractions, with combinations of air-drying, freeze-drying and
grinding the soil, on the extractability of pirimicarb and its
degradation products. Both methods of drying the soil enabled
further amounts of radioactivity to be extracted by repeated
acetone:water reflux. Grinding the soil caused a further but
small increase in extractability. After 15 weeks incubation, 54%
of the radioactivity present was extracted from the moist soil.
Subsequent sequential drying and refluxing (3 times) extracted an
additional 15% of the radioactivity. Although drying may have
released ^{14}C-material bound to soil surfaces, it is also a possibility that the pesticide and/or its degradation products can
accumulate inside microbial cells. These cells might not all be
ruptured without an extraction from the dried state. Studies are
in progress to examine this possibility.

Soils GFR and GFC contained approximately 14% and 21%,
respectively, of radioactivity derived from the ring-^{14}C and
carbamate-^{14}C radiolabeled pirimicarb originally applied. Extraction of GFC with 1.0 M HCl (shaken at room temperature) resulted
in a loss of 75% of the bound radioactivity from the extraction
vessel, and its recovery in the ethanolamine 'trapping' solution.
It was, however, observed that 98% of the radioactivity was also
evolved from soil treated with carbamate-^{14}C-pirimicarb and extracted with 1.0 M HCl after only 4 hours incubation. As 1.0 M
CH$_3$COONa (adjusted to pH 5 with glacial CH$_3$COOH) did not cause any
such 'loss' of radioactivity, soil GFC was also shaken with the
acetate buffer. During this extraction, 55% of the 'bound'
radioactivity was evolved and recovered in either ethanolamine or
NaOH solutions. Almost all of the radioactivity in the NaOH
'trap' was identified as Na$_2$^{14}CO$_3$. It was thus concluded that
approximately half of the 'bound' radioactivity in GFC was
probably present as ^{14}C-carbonate, derived from fixation of ^{14}CO$_2$
evolved after hydrolytic cleavage of the ^{14}C-carbamate ester
under flooded conditions. As expected, a similar amount of
radioactivity (approx 50%) remained in the soil residue after

NaOH extraction. The remaining radiolabeled residues were alkali-extracted, and after acidification were recovered from the fulvic acid fraction. Acetone:water refluxing extracted only 20% of the GFC 'bound' residues, of which three quarters was identified as pirimicarb.

During acid extraction of GFR, less than 2% of the radioactivity was evolved into the air stream. Instead, almost all of the 'bound' [14]C constituents behaved as those in BAR and GAR, being extracted with alkali and soluble in acids. From acetone:water extraction, 10% of the 'bound' radioactivity in GFR was identified as pirimicarb and 25% as 5,6-dimethyl-2-dimethylamino-4-hydroxypyrimidine. A further 20% was extracted but not characterized.

It has thus been possible to identify between 20% and 65% of what was originally believed to be the 'bound' radioactive residues of pirimicarb resulting from incubation of soil under 'aerobic' and flooded conditions.

The 'bound' ^{14}C-residues of pirimicarb remaining unidentified are at present under investigation using chromatographic and isotope dilution techniques.

26

Chloroaniline - Humus Complexes—Formation, Persistence, and Problems in Monitoring

R. BARTHA and T.-S. HSU

Rutgers University, New Brunswick, N. J. 08903

The microbial metabolism of some phenylamide herbicides results in reactive chloroaniline moieties. At field application levels, a small portion of these is metabolized further by dimerization and polymerization, but the bulk of the chloroaniline is immobilized in spontaneous interactions with soil. Their physical adsorption on clay and soil organic matter is a temporary mechanism of limited importance, but their chemical reactions with soil organic matter renders them stable and in-extractable by solvents.

In a test soil about 50% of the chemically bound chloro-aniline was recovered after alkaline hydrolysis. Model ex-periments with humic monomers indicate that this portion is held as anil or anilinoquinone. The remaining 50% could not be liberated, and considerations of bond stability indicate that their attachment involved heterocyclic ring closures and/or ether bonds.

$^{14}CO_2$ evolution from radiolabeled chloroanilines indicates a slow mineralization of the residues that is mediated by aerobic soil microorganisms. The degradation of purified chloro-aniline-humus complexes was studied using a *Penicillium frequentans* and an *Aspergillus versicolor* culture, and evidence for two distinct degradation mechanisms was obtained.

$^{14}CO_2$-evolution from the non-hydrolyzable fraction of purified chloroaniline-humus complexes indicates that this fraction has no absolute resistance to microbial attack. Nevertheless, in natural soil the proportion of the hydrolyzable complex decreased with time, while both the relative and the absolute amount of the nonhydrolyzable residue increased during 190 days of incubation. The most plausible explanation of this finding is that the nonhydrolyzable complex is degraded in natural soil only very slowly and, in addition, some of the hydrolyzable complexes shift to nonhydrolyzable ones with time. Chloroanilines immobilized as non-hydrolyzable humic complexes are not detectable with currently available analytical techniques other than radiotracer methods. This fact prevents any meaningful monitoring of chloroaniline residues in agricultural soils with a phenylamide treatment history. Novel analytical approaches designed to correct this situation need to be developed and applied.

Chloroaniline-humus complexes do not affect the respiration of soil microorganisms, and may be regarded as temporarily detoxified as well as immobilized. It appears that some soil microorganisms are able to gradually mineralize the bound residue, while others tend to re-mobilize it without an immediate mineralization. It remains to be established which of these two mechanisms predominates in natural soils.

Documentation and details of the above summary are available in the following references:

Hsu, T.-S., and Bartha, R. Soil Sci. (1974) 116:444-452.
Hsu, T.-S., and Bartha, R. Soil Sci. (1974) 118:213-220.
Hsu, T.-S., and Bartha, R. J. Ag. Food Chem. (1975), in press.

27

Determination of the Release of Bound Fluchloralin Residues from Soil into Water

GARY M. BOOTH and R. WARD RHEES
Brigham Young University, Provo, Utah 84601

DUANE FERRELL
Thompson-Hayward Chemical Co., Kansas City, Kans. 66110

J. R. LARSEN
University of Illinois, Urbana, Ill. 61801

The objectives of this study were: (1) to determine the total extractable and non-extractable residues from aged soil, (2) to identify the soil residues that were extractable in organic solvents, (3) to determine the rate of release of the total bound residues in the aquatic environment followed as a function of time, (4) to investigate the effects of sterile versus non-sterile soil and photoperiod changes on the above objectives.

Aged soil (18 months) which had been treated with ^{14}C-fluchloralin ($\underline{N}$-[2-chloroethyl]-2,6-dinitro-N-propyl-4-[tri=fluoromethyl]aniline) was supplied by BASF Wyandotte Corporation. Determination of total ^{14}C was obtained by a Harvey Biological Material Oxidiser (BMO). The BMO was used to calculate the total residues and then the aged soil was exhaustively extracted until no more residues were obtained. In practice, the aged soil was exhaustively extracted 3X with methanol (MeOH) and then repeated using hexane (1000 ml solvent/400 g soil/extract). The extracts were then subjected to thin layer chromatography analysis for a material balance.

The results of the combustion data showed that the $\bar{x}$ dpm/g = 4025 $\pm$ 584. No residues could be recovered using hexane. The MeOH extraction showed that 14.73% of the total residues could be extracted with MeOH leaving 85.27% as total ^{14}C bound residues (TBR). The residues from the MeOH extracts were examined by thin layer chromatography (tlc) and the parent compound, Fluchloralin accounted for 76.38% of the total MeOH extracted residues from the aged soil.

The determination of the release of the TBR from the soil into water was followed by combusting soil samples and analyzing water samples at selected intervals over a 128 day period. In general, the data showed that varying regimens of photoperiod and sterility had little effect on the release of the TBR. A possible exception to this might be the tanks that were subjected to 24 hours dark: non-sterile soil, which were consistently lower in ppm than the other treatments.

Analysis of the water over time showed that low concentrations of the TBR were released into water with the values remaining rather constant after day 16. Fluchloralin accounted for about 85% of the total residues released into the water.

28

Dinitroaniline Herbicide Bound Residues in Soils

CHARLES S. HELLING

Agricultural Research Service, USDA, Beltsville, Md. 20705

[Ring-^{14}C]Butralin, incubated 6 months aerobically and anaerobically in Chillum soil, was converted to 3.0 and 12.8% bound residue, respectively. Extraction of humic and fulvic acids was conducted by a classical procedure, i.e., extraction with 0.5 N NaOH and precipitation of humic acid with HCl. The balance of humic and fulvic acids was apparently influenced by soil aeration status during the course of original pesticide incubation. In aerobic Chillum soil, 61% of bound butralin residue was present in fulvic acid, and only 6% in humic acid. By contrast, in anaerobic Chillum, 18% and 33% were in fulvic and humic acids, respectively. The remainder, 33% and 49% in aerobic and anaerobic soil, was by definition classified as in humin. In the subsequently described Gascho/Stevenson procedure (Soil Sci. Soc. Amer. Proc. 32:117-119, 1968),ca. 70% would be in this "all other" (humin) category.

In a separate experiment, six ring-^{14}C-labeled dinitroaniline herbicides were aerobically incubated in Matapeake silt loam for 5 months (dinitramine) or 7 months (other compounds). Bound residues as measured by combustion of the extracted soil were: trifluralin--14, dinitramine--18.5, profluralin--21, butralin--24.5, chlornidine--28, and fluchloralin--37%. The relative distribution of bound ^{14}C (after using the classical fractionation procedure) was similar among the dinitroanilines. Generally, 50 ± 5% was in fulvic acid, 15-20% in humic acid, and 25-30% in humin. However, trifluralin and dinitramine contained ca. 40% of bound ^{14}C in the humin.

Butralin-treated, anaerobic Chillum soil was also fractionated by the procedure of Gascho and Stevenson. This involved pretreatment with 0.3 N HF to destroy hydrated silicates, followed by dialysis against 0.02 M Na$_4$P$_2$O$_7$ and 0.03 N NaOH. Relatively less fulvic acid is produced by these dilute alkaline reagents. Humic acid was precipitated at pH 1.25. The residual soil was further fractionated by centrifugation and sieving into additional mineral and organo-mineral components. Bound radioactivity was concentrated in the NaOH and Na$_4$P$_2$O$_7$ humic acids, in humin, and in undecomposed organic matter. On a total soil fraction basis, however, ^{14}C was present mainly in silt, followed by the afore-mentioned fractions. Little bound material was associated with fulvic acids or with the various dialysates. There was no difference in relative ^{14}C concentration among five humin fractions differing in size. It appeared that bound butralin was principally localized in more "highly humified" fractions, and <5% was associated with organic matter whose molecular weight was <12,000-14,000.

Biological availability of bound butralin was evaluated by incubating the anaerobic Chillum soil or several of its bound residue fractions (Gascho/Stevenson procedure) with biologically active Matapeake soil. Incubation was at 75% of field moisture capacity and ca. 24°C. $^{14}CO_2$ was trapped within the biometer flasks and measured by liquid scintillation counting. The soil was also periodically sampled for moisture and residual ^{14}C contents. A slow decline in residual ^{14}C occurred during the 21-week incubation. Only 1% of added radioactivity was recovered as $^{14}CO_2$, and this was from two samples (NaOH humic acid and humin I) only. Projected half-lives for bound butralin are: $Na_4P_2O_7$ humic acid--10-11 months; NaOH humic acid--15 months; whole anaerobic soil--2 years; and humin I--3 years). Bound residue associated with mineral fractions was lost most slowly.

Biological availability was further evaluated by growing soybeans, crabgrass, and pigweed in soil given the 21-week biometer flask incubation. Little or no plant uptake of ^{14}C was found 2,6, and 9 weeks after planting. The conditions of the experiment were intended to maximize the opportunity for uptake to occur.

29

Summary of Conjugate Papers

HANS GEISSBUHLER

Agrochemicals Division, CIBA-GEIGY Ltd., Basle, Switzerland

As a reviewer I might content myself with providing an abstract and mentioning some highlights of each of the papers presented. A more difficult but (maybe) more rewarding task would be to look at the presentations from a different angle. I shall attempt to follow the latter approach. The optics I have chosen are those of an industrial scientist whose daily managerial responsibility consists of providing answers to often reasonable, but sometimes not so reasonable biochemical and analytical questions raised by regulatory authorities all over the world.

Before coming to this conference and without knowing the contents of the presentations, I had decided that I should try to review the various conjugate papers according to the following question scheme:
(1) What are the type- or general conjugation reactions recognized with pesticides or pesticide moieties? How can they best be classified and in which organisms do they occur?
(2) What is the present stage of methodology for synthesis, purification/separation and identification of conjugates?
(3) What is the significance of conjugation reactions with regard to the behavior of pesticides in the environment, specifically in terms of:

- terminal residues
- elimination and/or transport mechanisms
- bioavailability.

Thus, I purposely neglect the various aspects of enzymatic mechanisms involved (for which theme I do not feel competent) and I also omit to evaluate the significance of conjugation reactions in the chain of events leading to the biocidal/bioregulating effects or to the selective behavior of pesticides in target organisms.

After having heard and read the papers I have to admit that the authors, who have approached their topics in quite different ways, have not exactly facilitated my task. If, in spite of this diversity in presentation, I adhere to my original question

scheme, I fully realize that my review will be no more than an
incomplete and sketchy mixture of facts, opinions, ideas and
questions.

I. Type, or General Conjugation Reactions Recognized with Pesticides

With the information provided at this meeting one would
hopefully like to be in a position to tabulate the major classes
of pesticides and/or their primary metabolites according to the
conjugation reactions to which they are subjected. However, I
assume that all of us have realized that we still have a long way
to go before we succeed in establishing such a comprehensive
listing. Even if we attempt to classify the chemical reactions
which lead to the described conjugates, we face what Dr. Hutson
has called the "xenobiochemical dilemma". Both systems, classification by functional moieties conjugated, and by groups transferred have been used at this meeting and I am in no position to
improve on this by presenting a unified scheme. If, in spite of
these limitations, I have tried two kinds of tabulations, it is
to demonstrate how the present or accumulating information might
usefully be collected and how the existing gaps might be
visualized.

Table I gives an overview of the type reactions which appear
to have been recognized with pesticides. This list is already
impressive. However, if one compares it with similar tabulations
in recent textbooks or reviews on drug metabolism ($\underline{1}$, $\underline{2}$), it is
to be expected that additional type reactions will be demonstra-
ted with pesticides.

Table II attempts to list the main conjugation reactions
described at this meeting in terms of groups of biosystems in
which they have been demonstrated or in which they may be expected
to occur. The table reveals that there is reasonable documentation on conjugate formation in mammals and plants, but a definite
lack of information on other vertebrates (birds, fish, etc.) and
non-vertebrates (possibly except for insects). Some of these
biosystems comprise species which may be important parts of the
"food chain" involved in the transport and distribution of pesticide moieties. Especially surprising is the observation that
there is apparently no more than limited demonstrated knowledge
on conjugate formation in microorganisms such as algae, fungi
and bacteria, all of which are prominent constituents of the
soil/water ecosystem. However, conjugates produced by these
organisms would be expected to be released into the surroundings
rather than being retained, accumulated or transported.

II. Stage of Methodology

In my personal judgement the information acquired under
this heading may have been the most rewarding of the present

Table I   Type of Reactions Recognized in Conjugate Formation of Pesticides or of Pesticide Moieties (list compiled from reactions discussed in various conjugate papers).

Glucuronides	O-(Ester/Ether) N- S-	Sulfuric Acid Esters	Aryl-S
Glycosides	O-(Ester/Ether) N- S-	Glutathione Conjugates	Alkyl (Aralkyl) Aryl (Epoxide) Alkene N-heterocycle
Amino Acid Conjugates	-glycine -glutamic -aspartic	Acylated Metabolites	Acetyl- Formyl- Malonic acid
		Alkylated Metabolites	Methyl-

Table II   Pesticide Conjugation Reactions Recognized or
            Presumed to Occur in Different Groups of Biosystems

	Mammals	Other Vertebrates	Non-Vertebrates	Plants	Micro-Organisms
Glucuronides	●	○		•	○
Glycosides	•	○	○	●	◑
Amino Acid Conjugates	●	○	○	●	○
Sulfuric Acid Esters	●	●	◑	○	○
Glutathione Conjugates	●	●	●	●	○
Acylated Metabolites	●	○	○	●	◑
Alkylated Metabolites	○	○	○	○	◑

Legend: ● Conjugate formation demonstrated with several
           pesticides or pesticide moieties

         ○ Conjugate formation presumed to occur from the
           presence of enzymes and/or natural conjugates

         ●○ Exceptional conjugation reaction

meeting. The authors have provided a number of useful personal experiences, references and suggestions regarding methodology and instrumentation. This information ought to assist us in coping with the various synthetic and analytical problems of conjugate chemistry, biochemistry and physiology. For this type of work the pesticide chemist needs the following:

- synthesis procedures for small-scale preparation of nonlabeled and labeled reference conjugates;
- information on physical/chemical properties of conjugates to assist in establishing analytical and physiological criteria;
- procedures for purification and separation of conjugates from extraneous materials of natural origin; and
- methods for identification of compounds at the submicrogram scale.

The part on purification/separation and identification has found the most attention at this meeting. Dr. Frear has clearly pointed out the three criteria which have to be fulfilled for adequate identification of conjugates:

- protect labile groupings;
- identify the conjugate in its entity; and
- identify both the pesticide moiety and the endogenous part after suitable hydrolysis.

An attempt to summarize the salient information on methodology in a short hand fashion is made in Table III. The most helpful and promising procedures as suggested by the authors themselves are listed to assist in the selection of appropriate techniques from among the bewildering variety of possibilities. Without going into the details of the table, it is evident that column chromatography (ion exchange, adsorption, gel permeation) and mass spectrometry are the most prominent tools at this time. However, new instrumentation in the µg-range, i.e. micro-IR and Fouier-transfer NMR appear to be very powerful aids for supplementing MS-data.

High pressure liquid chromatography (HPLC) has been mentioned as a possibility for speeding up and improving separations and according to experience acquired in our own laboratories, ought to be utilized without delay.

To acquire more detailed and definite quantitative data on conjugate formation under different environmental conditions (see section III), standardized and less elaborate procedures are required which allow clean-up and determination of major conjugates on a routine-scale.

III. Significance of Conjugation Reactions

To evaluate the significance of conjugation reactions with regard to the behavior of pesticides in the environment, I wish to demonstrate with two simplified models the various points which have to be considered:

Table III   Methods and Instrumentation Recommended for Purification, Separation and Identification of Pesticide Conjugates

Glucuronides	P/S:	LC (ion exchange, adsorption), GLC (derivatization), PC
	I:	MS (GLC/derivatization); for aglycone: NMR, IR
Glycosides	P/S:	TLC (preparative), LC (ion exchange, adsorption, gel filtration)
	I:	MS (GLC/derivatization), NMR (as supporting evidence)
Amino Acid Conjugates	P/S:	PC, TLC, LC (ion exchange), Electrophoresis
	I:	MS
Sulfuric Acid Esters	P/S:	LC (ion exchange, gel filtration)
	I:	IR
Glutathione Conjugates	P/S:	PC, TLC, LC (ion exchange)
	I:	MS (derivatization), Biosynthesis

Legend: P/S: Purification/Separation   I: Identification

LC:	Liquid Chromatography	MS:	Mass Spectrometry
GLC:	Gas Liquid Chromatography	IR:	Infrared Spectrometry
TLC:	Thin Layer Chromatography	NMR:	Nuclear Magnetic Resonance Spectrometry
PC:	Paper Chromatography		

- a model consisting of crop plants and mammals (Fig. 1); and
- a model comprising soil and an aqueous biotope (Fig. 2).

Provided the parent compound or any of its primary metabolites are significant terminal residues on a plant crop, the
study of their forming conjugates becomes part of the routine
mammalian metabolism investigation (Fig. 1, step A). If conjugation reactions do occur in treated plants, it has to be
established in quantitative terms if they represent a significant fraction of the terminal residue in the raw agricultural
commodity (B). Determination of their role in the formation of
non-extractable or "bound" residues (C) is not a requirement
but may provide some useful information on the structural identity of the latter fraction. Provided conjugates are significant terminal residues, their bioavailability has to be examined
in an appropriate mammalian organism (D).

The significance of conjugate formation in a soil/water
biotope can be visualized by an analogous sequence of potential
physiological and biochemical events (Fig. 2). However, polar
conjugates formed by microorganisms (B) in such a system would
normally appear to be too unstable to become part of the terminal residue fraction in soil or water. If they do, their bioavailability (D) to both plants and exposed soil or aqueous fauna
(fish, etc.) would need to be examined as described above for
terminal crop residues.

A large portion of the information provided at this meeting
has dealt with step A (Fig. 1), i.e. the different conjugates
formed from parent pesticides in mammalian systems. However,
this information has mainly been qualitative, and, with a few
exceptions, little has been said about the quantitative aspects
of conjugate formation. We need to distinguish between major
and minor pathways as this provides some essential indications
regarding the behavior of the examined compound in the mammal.
Thus, simple appearance of a particular conjugate in very small
quantities has to be placed in proper perspective.

As regards conjugate formation in plants (step B, Fig. 1)
we had an overview of the different type reactions which do or
can occur in young vegetative organs and tissues. Again, quantitative information appears to be minimal at this time. As with
mammals, we have to distinguish between major and minor pathways.
In addition, we do have to consider the physiological age of the
system. The discussed conjugates may indeed reflect important
metabolic routes in the early stages of plant growth. However,
what is finally significant is the presence of these conjugates
as measurable terminal residues in the raw agricultural commodity.

As regards steps C and D (Fig. 1), which refer to the implication of conjugates in the formation of non-extractable
plant residues and to their bioavailability in the mammalian
organisms, it is obvious from this meeting that we are at present
in the initial stages of experimentation. These experiments

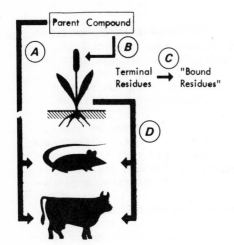

Figure 1. Scheme of simplified model consisting of crop plant and mammal(s) for demonstrating the significance of conjugate formation from pesticides applied to plants. Steps A to D described in text.

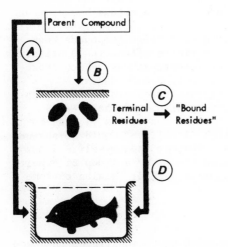

Figure 2. Scheme of simplified model comprising soil and an aqueous biotope for demonstrating the potential significance of conjugate formation from pesticides applied to or collected by soil. Steps A to D analogous to those of Figure 1 and described in text.

ought to lead us to an appropriate design of more definite
studies. The same remarks apply to the significance of conjugate formation in a soil/water biotope (Fig. 2). An adequate
design of experimentation is needed to properly evaluate the
role of conjugates in such a system.

At this point, and since it refers in part to the model
situations just described, I wish to make some comments on
Dr. Dorough's introductory paper which has dealt with certain
aspects of bioavailability and toxicity testing of pesticide
conjugates. According to my personal opinion it has set the
stage for some discussions which were beyond the purpose and
scientific competence of this meeting:

- a pesticide conjugate formed in plant leaves or stems,
  even if present in large quantities is not a significant
  residue unless it has been established that it represents
  a major portion of the terminal residue in the raw agricultural commodity as observed under practical application conditions.
- even if the bioavailability to mammals of such an identified or non-identified conjugate fraction has been demonstrated, there is no valid scientific reason to subject
  it as an entity to routine toxicity testing. Such testing
  may be indicated if the pesticide moiety or aglycone released from the conjugate has been identified and has
  structural features which set it definitely apart from
  the major primary mamalian metabolites of the parent
  compound.
- allusion to drug metabolism in this connection has to be
  made with care. Drugs are chemicals designed to have
  specific biological effects in mammalian systems and this
  in concentrations several orders of magnitude higher than
  those of pesticide terminal residues. Biochemists familiar
  with the mammalian metabolism of foreign compounds know
  that the dose applied has a definite effect in quantitative
  terms on the operation of different pathways which represent the transformation pattern of a compound.
- as regards the proposed mutagenicity/carcinogenicity testing of conjugates with microorganisms and the 90-days
  toxicity experiments with labeled conjugates, I wish to
  voice some reservations which, I beleive, would be shared
  by toxicologists much more competent than myself. I may
  refer to a recently published report by a group of experts
  convened by WHO who have expressed their opinion on some
  of these subjects (3).

IV. Concluding Remark

I do hope that the present meeting has demonstrated the need
for a major scientific effort by universities, government insti-
tutions and industrial laboratories. Once this effort is accom-

plished then the proper significance of conjugate formation in different biosystems can really be evaluated.

Acknowledgments

I gratefully acknowledge the assistance of Gino J. Marco in preparing this review and the help provided by the authors of conjugate papers in collecting the data for Tables I to III.

References

1. Foreign Compound Metabolism in Mammals (D.E. Hathway, senior rep.). Vols. 1 to 3. Specialist Periodical Reports, The Chemical Society, Burlington House, London, WIV OBN (1970, 1972, 1975).
2. Metabolic Conjugation and Metabolic Hydrolysis (W.H. Fishman, ed.). Vols. 1 to 3. Academic Press, New York and London (1970, 1973).
3. Assessment of the Carcinogenicity and Mutagenicity of Chemicals. Report of a WHO Scientific Group, Technical Report Series 546, World Health Organisation, Geneva (1974).

30

Summary of Soil Bound Residues Discussion Session

P. C. KEARNEY

Pesticide Degradation Laboratory, Agricultural Research Service,
U. S. Department of Agriculture, Beltsville, Md. 20705

The Discussion Section on Soil Bound Residues considered three major questions:
1. When should soil bound residue studies be initiated?
2. How should soil bound residues be measured?
3. What is the significance of a soil bound residue?

Each of these questions was discussed in length by the participants. Summarized below are the salient points of this discussion:

When should soil bound studies be initiated?

In the early planning of this conference, one of the ideas suggested was the development of criteria for determining when pesticide scientists need to initiate in-depth studies on soil bound pesticide residues. For example, if after completion of a residue or metabolism study, (a) < 10% of the applied pesticide was bound to soil particles after exhaustive extraction with both nonpolar organic and polar solvents, (b) the compound under study was not a persistent pesticide, and (c) the toxicology of the parent material or suspected metabolites was not a major problem, then in-depth studies on bound pesticide residues were not essential or needed. In contrast, if it were ascertained that binding was a major mechanism for a toxic compound (>10%), then detailed study on the nature, stability, and chemistry of the bound residue would be mandatory. Conference discussion of this proposal did not yield a favorable reaction. Some of the reasons given were as follows:

1. The methods and interpretation of results gleaned from current methodology are not now sufficiently developed to merit in-depth investigations. The general consensus was that detailed studies on soil bound residues are still in their infancy and considerable exploratory and theoretical research must be done before standard methods can be recommended and adequately interpreted.

2. There is still considerable confusion as to what exactly is being measured when radioactive materials are used. It is unclear whether one is measuring the intact bound parent material or one of several metabolites.

3. There is still considerable question as to the significance of soil bound residues in succeeding crops, on the soil ecology, and on other segments of the environment into which a soil bound residue might migrate. There may not be sufficient justification to merit in-depth investigations unless there is evidence for adverse environmental effects of soil bound residues.

Much of the subsequent discussion dealt with paraquat, a herbicide that is strongly bound to soil components soon after application. Because of the strong binding capacity of paraquat, and because of potential health-related problems associated with the unadsorbed molecule, this herbicide has been the most extensively studied from the standpoint of bound residues. It was pointed out that because of its high biological activity when unadsorbed, binding of paraquat represents a rapid and safe decontamination mechanism in soils. There was considerable debate on whether plant uptake of paraquat residues might be a problem in sandy soils, where binding is limited. It was pointed out, however, that the label clearly states that use of paraquat on sandy soil is not recommended. It was also pointed out that it is difficult, if not impossible, to extend findings on paraquat in soils to other pesticides because of the unique chemical structure of this herbicide. Paraquat and the structurally related analog diquat, are both di-cations, and represent a unique class of compounds when compared to most other herbicides which are either neutral or anionic. Because of the unique chemistry of paraquat, it was cautioned that extrapolation of findings with this compound to other herbicides or insecticides could be misleading. It was the general opinion of the participants that we will continue to conduct bound residue studies on most new pesticides considered in the future. The depth of the study will depend on the compound being considered for registration.

How should soil bound residues be measured?

The second critical question on the chemistry of soil bound residues is the adequacy of current methods. Basically this question resolves into two more specific questions, i.e., (a) Is the method recommended by the American Institute Biological Sciences Environmental Chemistry Task Group satisfactory? and (b) Are there other better methods? The classical method recommended by the AIBS committee, which subsequently appeared in The Environmental Protection Agency's "Guidelines for Registering Pesticides in the United States" (Federal Register 40 [123]:

26802-26928, June 25, 1975) involves extraction of the organic
matter from soils with caustic alkali and the further subdivision
of the extracted material by partial precipitation with mineral
acids. The proposed scheme is outlined below:

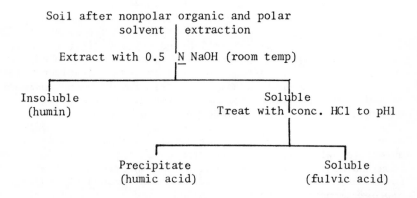

Soil after nonpolar organic and polar
solvent | extraction

Extract with 0.5 N NaOH (room temp)

Insoluble
(humin)

Soluble
Treat with conc. HCl to pH1

Precipitate
(humic acid)

Soluble
(fulvic acid)

The method as outlined is approximately 100 years old and yet it
is a method that has been widely accepted by soil scientists for
the fractionation and characterization of soil organic matter.
The method is reasonably simple and conducive to routine analytical techniques. It suffers the disadvantage of employing
rather strong agents to separate the humic and fulvic acid fractions. Consequently, there has been criticism as to the integrity of the extracted components. These rigorous conditions may
also alter the bound pesticide moiety to the point where it no
longer resembles what is initially bound in the undisturbed soil
organic matrix. Nevertheless, due to its extensive use in the
past as a method of characterizing organic matter components in
soils, it was recommended this method be employed until better
ones are developed and verified.
      The second specific question discussed under the broad
general topic of methodology dealt with new or improved methods
of fractionating soil bound residues. It was pointed out that
there are many methods for extracting soil organic matter, some
more chemically facile than the method outlined above. For example, a DMF/oxalate method might yield valuable fractions which are
less distorted than those from strong acid and base extraction.
It was also noted that pyrophosphate extractions offer more
facile methods of cleaving linkages in the complex soil humic
acid fraction, again yielding fractions that are more characteristic of the natural soil matrix. It was generally concluded
that methodology needs considerable future attention. A fruitful area of research would be to examine several methods of
extracting soils and determine the magnitude and distribution

of bound pesticides yielded by these methods. A very important
point was made that the living plant might be one of the best
bioassays for soil bound residues in the unextracted soil. It
was further recommended that methodology for bound soil residues
proceed both with chemical and biological assays that would shed
some light on the overall question of significance.

What is the significance of a soil bound residue?

The question of significance of soil bound residues is one
of the major imponderables at the current time. In an attempt
to assess the environmental significance of soil bound residues,
a number of environmental components were considered. Those
components that would be least affected by soil bound residues
are probably higher animals, including man. Those components
that may be directly impacted by soil bound residues because of
their proximity would include agronomic plants, aquatic organisms, soil processes, and soil organisms. Specifically, there
is concern that changes in cultural practices may liberate bound
residues, reintroducing them into the soil solution and subsequent uptake and translocation into the economic portions of
plants. It is conceivable that soil bound pesticide residues
may enter into the aquatic environment, be released, and subsequently be accumulated in aquatic food chains. Likewise, there
is some concern that continued buildup of bound residues in soils
may affect important physical and biochemical processes such as
water holding capacity, soil structure, and the processes of
nitrification, ammonification, cellulose decomposition, and a
host of related processes. Finally, there is some concern that
bound residues may adversely affect soil organisms such as the
earthworm. Presently there appears to be little foundation for
these concerns, based on the current available literature. Nevertheless, it was pointed out that experiments should be designed
to ascertain the significance of each of these potential problems.
In summarizing the overall philosophy of the significance of
soil bound residues, two opposing viewpoints were presented. On
the negative side, it can be argued that bound residues are
really hidden residues that keep an intact molecule capable of
subsequent release and exertion of long-term biological effects.
On the positive side, it can be argued that binding of soil residues represents the most effective and safest method of decontamination by rendering the molecule innocuous and allowing slow
degradation in the bound state to products that pose no short-
or long-term problems. In defense of the second argument, several
examples were cited where soil disposal represented a reasonable
option for disposal of hazardous materials. In these cases,
binding of the toxic residue represents a mechanism of immobil-
izing the toxicant so that other soil processes can degrade the
molecule.

In summary, it was recommended that the Division of Pesticide Chemistry, American Chemical Society take no strong position
on soil bound residues at the current time. A series of recommendations now might be counterproductive to the development of
exploratory research needed to elucidate the true significance
and nature or soil bound residues. The regulatory agencies, particularly EPA, expressed a willingness to remain
flexible on requirements on bound residue studies. It
will be the policy of these agencies in the future to consider these compounds on their individual merits as to
what additional information is needed on soil bound residues.

INDEX

INDEX

A

3-Acetamido group 287
Acetonitrile–water extraction analysis 291
2-Acetylaminofluorene 23
Acetylation 133, 261
N-acetylation 133
Acetyl-coenzyme A biosynthesis 132
Acetyl transfer 136
Acid(s)
   biosynthesis of fatty 139
   extraction 361
   strength of some clay minerals 212
Acidities of fulvic acids 182
Acidity, increased 4
Actinomycetes 249
Activation 22
   of the carcinogen, metabolic 23
   step for GSH conjugation 116
Acylated, metabolites 371
Acylation, reactions involving 4
Acylation of xenobiotics 132
Adenosine-3′-phosphate-5′-phospho-
   sulfate (PAPS) 86
Adenosine-5′-phosphosulfate (APS) .. 86
Adsorbed pesticides, surface chemis-
   try of 216
Adsorbent, organic matter as 184
Adsorption
   capacities of soils, strong 315, 316, 328
   by direct protonation 200
   equation, Freundlich 188
   exchange sites for 200
   forms of paraquat in soil according
      to strength of 323
   of herbicides by soil organic matter 203
   mechanisms 199, 211
      for pesticides 7
   for retention of nonionic polar
      pesticides 201
   by organic matter 180
   parameters 214
   organic matter, clay, and other
      soil properties correlated
      with 189
   of soils, strong 331
   of the s-triazines 200
Aerobic soil microorganisms 362
Aerobically incubated soils 358
Aglycone 24, 63
Alcohols, polymers of the lignin 246

Aldehydes 267
Alfisols 180
Aliphatic acids 192
Aliphatic glucuronides 63
Alkene transfer 112
Alkene transferase, GSH 113
Alkyl
   halides 107
   mercapto transfer 112
   sulfonates 106
   thiocyanates 115
   transfer 106, 150
   transferase, GSH 108, 115
Alkylated metabolites 371
Alkylation 142
   reactions involving 4
   of xenobiotics 132
Alkaline hydrolysis 148
Amadore rearrangement 193
Amerlite XAD-2 55
Amiben 191
Amino acid(s)
   condensation of 194
   conjugates 68, 370, 371
      biological properties and metabo-
         lism of 78
      chemical and physical properties
         of 69
      of 2,4-D 26, 72, 73, 79
      isolation, purification and identi-
         fication of 76
      TMS derivatives of 81
   diagenesis of 197
   nonbiological reactions involving 197
   reactions with 4
   of sediments 197
m-Aminophenol 136
Ammonia complexes, lignin– 244
Ammonia, reactions of 4-methylcate-
   chol and p-quinone with 269
Ammonium chloride 320
Aniline(s) 39, 139
   N-acetylation of 133
   derivatives with malonic acid 142
   microbial transformation of 139
   mustard 25
   non-hydrolyzable attachment of 268
   residues, covalently bound 269
Animal bioavailability experiments 153
Animals, fate of pesticide conjugates
   in 26

Anion
    exchange cellulose 57
    exchange paper chromatography,
        liquid 519
    exchangers, liquid 56
Anthelmintic 178
Anthranilic acid 139
Aphicide 358
Apocrenic acids 181
Apple 173, 175
APS (adenosine-5'-phosphosulfate) .. 86
Aralkyl
    halides 106
    sulfates 109
    transfer 106
    transferase, GSH 110
Argentine Basin sediments 198
Aromatic glucuronides 63
Aromatic hydrocarbons 111
Aryl
    acylamidase 156
    sulfatases 89
    sulfate esters 92, 94
    transfer 109
    transferase, GSH 110
Arylhydroxylamine 136
Aspartic acid conjugates 72, 82
ATP:sulfate adenyltransferase 86
Atrazine 185, 215, 223, 224
Auxins 44
Avena coleoptile sections 79
Azinphos 166, 167
Azobacter spp 342

B

Bacterial mutagenesis of pesticides 30
BAM (2,6-dichlorobenzamide) 173
Barley 319
    seed 333, 339
Basidiomycetes 245
Bean plants 27, 28, 166
Benzoxazole radical ion 136
Bile 105
Binding of pesticides 193
Binding radiocarbon 297
Bioavailability experiments, animal 153
Bioavailable conjugates 17
Biodegradation 122
Biological
    activity 18
    of adsorbed organic compounds .. 213
    of pesticide conjugates 11
    availability of bound butralin 367
    properties of amino acid conjugates 78
    unavailability of bound paraquat
        residues in soil 201
Biologically
    active 2,4-D amino acid conjugates 26
    active pesticide metabolites 15
    available soil-bound pesticides 354

Bioregulation of phenol intermediates 37
Biosynthesis 86, 117, 122
    acetyl–coenzyme A 132
    of fatty acids 139
    glucose ester 43
    mercapturic acid 103
    of purines 136
Biosystems 371
Bio-unavailable conjugates 17
Bipyridylium 186
    herbicides 264
Birds 124, 146
Bleidner distillation/extraction
    head 159, 160
Blood and blood rich organs, residues
    in 178
Bond formation, covalent 266
Bonding
    hydrogen 201
    hydrophobic 202
    mechanisms for adsorption of herbi-
        cides 203
    mechanisms for the retention of pes-
        ticides 199
Bound
    butralin 367
    capacity of soils, tightly 328, 331
    conjugates 17
    fluchloralin residues 364
    material 283
    in rats, conjugated 27
    metabolites 14
    of dichlobenil in field crops 173
    pesticides, fixed and biologically-
        available soil- 354
    pesticide residues 1
    to plant material, pesticides 166
    radioactivity 359
    residue(s) 5, 208
    analysis 258
    azinphos-^{14}C 166
    characterization of 358
    3-4-dichloroaniline 164
    from 3,4-dichloroaniline-^{14}C and
        propanil-phenyl-^{14}C treated
        rice root tissues 156
    in erythrocytes 178
    measurement, soil 379
    of nitrofen 166
    reservoir 232
    soil 378, 381
    in soil organic matter 272
    in soils, dinitroaniline herbicide .. 366
    (TBR), total ^{14}C 364
Breakdown products in apple leaves
    and fruits 175
Broadricks soil 335
p-Bromoaniline 133
Bromobenzene 104
Butralin 366, 367

C

California soil 227, 229, 327
Callus tissues 74, 77–80, 82
Carbaryl
  in cows, metabolism of 146
  indirect conjugative deactivation of   19
  metabolism by rat liver microsomes
    + NADPH 20
  naphthyl-^{14}C in bean plants 27
Carbohydrates 3
^{14}Carbon
  -benzoic acid 179
  bound residues (TBR) 364
  -dichlobenil granules 175
  -labeled dinitroaniline herbicides,
    ring- 366
  -metabolites in bean plants, ethio-
    fencarb 28
  nitrofen 170, 171
  nuclear magnetic resonance spectra 121
  paraquat 325
Carbonyl position, azinphos ^{14}C-
  labeled in the 167
Carcinogen, metabolic activation of
  the 23
Carcinogenic potential, screening for 29
Carcinogenicity 22
  testing of conjugates 376
Cardiac glycoside digitalis 24
Carrots 319, 356
Catabolism of a sym-triazine conju-
  gate in plants 115
Catechol 194
Cation exchange capacity 210
Cation-exchange resin chroma-
  tography 58
Cationic herbicides 202
Cellulose, anion exchange 57
Cereal grain, residues of nitrofen in 170
Charge transfer, attachment of s-
  triazines to humic acid by 265
Charge transfer complex of ciquat
  with humic acid 264
Chemical
  accumulation and dissipation of
    residues from annual addition
    of 235
  analysis of soils 322
  binding of pesticides 193
  defense 4
  designations of organics 186
  nature of soil organic matter 180
  reaction involving pesticides and
    organic substances in soil 192
  requirements for glutathione conju-
    gation 105
Chemotherapeutic conjugates 24
Chillum soil 366
Chitosan 251
Chloramben, N-glucosyl 42

Chlordimeform 139
Chlornidine 366
Chloroaniline–humus complexes 362
Chloroaniline residues 266
p-Chloroaniline 133
p-Chloroanisole 144
N-(2-chloroethyl)-2,6-dinitro-N-
  propyl-4-(trifluoro-methyl)ani-
  line 364
4-Chloro-2-hydroxy-acetanilide 133
Chloroneb 146
4-Chloro-o-toluidine 139
Chloro-s-triazines 192, 194
Chromatography
  cation-exchange resin 58
  gas 93
  gas–liquid 59
  gel 274
  high pressure liquid 372
  ion exchange 93
  liquid anion-exchange paper 59
  thin-layer 286
CIPC 191
Cisanilide 38
Clay(s)
  as adsorbent 184
  adsorption of paraquat on 303
  correlated with adsorption parame-
    ters 189
  –metal–organic matter complex 185
  minerals
    acid strength of some 212
    cation exchange capacity and sur-
      face area values for 210
    structure and properties of 208
    swelling 210
  –pesticide interactions 208
Coefficient distribution 222
Colloids, soil 354
Condensation of polyphenols and
  amino acids 194
Conjugates 3
  amino acid
    (see Amino acid conjugates)
  aniline mustard and its protein 25
  in animals, pesticide 26
  aspartic acid 72, 82
  bioavailable 17
  biological activity of 11
  bound or bio-unavailable 17
  ^{14}C nuclear magnetic resonance
    spectra of 121
  chemotherapeutic 24
  of 2,4-D, amino acid 26, 72, 73, 79, 82
  of 17 β-estradiol, glutathione 126
  exocon 21
  formation from pesticides 370, 375
  formation in plants 374
  glucuronide 55
  glutamic acid 72

Conjugates (*continued*)
glutathione 103, 117, 119, 122. 371
GSH 120, 126
miscellaneous 132
nomenclature 16
papers, summary of 368
from parent pesticides in mam-
    malian systems 374
pesticide (*see* Pesticide conjugates)   35
sulfate ester 86
Conjugated
materials to rats, fate of 27, 28
metabolites 13, 173
pesticide residues 1
Conjugation 132
and biological activity 18
glucuronide 20
glutathione 105
glycoside 22
GSH 116, 125
malonic acid 139
reaction(s)
    classes of 116
    pesticide 369, 371
    significance of 372
    types of 114
Conjugative deactivation of meproba-
    mate and carbaryl by glucuroni-
    dation, indirect 19
Contaminates, environmental 13
Coordination 201
Cotyledon callus tissue, soybean 79
Counter current distribution 57
Covalent bond formation 266
Covalently bound aniline residues 269
Cows 146
Crenic acids 181
Crop(s)
    dichlobenil in field 173
    paraquat residues of 319, 327, 329
    plant 375
Cyclohexylamine glucuronide 22
Cysteine conjugates 117
Cytokinin metabolism 42

D

2,4-D 78, 144, 191, 248
amino acid conjugates 26, 72, 73, 79
-Asp 78
-1-^{14}C, metabolites of 77
conjugates from soybean callus
    tissue 82
Glu 78, 80
-Ile 75
metabolites 80, 81
DCA (*see* 3,4-Dichloroaniline)
DCPA 191
Deactivation
direct 21

Deactivation (*continued*)
indirect 19
of meprobamate and carbaryl by
    glucuronidation 19
DEAE-sephadex 58
Decomposition
ditalimfos 273
microbial 244
pesticide 193
phenol 247
rates of 223
soil 221, 223, 247
Degradation
of atrazine and simazine 224
of a dialkyl phosphate 110
of dichlobenil 173, 174
humin 282
of pesticide residues in soil 248
of pirimicarb 358
Derivatization procedures 93
Desorption studies, soil 214
Destun 285
Detoxication of diazinon 116
Detoxication of exocons, direct 22
Diagenetic changes in the amino acids 197
Dialkyl phosphate 110
Dialysis of soil extracts 275
Diazinon 116
Dichlobenil (2,6-dichlorobenzo-
    nitrile) 173, 174
3,4-Dichloro-acetanilide 156
3,4-Dichloroaniline (DCA) 267, 268
with aldehydes and *p*-quinone reac-
    tions of 267
bound residues 164
humic acid, IR spectra of 267
liberation of 159
solubilization of bound residues
    from 156
2,6-Dichlorobenzamide (BAM) 173
1,2-Dichlorobenzene 162
5,6-Dichlorobenzofuran (5,6-dichloro-
    benzo-2,1,4-benzoxodiazole) 162
2,6-Dichlorobenzonitrile (dichlo-
    benil) 173
Dichloromethoxyphenol 146
2,5-Dichloro-4-methoxyphenol 146
2,6-Dichloro-4-nitroaniline 142
1,2-Dichloro-3-nitrobenzene 162
1,2-Dichloro-4-nitrobenzene 162
2,4-Dichloro-1-(4-nitrophenoxy)
    benzene 170
N-3,4-Dichlorophenylcyclohexyl-
    amine 156
N-3,4-(Dichlorophenyl)-furfuryli-
    dimine 159
3,4-Dichloropropionanilide 153
O,O-Diethylphthalimido-1-^{14}C-
    phosphonothioate 273
Differential equations, solution of 240

Digitalis 24
Digitoxigenin 24
2-Dimethylamino-5,6-dimethylpyrimi-
    din-4-yl dimethylcarbamate 358
N,N'-Dimethyl-N'-(2-benzthiazoyly)-
    urea 356
Dinitramine 366
Dinitroanilines 355, 366
4,6-Dinitro-o-cresol 133
Diphenamid 46, 49, 191
Diphenyl ether 111
Diquat 199, 264, 354
Disappearance curve 225
Dissipation of residues 234–236
Distillation/extraction head,
    Bleidner 159, 160
Distribution coefficient, Kd ...188, 222
Distribution, counter current 57
Ditalimfos 273, 282
DNPB 191
Drug meprobamate 19
Drug metabolism 376

E

Earthworms 330, 331
Electron
    impact mass spectroscopy 50
    paramagnetic resonance (EPR) 262
    spin resonance (ESR) spectra 262
Elution diagrams 278, 280, 281
Endocon 16
Environmental contaminates 13
Enzymes, role of the 125
Enzymes, sulfatase 89
Epoxide transferases 109, 113
EPR (electron paramagnetic reso-
    nance) 262
Erythrocytes 178
ESR spectra (electron spin resonance) 262
17 β-estradiol 126
Ethiofencarb ^{14}C-metabolites 28
Exchange sites for adsorption 200
Exocon 16, 21, 22
Extractability, radiocarbon 297
Extraction
    acid 361
    analysis, acetonitrile–water 291
    of glucuronides 55
    head, Bleidner distillation/ 160
    of humic and fulvic acids 366
    scheme 380
    of trifluoromethanesulfonanilide
        pesticides 285

F

Fatty acids, biosynthesis of 139
Field trials 316, 324, 326
Fluchloralin 364, 366
Fluoridamid 285, 293

Forest soils (alfisols) 190
Formanilides 139
Formylation 136
Formyl-CoA 136
N-Formyl-L-kynurenine 139
N-10-Formyl-tetrahydrofolate 136
Fractionation
    of glucuronides 57
    plant 153
    of soil extracts 275
    of soil organic matter 298
    of the unextractable soil radio-
        carbon 299
Fragment ions 61, 63
Free radicals 262
Freeze-dried 359
Frensham soil 317
Freundlich adsorption equation 188
Fruits, mean residues of breakdown
    products in 175
Fulvic acid(s) 181, 191, 259, 272
    extraction of 366
    fractionation of the unextractable
        soil radiocarbon into 299
    incorporation of nitrogen into 197
    nitrogen containing compounds
        from 250
    radioactive 281
    relationships, humic acid– 190
    total acidities of 182
Fungi 245, 249, 250, 341
Fungicide 273
Fusarium oxysporum 133

G

Gentiobioside 46, 49
Glasshouse studies 306, 320
Globin 179
Glucosamine 251
Glucosazones 171, 172
Glucose esters 41, 42, 43
Glucosides 40, 42, 43, 45, 46, 371
N and O Glucosides 36, 39
Glucosinolates 43, 45
N-Glucosyl chloramben 42
Glucosyltransferase 43
Glucuronic acid moiety 61
Glucuronidation 19
Glucuronide(s) 370, 371
    aliphatic and aromatic 63
    conjugates 55
    conjugation, reduced 20
    extraction of 55
    fractionation of 57
    perTMS and TMS-methyl deriva-
        tives of 60
Glutamic acid conjugate 72
Glutathione
    ^{14}C nuclear magnetic resonance
        spectra of 121

Glutathione (*continued*)
  conjugates 103, 117, 122, 371
    of 17 β-estradiol 126
    identification of a 119
  conjugation, chemical requirements
    for 105
  s-transferases 123
Glycine 194
Glycoside(s) 35, 46, 370
  conjugation 22
  digitalis 24
  formation 3
  1-naphthol-¹⁴C and its 29
  in pesticide metabolism 37
Glycosidic linkages, labile 48
Glyoxylate cycle 139
Grain, cereal 170
Grapes 177
Grassland (mollisols) 190
Group transfer 105
GSH
  alkene transferase 113
  alkyl transferase 108, 115
  aralkyl transferase 110
  aryl transferase 110
  conjugates 117, 120, 126
  conjugation 116, 125
  -dependent pathways for dialkyl
    phosphate degradation 110
  -dependent cleavage of a diphenyl
    ether 111
  epoxide transferase 113
  transferases 124
  triazinyl transferase 115

H

Hammett indicators 212
Heme 179
Herbicide(s) 285, 356
  bipyridylium 264
  bound residues in soils, dinitroani-
    line 366
  ring-¹⁴C-labeled dinitroaniline 366
  cationic 202
  contact 301
  decomposition and constituents of
    soil organic matter 196
  dichlobenil (2,6-dichlorobenzo-
    nitrile) 173
  paraquat 379
  phenylamide 362
  by soil organic matter, adsorption
    of 203
  for soil organic matter surfaces,
    affinities of 203
  s-triazine 354
HOBAM metabolite (3-hydroxy-2,6-
  dichlorobenzamide) 173
Homocysteine 144

Humic
  acid(s) 181, 182, 259, 272
    attachment of s-triazines to 265
    charge transfer complex of diquat
      with 264
    core 260
    extraction of 366
    fractionation of the unextractable
      soil radiocarbon into 299
    –fulvic acid relationships 190
    incorporation of nitrogen into 197
    IR spectra of 262
    nitrogen containing compounds
      from 250
    nonhydrolyzable attachment of
      DCA to 268
    radioactive 280
    structure for 183
  compounds 252, 253
  substances 183, 272
    formation of 193
    metal ion interactions with 261
    microbial synthesis of 244
    origin, classification, and compo-
      sition of 258
    polymeric nature of residues in 272
    in soil, formation of 282
    spectroscopic characterization of . 263
    spectrometry of 260
Humification 300
Humin 259, 272
  degradation, rate of 282
  fractionation of the unextractable
    soil radiocarbon into 299
Humus 181
  complexes, chloroaniline– 362
  interactions, spectrometric studies
    on pesticide residue 264
Hydrocarbons, mercapturic acid formation from aromatic 111
Hydrogen bonding 201, 265
Hydrogenolysis 118
Hydrolysis
  chemical 118
  degree of 217
  of globin 179
  of sulfate ester conjugates 86
Hydrophobic bonding 202
Hydroxamic acid 136
m-Hydroxyacetanilide 136
p-Hydroxybenzoic acids 249
4-and-5-Hydroxycarbaryl glucoside 22
p-Hydroxy-cinnamic acids 249
3-Hydroxy-2,6-dichlorobenzamide
  (HOBAM), metabolite 173
Hydroxylation of the chloro-s-triazines 192

I

IAA 44
Illinois soil 228, 230, 234

Illite 184
Incubated soils, aerobically 358
Indirect deactivation 19
Infrared (IR) spectra 260
Insects 45, 125, 127
Insecticide
  azinophos in bean plants 166
  carbaryl 19
  organophosphate 106
  pirimcarb 358
Ion(s)
  exchange 199
    chromatography 93
  fragment 61, 63
  from fragmentation of 2,4-D-Ile,
    mass spectral 75
  interactions with humic substances,
    metal 261
Isolation 55, 76, 117
Isopropyl 3-methoxy-4-hydroxy-
  carbanilate 146

K

Kaolinite 184, 212
Kd, distribution coefficient 188

L

Labeled benzoic and cinnamic acids .. 247
Labile pool 220
Laboratory synthesis of sulfate esters   90
Lake, distribution of pesticides in a 210
Laser ionization mass spectrometry 95
Layer lattice silicates 209
Leaves, mean residues of breakdown
  products in apple 175
Lemna polyrhiza 306
Lignin
  alcohols 246
  isolation of 170, 171
  –protein of lignin–ammonia
    complexes 244
  purification of 172
Liquid anion exchangers 56
Liquid–liquid partition 93
Liver microsomes, carbaryl metabo-
  lism by rat 20
Loam 273
Lysimeter experiments 356

M

Maillard reaction 193, 196
Malonanilic acid derivative 139
Malonic acid, aniline 142
Malonic acid conjugation 139
Malonyl-CoA 139
Mammal 124, 127, 375
Mammalian systems, conjugates from
  parent pesticides in 374

Mass
  spectral ions 75
  spectrometry 60, 120
    laser ionization 95
  spectroscopy, electron impact 50
MBR 12325 285
Mechanical analysis of soils 322
Melanins 244
Meprobomate 19
Mercapturic acid(s) 117
  biosynthesis 103, 104
  from bromobenzene 104
  formation from aromatic hydrocar-
    bons 111
Metabolic pathway of pesticides 173
Metabolism
  of amino acid conjugates 78
  carbaryl 20
  diphenamid 49
  drug 376
  of 3-hydroxy-5-methylisoxazole in
    plants 42
  of pesticides, bioregulation of phe-
    nol intermediates in the 37
  in plants, cytokinin 42
  propanil 157
  role of glycosides in pesticide 37
Metabolite(s)
  acylated 371
  alkylated 371
  in the bile 105
  biosynthesis of 49
  bound or unextracted 14
  of cisanilide, phenolic 38
  conjugated 13
  of 2,4-D and 2,4-D-Glu 80
  of dichlobenil 173
  fate of p-toluoyl chloride phenyl-
    hydrazone 178
  3-hydroxy-2,6-dichlorobenzamide
    (HOBAM) 173
  pesticide (see Pesticide metabolite)
  significance 12
  soil 173
Metal ion interactions with humic
  substances 261
Metal–organic matter complex, clay–   185
Methabenzthiazuron 356
Methionine activation 144
Methodology 370
1-Methoxy-5-(methylcarbamoyloxy)-
  2-napthyl sulfate 146
N-5-N-10-Methyenyl-tetra-hydrofolate 136
Methyl transfer 146
Methylation 142, 144, 148
4-Methylcatechol 269
Mice 21, 25
Microarthropods 336, 338
Microbial
  decomposition of the original plant
    constituents 244

Microbial (*continued*)
  synthesis of humic materials 244
  transformation of anilines 139
Microorganisms, effect of paraquat
  on 341, 343
Microorganisms, unavailability of
  bound paraquat soil residues to .. 340
Microsomes, rat liver 20
Minerals, clay 208, 210, 212
Mohole sediments, experimental 198
Molecular ion, mass of the 63
Molecular weights 276
Mollisols 180
Montmorillonite 184, 212
Muck soil 190
Mutagenesis of pesticides, bacterial 30
Mutagenicity, testing 29, 376

N

NADPH 20
1-Naphthol-^{14}C 29
Naphthyl-^{14}C, carbaryl 27
Naphthylamine 139
NH$_2$ group 288, 299
NHR groups 287
Nitrobacter 342
Nitrofen 170
Nitrogen
  containing compounds from humic
    and fulvic acids 250
  heterocycles, transfer of 112
  into humic and fulvic acids, incor-
    poration of 197
  transformations in soil 342
Nitrogenous polymers, brown 194
p-Nitrophenyl-6-O-malonyl-β-D-
  glucoside 49
Nitrosomonas 342
Nomenclature, conjugate 16
Nonbiological reactions involving
  amino acids 197
Nonhumic substances 272
North Dakota soil 278, 280, 281
Nuclear magnetic resonance (NMR)
  spectroscopy 50, 94, 120, 263

O

Organs, residues in blood and blood
  rich 178
Organic(s)
  chemical designations of 186
  compounds, biological activity of
    adsorbed 213
  matter
    as adsorbent 184
    adsorption by 180
    adsorption of paraquat on 303
    affinities of pesticides for 202

Organic matter (*continued*)
  complex, clay-metal- 185
  correlated with adsorption
    parameters 189
  of natural soils 188
  soil (*see* Soil organic matter)
  in sediments 197
  substances in soil, bonding mecha-
    nisms for the retention of
    pesticides by 199
  substances in soil, potential chemi-
    cal reactions involving pesti-
    cides and 192
Organophosphate substrates for GSH
  alkyl transferase 108

P

PAPS (adenosine-3′-phosphate-5′-
  phosphosulfate) 86
Paraquat herbicide 191, 199, 354, 379
  adsorption and deactivation 307
  application
    earthworms after 331
    long term consequences of re-
      peated 346
    microarthropods after 338
    rates and soil residues 347
  capacities of soils to deactivate 312
  capacity of soils to bind 320
  capacity of soils to reduce 305
  effect of microorganism on 343
  high rate soil trials, crop growth in 327
  on microorganisms, effect of 341
  residues
    availability of bound 324
    in Broadricks soil 335
    of crops 319
    in crops from high rate soil trials 329
    in soil, biological unavailability of
      bound 301
    in worms 334
  sensitivities of roots and shoots to .. 308
  soil absorption isotherms 309
  on soils, clays, and organic matter,
    adsorption of 303
  in soil, forms of 323
  soil residues 348
    to earthworms, unavailability of
      bound 330
    to microarthropods, unavailability
      of bound 336
    to microorganisms, unavailability
      of bound 340
    nature and amounts of 302
    in outdoor pot experiment 345
    to plants, unavailability of bound 304
  wheat bioassay of soils treated with 311
Parathion 223, 224
Partition, liquid–liquid 93

Peanut 49
Pentachloroanisole 144
Pentachloronitrobenzene 148
Pentachlorophenol 144
Pentachlorothioanisole 148
Pepsin 179
Perfluidone 285, 292
perTMS derivatives 60
Pesticide(s)
  N-acetylation of a 133
  adsorption mechanisms and sites
    for 7, 201
  bacterial mutagenesis of 30
  bioregulation of phenol intermedi-
    ates in the metabolism of 37
  bound to plant material 166
  chemical binding of 193
  chemical reactions involving 192
  conjugates 35
    in animals 26
    biological activity of 11
    formation from 375
    formation of 370
    purification, separation, and iden-
     tification of 373
    screening 30
    significance of 15, 31
    synthesis 12
  conjugation reactions 369, 371
  interactions, clay– 208
  in a lake 210
  in mammalian systems 374
  metabolic pathway of 37, 173
  metabolites
    biologically active 15
    characterized as N-and-O-gluco-
     sides 40
    common sources of 13
    phenolic 36
  of nonionic polar 201
  for organic matter, relative affinities
    of 202
  by organic substances in soil, reten-
    tion of 199
  phenolic 36
  residues
    bound and conjugated 1
    –humus interations 264
    plant availability of soil bound 356
    in soil, degradation of 248
    in soil, turnover of 219
    soil bound 354
  in soil, organic matter reactions
    involving 180
  surface chemistry of adsorbed 216
  toxicity, role of glycosides in 37
  tracer experiment with 356
  trifluoromethanesulfonanilide 285
Phenanzine type compounds, phen-
  oxazine 269

Phenol(s)
  decomposition rates of the 247
  gentiobioside 46
  glucoside 46
  intermediates 37
Phenolic
  constituents, plant 246
  metabolite of cisanilide 38
  pesticides or pesticide metabolites .. 36
Phenoxazine type compounds 269
Phenylamide herbicides 362
Phenyl carbamates 186, 189
Phytotoxic levels 305
Pigments, plant 47
Pirimicarb 358
Plant(s) 125, 127
  availability of soil bound pesticide
    residues 356
  callus tissues 74
  conjugate formation in 374, 375
  constituents, microbial decomposi-
    tion of the 244
  crop 375
  cytokinin metabolism in 42
  dichlobenil in 174
  effects on 25
  fractionation 153
  s-glucosides in 45
  growth regulators 44, 285
  material, pesticides bound to 166
  phenolic constituents 246
  pigments 47
  rice 153, 170
  samples, propanil-phenyl-14C
    treated 154
  soybean 77
  sym-triazine conjugate in 115
  wheat 170
Polarity 12
Polyelectrolytes, high molecular
  weight 181
Polymer(s) 259
  brown nitrogenous 194
  dark colored 250
  glucosamine and chitosan as 251
  of the lignin alcohols 246
  radioactive 277, 278, 280, 281
Polymeric nature of residues in humic
  substance 272
Polyphenols (quinones) 193, 194
Pool, labile 220
Porapak Q 55
Pot experiment, outdoor 344, 345
Precursor molecules, pool of 193
Profluralin 366
Pronase 179
Propanil
  -bound residues in rice plants 153
  metabolism in rice 157

Propanil (*continued*)
-phenyl-^{14}C treated plant samples .. 154
-phenyl-^{14}C treated rice root tissues 156
Propham treated birds 146
Protein conjugates, aniline mustard
    and its 25
Protein of lignin–ammonia complexes,
    lignin– 244
Protonation 199, 200, 217
Purification 118
    of amino acid conjugates 76
    or aryl sulfate esters 92
    of glucosazones 172
    of pesticide conjugates 373
    of sulfate ester conjugates 86
Purines 136

Q

p-Quinone 267, 269
Quinone structures 250
Quinones, polyphenols 193

R

Radioactive
    ditalimfor fungicide 273
    fulvic acid 281
    humic acid 280
    polymers 277, 278, 280, 281
Radioactivity 154, 172
    bound 359
    of glucosazones and lignin 171
    from soils, extraction of 275
    in the soil organic matter 272
    taken up by the soybean callus 72
Radiocarbon
    binding 297
    extractability 297
    unextractable soil 299
Radiotracer technology 166, 173
Rate model for soil decomposition 223
Rats 20, 27–29
Reactivation, direct 21
Residence time of humic compounds
    in soil 253
Residue(s)
    accumulation of 231
    aniline 269
    in blood and blood rich organs 178
    bound (*see* Bound residues)
    of breakdown products in apple
        leaves and fruits 175
    chloroaniline 266
    fluchloralin 364
    in humic substance 272
    paraquat (*see* Paraquat residues)
    pesticide (*see* Pesticide residues)
    soil (*see* Soil residues)
    to soil, accumulation and dissipation
        of 236

R$_f$ values of amino acid conjugates 73
Rhizobium spp 342
Rice plants 153, 156, 157, 170
[Ring-^{14}C] butralin 366
Ring structures of the humic acid core 260
Rye 356

S

Samonellao typhimurim 30
Screening techniques 16, 29, 30
Sediments
    amino acids of 197
    Argentine Basin 198
    experimental mohole 198
    fate of organics in 197
Separation of aryl sulfate esters 92
Separation of pesticide conjugates,
    purification 373
Sephadex
    DEAE- 58
    gels 273
    LH-20 ....56, 58
Sheep 178
Silicates, layer lattice 209
Simazine 223, 224
Soil
    accumulation and dissipation of
        residues to 236
    aerobically incubated 358
    aged 364
    analysis 290
    atrazine adsorbed by 185
    bound
        pesticides, fixed and biologically-
            available 354
        pesticide residues, plant availa-
            bility of 356
        residues 378, 379, 381
    Broadricks 335
    California 229, 277
    chemical extraction of pesticides
        from the 285
    chemical and mechanical analysis
        of 322
    chemical reactions involving pesti-
        cides and organic substances in 192
    chillum 366
    colloids 354
    conjugate formation from pesticides
        applied to 375
    degradation and transfer routes of
        dichlobenil in 173
    decomposition 221, 223
        of labeled benzoic and cinnamic
            acids in 247
    degradation
        of atrazine and simazine applied
            to 224
        of the insecticide pirimicarb in 358
        of pesticide residues in 248

Soil degradation (*continued*)
and transfer routes of dichlobenil
in 174
desorption studies 214
dinitroaniline herbicide bound resi-
dues in 366
in ditalimfos decomposition 273
extraction of radioactivity from 275
extracts 275
forest 190
frensham 317
from high rate field trials 326
humic substances in 282
Illinois 230, 234, 278
metabolism 8
metabolite 173
microorganisms, aerobic 362
nitrogen transformations in 342
North Dakota 278, 280, 281
organic matter
adsorption of some common
herbicides by 203
bound residues in 272
chemical nature of 180
fractionation of 298
intermediate products of herbi-
cide decomposition and
constituents of 196
radioactivity in the 272
reactions involving pesticides 180
specific activities for 282
spectroscopic characterization of 258
surfaces, affinities of herbicides
for 203
in outdoor pot experiment 344
parathion in 224
pH 200
properties correlated with adsorp-
tion papameters 189
qualitative differences in the organic
matter of natural 188
radiocarbon 299
residence time of humic compounds
in 253
residues, paraquat
(*see* Paraquat soil residues)
retention of pesticides by organic
substances in 199
samples, preparation of 290
solution 200
strong adsorption and tightly bound
capacities of 315, 328, 331
treatment with ^{14}C-dichlobenil
granules 175
trials, high rate 327, 329
triclopyr in 227–230
turnover of pesticide residues in 219
into water, release of bound fluchloralin residues from 364
wheat bioassay of 311

Solubilization methods 156, 159
Solution, soil 200
Sorption-catalyzed hydrolysis of
chloro-*s*-triazines 194
South Africa 347–348
Soybean
callus tissue 77, 79
2,4-D-conjugates from 82
metabolites of 2,4-D and 2,4-D-
Glu incubated with 80
radioactivity taken up by the 72
2,6-dichloro-4-nitroaniline in 142
plant 77
Specific activities 282, 289
Spectral analysis of aryl sulfate esters 94
Spectral properties of sulfate ester
conjugates 86
Spectrometric studies on pesticide
residue–humus interactions 264
Spectrometry of humic substances 260
Spectrometry, mass 60
Spectroscopic characterization of
humic substances 263
Spectroscopic characterization of soil
organic matter 258
Spectroscopy 120
electron impact mass 50
infrared 216
nuclear magnetic resonance (NMR) 50
raman 217
uv 217
Starch 170, 172
Straw 170
Structural identification 118
Substituted ureas 189
Sulfatase enzymes 89
Sulfate
adenyltransferase, ATP 86
ester(s)
aryl 92, 94
conjugates 86
of isopropyl 3-methoxy-4-
hydroxycarbanilate 146
laboratory synthesis of 90
metabolism of 88
group 93
Sulfotransferase 87
Sulfoxidation 116
Sulfur, reactions involving 4
Sulfuric acid esters 371
Surface
area values for clay minerals 210
chemistry of adsorbed pesticides 216
tension 4
Sustar 285
Sym-triazine conjugate in plants,
catabolism of a 115
Synthesis 117
of GSH conjugates, chemical 120
of humic materials, microbial 244

Synthesis (continued)
pesticide conjugate 12
of sulfate esters 86, 90

T

Talc 212
TBR (total ¹⁴C bound residues) 364
TCPH (p-toluoyl chloride phenyl-
hydrazone) 178
TDA, radiolabeled 296
Terminal analyses 120
Thiophenol methylation 148
Tissue(s)
soybean 80
callus 82
cotyledon callus 79
plant callus 74
rice root 156
TLC analysis, two-dimensional 289
TMS derivatives 81
TMS-methyl derivatives 60
p-Toluoyl chloride phenylhydrazone
(TCPH) 178
Tomato 49
Toxicity
of aniline mustard 25
of exocon and conjugate 21
pesticide 37
Tracer experiment with pesticides 356
Transfer
group 105
of larger alkyl groups 150
rates of 223
routes of dichlobenil in soils and
crops 173
Transferase(s) 107
epoxide 109
GSH
alkene 113
alkyl 108, 115
arlkyl 110
aryl 110
epoxide 113
triazinyl 115

Transmethylation 146
s-Triazine herbicides 186, 189, 202, 265, 354
Triazinyl transferase, GSH 115
Triclopyr (2,3,5-trichloro-2-pyridyl-
oxyacetic acid) 226–230, 234
Trifluoromethanesulfonanilide pesti-
cides, chemical extraction of 285
Trifluralin 191, 366
Trypsin 179
Tumor inhibitory potency of aniline
mustard 25
Two-dimensional TLC analysis 289

U

UDP-glucosyltransferase mechanism .. 43
Ultraviolet spectra 260
Unavailability of bound paraquat soil
residues 304, 330, 336, 340
Unextracted metabolites, bound 14
Uptake of ¹⁴C paraquat from treated
soils 325
Ureas, substituted 186, 189

V

Van der Waals forces 201
Vines 347, 348
Visible spectra 260

W

Water
extraction analysis, acetonitrile– 291
liberation of 3,4-dichloroaniline by
boiling 159
release of bound fluchloralin residues from soil into 364
Weed control 347, 348
Wheat 170, 309, 311
Worms 334

X

Xenobiotics 132